CITIZEN TURNER

ALSO BY ROBERT GOLDBERG AND GERALD JAY GOLDBERG:
Anchors: Brokaw, Jennings, Rather and the Evening News (1990)

ALSO BY GERALD JAY GOLDBERG:
Heart Payments (1982)
126 Days of Continuous Sunshine (1972)
The Lynching of Orin Newfield (1970)
The National Standard (1968)

ALSO BY ROBERT GOLDBERG:
Getting the Talk Right: The CEO and the Media (1993)

ROBERT GOLDBERG
and GERALD JAY GOLDBERG

CITIZEN

THE WILD RISE OF
AN AMERICAN TYCOON

TURNER

Harcourt Brace & Company

New York   San Diego   London

Title page photograph is by John Chiasson / Gamma Liaison.

Library of Congress Cataloging-in-Publication Data
Goldberg, Robert.
    Citizen Turner/Robert Goldberg and Gerald Jay Goldberg.—1st
ed.
        p.   cm.
    Includes index.
    ISBN 0-15-118008-3
    1. Turner, Ted. 2. Cable News Network—Biography.
3. Businessmen—United States—Biography.
4. Telecommunication—United States—Biography.
I. Turner, Ted.   II. Goldberg, Gerald Jay.
III. Title.
PN4888.T4G645   1994
338.55'5'092—dc20   94–43761
[B]

Designed by Lydia D'moch

Printed in the United States of America

First edition

B C D E F

For Nancy and Colleen
and for
James Brendan

CITIZEN TURNER

ON A SUNNY NOVEMBER AFTERNOON in 1991, in the large, gleaming, Upper West Side loft of artist Peter Max, the crowd of journalists, celebrities, and businessmen in power blue suits is packed shoulder to shoulder, sipping cocktails and imported mineral water. Polished hardwood floors and whitewashed walls frame a row of the artist's rainbow-colored prints. But there is only one focus of attention. Five deep, the photographers press closer, shoving their cameras forward, elbowing each other for that better angle, that perfect shot. "Ted!" "Look this way, Ted!" "Hey!" A brief fracas breaks out as two of them jostle for position. The guest of honor turns again, and the flashes go off to the cries of "Ted! Ted!"

At the center of the crowd, tall and slim and slightly rumpled in a charcoal-gray suit, stands the man everyone has come to see. He looks surprisingly distinguished, with his craggy face, trim mustache, and silvery-white hair. His expression is unusually somber. But then he flashes his trademark grin, that gap-toothed, fun-loving, hey-what-the-hell grin, and there is the tabloid Ted Turner, the wild man of American business. He's come to talk about his latest passion, the environment, and to announce the Turner Broadcasting System campaign called "Save the Earth," a series of television programs geared to culminate at the June 1992 Earth Summit in Rio de Janeiro. And as always, where Ted Turner goes, he makes news.

Half visionary, half crackpot, and all-American character, Ted

Turner is a genuine original. Some see him as a manic genius, others as an idiot savant. Playboy, sportsman, entrepreneur, risk taker, thrill seeker, hard-charging winner, lover of fast boats and even faster women, Turner is all of these.

But despite the sloppiness of his high-visibility escapades and his high-decibel notoriety, Turner is a man of impressive achievements, a man who—almost in spite of himself—must be taken seriously. By *Fortune*'s most recent estimate, he is one of the hundred richest men on earth. More than luck has made him a multibillionaire.

Like Carnegie and Rockefeller, the robber barons of the last century, like Ford and the assembly-line princes of American industry in the first decade of this one, Turner, founder of seven television networks, is *the* magnate of his age, the media age—a televisual tycoon. Herbert Schlosser, former president of NBC, has said that along with David Sarnoff (who built the box) and William Paley (who put programming into it), Ted Turner is one of the three most important men in the history of television.

Turner has muscled his way into our living rooms with his Cable News Network and, quite literally, changed the way we look at the world. Today, news is available when it actually happens, not when it's convenient for the three broadcast networks to carry it. Whether presenting Tiananmen Square or the Berlin Wall, the *Challenger* disaster or the Anita Hill/Clarence Thomas morality play, his CNN—the first worldwide, round-the-clock, all-news television network—has become the place to watch history being made . . . live. In times of political crisis, in times of starvation and flood, at any time when things are happening, people everywhere are watching the world through the eyes of CNN. And Ted Turner owns those eyes.

If there was a moment that confirmed Turner's power, it occurred on Wednesday, January 16, 1991, at 6:35:26 P.M., EST—the beginning of the battle in the Persian Gulf, the first electronic war. At that moment, CNN came of age. As the allied aerial bombardment swept over Baghdad and all the other networks shut down, the CNN correspondents—linked to their satellite by a special "four-wire" phone—peered out the window of the Al-Rashid Hotel and described the scene—the bombs erupting, the tracer fire strung out across the night sky like a silvery web.

In the White House, in the Vatican, even in the bunkers of Iraq, transfixed viewers watched, fascinated by the strange phenomenon of a live, instantaneous, real-time war, a war that unfolded on their screens as it happened, like some sporting event, complete with commentators providing the play-by-play. And strangest of all was the worldwide linkage: in the global living room, everyone around the planet was watching the same show at the very same moment, including world leaders. Secretary of Defense Richard Cheney said that his best source of intelligence was neither the CIA nor the NSA but CNN. Likewise, "the best source of information Saddam Hussein may have," as one military expert commented, "is American television"—that is, the American television that the Iraqi leader could receive on his newly installed CNN satellite dish.

If the power brokers in the boardrooms and the war rooms were depending on CNN for the latest news, so, too, were those in the pressrooms. On any given day, most of the world's reporters who had come to cover this event were in fact huddled around a TV set with the dial turned to the Cable News Network. Though weak on analysis and insight, CNN was bringing raw information directly from where it was happening to home television screens with microchip speed, and in so doing established itself as the video wire service of record—the first draft of the rough draft of history.

Most significantly, in the Gulf War, TV was more than just a recorder of events; it became a player in those events. In place of the Oval Office red phone, diplomacy was being conducted publicly by television interview: Israel's Benjamin Netanyahu, the Iraqi ambassador to France, Jordan's King Hussein, all peered into the cameras and floated their separate trial balloons concerning possible negotiations. It was a nonstop video feedback loop, where the protagonists acted, then studied themselves on TV, then revised their act, with each performance inspiring a dozen commentaries.

And since the Gulf War, the astonishing power of Turner's news channel has continued to grow. According to Anthony Lake, Assistant to the President for National Security Affairs, "Our foreign policy seems to be increasingly driven by where CNN places its cameras."

Interestingly enough, given his creation, Ted Turner has never been a man with a huge passion for news. For him, news has always been just another way to fill airtime, like sports or situation comedies. Not that these especially fascinate Turner either. He built his

WTBS Superstation on sports—Atlanta Braves' baseball and Atlanta Hawks' basketball—but even though he owns both teams,
when he bought them he could barely tell the difference between
a balk and a slam dunk. Nor does he have a particularly sophisticated
taste in movies or any of the other programming that runs on his
networks. His passion has always been to build an empire.

Using second-string producers, third-rate correspondents, and
recycled network programming, he has gone about his task with a
single-minded determination. The knowledgeable people around
him once considered his prospects for success ludicrous. "I literally
thought Ted would crap out," one of his earliest financial advisors
acknowledged, having watched him repeatedly place all his chips
on the table and roll the dice. Yet Turner—who as a child dreamed
of equaling the exploits of the boy king Alexander the Great, and
as an adult sat through Orson Welles' *Citizen Kane* dozens of
times, captivated by the thinly disguised life of William Randolph
Hearst—has succeeded quite remarkably in building himself an empire of his own.

Standing at the podium on this November afternoon, a video
wall behind him six screens wide, Ted Turner represents something
new—a media baron who is not just multinational, like Rupert
Murdoch or Silvio Berlusconi, but global. And in keeping with the
scope of his new vision, he has decided that his next project is to
save the planet. Once again, he is showing the same stunning combination of hubris and sheer naïve enthusiasm that helped him build
his fortune. And so on this sunny fall day, when even the New
York City streets outside seem to sparkle, he is holding a press
conference to bring an environmental warning of doom and gloom,
and to point the way to salvation.

Turner's high-pitched Georgia drawl comes crackling with electricity through the room. "With the increasin' numbers of people,
plus the fossil-fuel use"—he chews his words as if he still has a
plug of Red Man in his cheek left over from his old tobacco-chewing
days—"we're puttin' tremendous pressure on our little home here."
Scanning the faces before him, he speaks with the absolute confidence of one who enjoys being the center of attention and has been
in front of audiences all his life. He has no prepared text, no notes.

"The numbers have tripled durin' my lifetime, and the size of
the planet hasn't," he points out. Then he launches into the day's
theme—the new TBS programs on the environment. "We have to

plan for our futures," Turner urges his listeners. But suddenly the cameras turn away as if they've lost interest. Who's that? The audience cranes to look, ignoring the speaker. Up in the front row, trying to slip in unobtrusively, Turner's wife-to-be, Jane Fonda, has started a stampede of photographers. "It's Jane! Jane!!"

Resplendent in a lioness-beige pantsuit, shades, and windswept blond hair with windswept blond highlights, every inch aerobically toned, Fonda looks extraordinary for her fifty-three years.

"We *have* to think globally!" Turner insists, reminding them why they're there. "It's not just fortress U.S.A. I mean, we can't just pull up the drawbridge. We all breathe the same air." He punches the air for emphasis.

And then, without warning, his speech starts to ramble, passionate but disjointed, as if his speeding train of thought has gotten ahead of his words, or fallen behind them, or the two have taken totally different tracks and derailed. He lurches from toxic waste to Haitian refugees. "What are we gonna do with 'em?" he demands to know. The members of the audience look sheepishly at one other, perplexed. This is the other side of Ted Turner, the erratic side, the side that makes people wonder, How did *this* man ever get to be *so* important, *so* rich?

But before anyone has time to ponder the question, Turner is back to the global conference again, and the crowd seems to breathe a sigh of relief. "You know, we want to make everyone here aware. . . . The U.S. is draggin' its heels." Revved up, this is the Turner they all know—Turner the motivator, Turner the firebrand, Turner the leader, who wanted everyone else to follow or get out of the way as he defended the America's Cup in 1977 and sailed to victory aboard *Courageous,* or as he attempted to swallow up CBS in 1985, or a year later bought MGM, or established the Goodwill Games in partnership with the USSR, or helped to found the Better World Society "for the production of television programming on issues of critical importance to the survival of the planet." This is the megavolt Turner—the subject of more ink, more video, more *gossip* than any other businessman alive.

Later, under the strobes, they look a charmed pair—Jane, the Oscar-winning sex goddess of the sixties (*Barbarella, Klute*), the political firebrand of the seventies ("Hanoi Jane"), the exercise queen of the aerobic eighties, and now, for the nineties, a retro wife, teamed up

with the Southern cracker turned presidential pal and global ambassador. Labels like left-wing and right-wing don't fly here. What
the two have in common is money, muscle, and love. To judge by
the fuss, not since the Medicis married off a child to the Bourbons
have two such powerhouses teamed up—Ted and Jane, the great
celebrity couple of the *fin de siècle*.

"Ted!" "Jane!" "Ted, Jane!"

A phalanx of security guards with walkie-talkies plows a wedge
through the crowd, sweeping the two of them past the cameras,
past the reporters, and out through a side door onto a landing and
into a waiting elevator.

Down on the street, Ted packs Jane into the long black limousine that waits curbside. "I'll seeya later," he says, beaming a
huge smile, and, with a gallant flourish, closes the door of her limo.
They're staying at the Waldorf, and Ted, eager for some exercise,
will walk back to the hotel across Central Park. He sets off down
the street, his long legs moving at a brisk clip.

Within minutes, he reaches the edge of the park, striding past
the ball fields and pathways and the old bowling green. Joggers,
skaters, even a horse-drawn carriage driver turn to look. "Ted," a
guy with a baseball glove yells. "Hey, Ted!" He waves at Turner
and flashes a big thumbs-up. It isn't just Turner's celebrity that
brings out the smiles. They really seem to like him and what he
stands for, the sheer outrageousness of the guy. After all, Turner
is a competitor who has put himself on the line again and again.
He's tweaked the nose of authority—venerable institutions like the
New York Yacht Club, major-league baseball, CBS. And most
quixotically of all, he's been willing to look like a fool, a loudmouth,
a boor, and to endure ridicule, even huge monetary losses, for those
simple, Jiminy Cricket, Save-the-Earth ideals of his—clean air,
world peace, everybody living happily ever after.

Whatever it is that makes these people smile and wave at Ted
Turner as he goes by, one can't imagine them doing it for any other
billionaire or media mogul—not Larry Tisch, not John Malone,
not Rupert Murdoch. No one is going to grin and shout, "Hey
Rupert!" The guys standing around the baseball diamond don't even
know who he is.

Warts and all, Ted Turner is clearly a remarkable character. Like
some two-fisted Indiana Jones of the boardroom, he's not simply
a businessman; he's an adventurer. Jane Fonda calls him "my buc-

caneer." He's a man who can state his business goal in six words or less: "I'd like to own everything." His story, much of it lived in boldface, is on an operatic scale and filled with roller-coaster ups and downs. It's a tale of sailing triumphs and marital failures, of bold and outrageous deals, each one bigger and riskier than the last. It's the story of a man who took a small, shabby, ultrahigh-frequency station in Georgia with unreliable, broken-down equipment and built it into an $8-billion global media conglomerate—and along the way just happened to transform himself from a provincial bad boy into *Time*'s Man of the Year.

But the public Turner is just part of the portrait. There is another Turner story as well, a deeper and in some ways much more interesting one. It concerns Ted's strange, complex, intensely competitive, love-hate relationship with his father.

Throughout his life, Ed Turner was the yardstick of his son's success. And after Ed died, his ghost continued to peer over Ted's shoulder—shoving him, prodding him, demanding to know "Why in hell ain't you doin' any better?" Ed was a charmer, an alcoholic, a woman chaser, a shrewd and driven businessman, and those traits were all part of his legacy to his son. Perhaps the most painful part of that legacy was Ed's sudden death shortly before the age of fifty-three.

As he walks through Central Park, Ted Turner has just passed his own fifty-third birthday, a milestone he never thought he'd reach. All through his life, he told friends that he expected to die young, that death was hounding him as it had his father. But here he is at fifty-three, reasonably healthy, mind-bogglingly rich, and about to be married to Jane Fonda.

At the corner of Park Avenue, engrossed in his thoughts, he fails to look up at the stoplight and strides into the street. Suddenly from the right comes a screech of brakes, a flash of chrome. The car, with no room to stop, swerves on by. Turner, astonishingly, doesn't even seem to be aware that he's nearly been run over. He just keeps on walking across the street toward the entrance to the Waldorf, apparently oblivious. It's like the time some twenty years earlier when he was running down the main street of Anderson, South Carolina, on his way to a business appointment and was struck by a car. Preoccupied with the sales pitch ahead, he just bounced off and kept on running.

Two decades later, Ted Turner is still a survivor.

IT WAS UNUSUALLY WARM when Ed Turner got out of bed, and the dove-gray clouds hung ominously low over the old plantation. Rolling down the Carolina coast from Charleston to Savannah, the humidity was oppressive that March 5, 1963, the air heavy, pollen-sweet, and melancholy.

As Ed walked inside for breakfast, the floorboards of the plantation house groaned under his weight. Over six feet tall and about two hundred pounds, Ed was thickset, outsize, unwieldy. He gave the impression of being as big as the billboards he owned all over the Southeast. And his appetites matched his size. There was a time when he could polish off a fifth of liquor a day and, despite emphysema, puff his way through three packs of cigarettes. Rather than stop drinking when he discovered he was developing an ulcer, he chose to cap his scotch with a buttermilk chaser. But that was ages ago now.

As Ed strolled into the dining room at the back of the house, cheerfully greeting his surprised wife and sitting down across from her, he seemed to be a new man. He eyed his breakfast with pleasure and began to wolf it down. He and Jane Turner were waited on by Jimmy Brown, a black servant with a crippled left arm, who often cooked their morning meals. According to Jane, Jimmy was a good cook. He had been working for her husband ever since he was a teenager, and to hear Jimmy talk, there was nobody like Mr. T. To Jimmy, he was a handsome man and "tall, always well dressed," with "Irish linen suits, white shirts, and green ties. And Panama hats. Real ones . . . a Southern gentleman."

If Jimmy Brown had been worried about his employer lately, he was heartened by the way he now went at his food. Mr. T seemed to be his old self again.

Polishing off his breakfast, Ed got up, leaving his wife to finish her coffee in peace. As he passed the window, she saw him stop to stare for a long moment. He loved the old hunting lodge, the creek outside the front door, the broad trees hanging heavy with Spanish moss. His 800-acre plantation in Beaufort County, South Carolina, stretched as far as the eye could see, from the long entrance drive flanked by the pines he had planted to the thick woodlands beyond. Everywhere it was full of wildlife—duck, dove, quail, raccoon, possum.

He and Jane had remade Binden, converting the old hunting lodge into a gracious country home. In Jane's photo albums, there are pictures of the two of them enjoying themselves out on the porch with friends, sipping cocktails on long, languid Southern afternoons. To Ed, the plantation had become a symbol of his success, although, as he acknowledged to his wife, it was odd that a man who had made his money in advertising—a business so inextricably linked with the twentieth century—should measure his success by such a nineteenth-century standard. But they both understood why. Ed's father, the first Robert Edward Turner, had been a dirt farmer in Sumner, Mississippi, and had lost almost all his land and everything on it during the Great Depression. For Ed, Binden was redemption—the tangible proof that he had really made it.

Turning away, he strolled through the kitchen, climbed the wooden stairs to the second floor and went into the guest bathroom. Closing the door behind him, he took out his .38-caliber pistol, placed it in his mouth, and pulled the trigger.

Jimmy Brown heard the shot at 9:40 A.M. just as he began clearing away the breakfast dishes. It echoed from upstairs like a thunderclap. Rushing up the stairs, Jimmy called, "Mr. Turner! Mr. Turner!" He found his employer crumpled on the bathroom floor. Ed had blown the side of his head off, but he was still breathing, his body twitching on the floor.

Downstairs, Jane sat stunned, immobile, frozen in her seat. Though she realized at once what had happened, she couldn't bring herself to move out of her chair. "I was afraid of what it was," she would say later. "I couldn't go up there and look."

In a few seconds, Jimmy Brown came hurtling down the stairs. "There's been an accident," he told her. "Call the doctor. Mr. T's still alive." Jane, galvanized into action, snatched the phone and quickly dialed Ridgeland, the nearest hospital. "I'm afraid my husband has damaged himself," she told the doctor delicately. "Could you please come right away?" But at a small, backcountry hospital like Ridgeland there was only one doctor on duty. "I'm sorry," he said. "I'm the only doctor here." Frantically, Jane dialed another number.

Sixty miles away in Savannah, Ed's friend, Dr. Irving Victor, chief of staff at Candler General Hospital, was making his hospital rounds when he was informed that there was an emergency call for him. It was Jane Turner, her voice trembling.

"Don't try to bring him to Savannah," the doctor ordered. "Take him to Ridgeland as quickly as you can. I'll drop what I'm doing and get right there."

Ridgeland was about midway between Savannah and Binden, and Victor covered the distance in record time. When he arrived, Ed was not yet there, but there was a call for Dr. Victor from the plantation. The sheriff was on the line.

Twenty-four-year-old Ted was stunned when he got the news that his father was dead. He couldn't believe it. He and his wife Judy both knew that Ed had been having problems, but, as she says, "He was such a vibrant person. I mean, we knew he'd get better. He had handled everything before. I don't think Ted ever believed he would do this." The two of them hurried down to Savannah.

Just six months earlier, in September of 1962, Ted had been thrilled when his father had made what they both thought was the deal of a lifetime—a $4-million purchase of three valuable divisions of General Outdoor Advertising (Atlanta, Richmond, Roanoke). In one stroke, Turner Advertising had been transformed from a middling, comfortably successful firm into the largest outdoor advertising company in the South, one of the ten largest in the United

States. And Ed Turner had gone from being a millionaire—he often said to his wife, "Having a million dollars is nothing"—to becoming one of the giants.

Yet almost as soon as he bought the three new plants, Ed began to have doubts about what he had done. He was afraid that he had neither the capital nor the people to make the deal work. After fretting about it for an agonizing month and a half, he picked up the phone and called his trusted accountant, Savannah CPA Irwin Mazo.

"I've made a big mistake, Irwin," he said. "This thing is going to break me."

Mazo could hear that his caller was upset. "It's the best deal you've ever made, Ed," he reassured him. "You can afford it. Believe me, it's *not* going to break you."

Mazo knew that in order to come up with the $800,000 down payment, Turner had sold his Savannah plant and other properties, but they were tapped markets. The potential of the new Atlanta plant alone was enormous. The interest he was paying on the $320,000 loan he took from the Citizens & Southern National Bank of Savannah was high, but the deal had been mostly financed by the sellers, and the cash flow out of Atlanta was strong. Furthermore, his other divisions were all doing fine.

"There's nothing to worry about," Mazo repeated again.

For a few seconds there was silence at the other end of the line. Then Ed suddenly shouted, "It's ruining my health! I've gotta sell it," and slammed down the receiver.

Shortly thereafter, Mazo received a second call from Atlanta. This time it was Ted's stepmother Jane, and she was plainly worried. She had to see him, talk to him, find out if there was any truth to her husband's fears that he might lose the business.

Mazo drove up to Atlanta to meet Jane and explained the situation to her: Ed had big lines of credit at the bank, all of his divisions were doing well, and despite the debt, he didn't have any long-range financial problems. Jane was relieved. "Don't let him sell it," she told Mazo. "It's something he's dreamed of all his life."

But once Ed's mind was made up, he was impossible to deal with. Blunt and uncompromising, he preferred candor to subtlety, enjoyed controlling events rather than muddling through. Even Ted, who had worked in the family business since he was a boy, had little influence on his father.

To Ted, inordinately proud of his father's success and thrilled by their company's recent expansion into the big time of Atlanta, the future looked rosy. To Ed, it promised only disaster. Ted didn't know what was going on, couldn't explain his father's change of heart. "Looking back," Ted says, "for about six months he had this terrible state of depression." Ted had tried to make him see reason. Their arguments about the business became fierce. Ed was sure he was about to lose everything he had ever worked for. His son talked a good game, but what did the boy really know about the future prospects of the business? Ed was convinced that he would end up destitute, just as his father had. He had to sell. He had to get some peace.

For years, Ed Turner had been battling a drinking problem. He had tried vitamin shots, Antabuse, anything that might help. In July 1961—almost two years before the March morning when he shot himself—he decided to try Silver Hill, a posh psychiatric hospital in New Canaan, Connecticut, specializing in the care of well-to-do people suffering from addictions (like Judy Garland and Truman Capote). Before heading north, Ed went to see Tom Adams, his Savannah lawyer, and made out his will. Giving up booze must have seemed like dying to him.

He came home in September, a recovering alcoholic. His friend Dr. Victor insists that Ed had accomplished a cure. But it wasn't necessarily a cure for the better. "You know," Victor says, "some people have an escape mechanism. Alcohol could have been his escape. If I had to say, he probably did better with alcohol than without it."

Struggling to keep off alcohol during the next twelve months, Ed became more intense, more driven, and finally, in September of 1962, when confronted by what appeared to be his great opportunity, utterly depressed. And so, three months later, unable to cope with the fast-approaching holiday season, the recovering alcoholic returned to Silver Hill.

It was from there that he put in a long-distance call to Bob Naegele, a business friend who ran a large outdoor-advertising company in Minneapolis. Naegele had been enthusiastic about Ed's recent purchase, and he was astonished when Ed told him that he could have the new plants—Atlanta, Richmond, Roanoke—for what they cost. All Ed wanted was to get his money back.

Naegele tried to convince him to think it over, but Turner refused to listen. Before hanging up, Ed said to him, "It's a long way from the master bedroom to the cellar." The voice was that of a frightened man. Naegele would never forget it.

Though he had no intention of exploiting a sick friend, Naegele was a businessman who knew a good deal when he saw one. He called Irwin Mazo in Savannah. "Look," he said, "not only will we give him his money back, but we'll give him a $50,000 profit he hasn't asked for, just to show that we didn't take advantage of him."

"Well . . . ," the accountant said, resigned to the inevitable, "if it's ruining his health." There was nothing more to be done.

Turner left Silver Hill in January of 1963 and returned home to Binden, where, according to Ted, he was still badly depressed and taking all kinds of prescribed medicines. The fact that Naegele had agreed to buy him out had only temporarily reassured him. His wife noticed that he had started to worry again. He feared he was being taken advantage of: the business was probably worth much more than he was being paid.

At this point, the arrival of two of Naegele's aides with a check for $50,000 as a down payment struck him as positively alarming. He paced the floor, appearing visibly upset to his visitors. He wasn't interested in piecemeal payments. He wanted *all* the money. After speaking on the phone to Naegele and coming to an agreement about the rest of the money, he seemed to feel a little easier.

His son felt just the opposite. Ted couldn't believe his father was selling out, and selling out *now,* just when all their plans were coming to fruition. Ted called to try to persuade him not to do it. He wanted his father to be the Ed Turner he had always known— dynamic, flamboyant, a force of nature, a pain in the ass at times, but always a winner. He tried to jolt him back into his old self, bluntly accusing his father of being a coward, a quitter, afraid to take risks. Jane, listening in on the conversation at Ed's request, was so shocked she started to cry. But Ed just listened. Normally volatile, he didn't seem to be at all angry at what his son was saying to him. In fact, he was eerily calm, as if all the fight had gone out of him.

A day or two later, while Ted was back at work in Atlanta, Ed was in Savannah at the DeSoto Hotel. He had finally gotten up the courage to write the letter to his wife that he had planned. It was probably the hardest thing he had ever done in his life. He gazed down at the DeSoto letterhead for a long time before finally finding the words. "I don't want to hurt anybody," he began. "I just feel like I've lost my guts." But after a few more sentences, he couldn't go on, and put the letter aside.

The following Sunday, at Binden, as Jane left for church, she asked Ed if he wanted to join her. Now more than ever, she was worried about him. The Antabuse was making him sick. He wasn't eating well. He rarely slept through the night. But he didn't want to go to church with her.

Sometime after she had left, he pulled out the note to her that he had started and picked up his pen. "While you were at church praying for me," he wrote, "I made up my mind what I was going to do." Forty-eight hours later, he was dead.

Promptly at 4:00 P.M. on Wednesday, March 6, funeral services were held at the chapel of Sipple's Mortuary, in Savannah. The Beaufort County coroner had demanded neither autopsy nor inquest. The Reverend Ernest Risley officiated at the obsequies, and a surprisingly large crowd turned out at such short notice—mainly business associates and sailing friends. The Episcopal minister knew the Turner family well and conducted a simple ceremony, just as Ed himself might have preferred it. Dr. Victor and Jane's father had taken care of the arrangements. In addition to Irving Victor, the pallbearers included Tom Adams and Irwin Mazo. Given the extent of Ed's injuries, the coffin remained closed.

He was buried in Savannah's Greenwich Cemetery. Gathered at the graveside near a narrow bend in the Wilmington River, Jane and Ted and Judy and the other mourners stood beneath huge oak trees covered with pale, ghostly Spanish moss. Ted, according to Judy, seemed to be in a state of shock. It was Dr. Victor who suggested to the widow the unusual epitaph that she would have inscribed on her husband's white marble headstone. ESSE QUAM VIDERI. To be rather than to seem—or, more colloquially, what you see is what you get.

People who knew Ed Turner were astonished by the news of his suicide. "R. E. TURNER DIES OF GUNSHOT WOUND" was the headline in that morning's *Savannah News*. "I never will understand why he killed himself," declared Mrs. Carl Helfrich, a close friend of the family. "He was so very successful." Not far away, at the Wessels' house, the reaction was much the same. Charlie Wessels, a former schoolmate of Ted's, recalled that he had seen Mr. Turner only a short time before. Flying home from Atlanta, he had run into Ted and his father on the plane. Aware of their past feuding, he had been genuinely surprised at how well the two of them seemed to be getting on. "You know," Ted's father said to both of them, "you-all have the opportunity to do anything that you want to. You just have to take time to think about things, and then you have to do 'em."

Though news of the death startled Savannah, Ed had posted warning notices of his grim intentions. Irwin Mazo's wife would later remember Ed's telling her that he had lost confidence in himself. Ted's friend Peter Dames recalled that he said, "Nothing has been worthwhile." Dr. Victor, who saw Ed only a few days before his death, had thought that he seemed seriously depressed. He warned Ed that he really should be in the hospital.

The one person who appears to have gotten an unequivocal message of Ed's suicidal plan was Ted's mother—Ed's first wife, Florence, with whom he had continued to maintain cordial relations even after their divorce in 1957. Shortly before that fateful Tuesday, he phoned her to say that he had lost his nerve in a business deal, that he had been humiliated, that he was going to kill himself. If Florence believed him, she apparently felt there was nothing she could do. Even when they were married, she had never had any control over him.

Ted had felt equally helpless. He didn't know what to make of his father's recent behavior. "I thought he was having a nervous breakdown" is the way he remembers those last terrible weeks. "I begged him to stop working, to take some time off, but he wouldn't."

Years later, discovering that his father had been taking prescribed tranquilizers and quaaludes to handle his drinking and smoking addictions, Ted attributed his suicide to them. But he also realized that nobody could have stopped him. "Nobody had any control over him. He was his own man. And I didn't even know what the problem was at the time."

What Ted did know was that he loved his father, even though Ed Turner had been a hard, demanding parent, sometimes distant, sometimes even cruel. Ted had tried to please him, and often he had failed. On this awful occasion, what he felt was guilt, as if he were in some way to blame for what had happened. As Ted would later tell the story, it was right after one of their many terrible fights that his father "blew his brains out." In this rather pathetic version of the event, not only does Ted fail to save his father, he holds himself directly responsible for his death. Heightening his feeling of complicity was the suicide weapon his father had chosen. It was, according to Ted, "the same gun that he taught me to shoot with."

In a sense, the sound of the bullet that Ed Turner sent crashing into his brain would echo throughout his son's life. Once again, Ted was being deserted by his father, as he had been on many occasions while growing up. Abandonment and exile loomed large for Ted Turner in his early years.

And it wasn't only his father who deserted him. When Ted was fifteen, his twelve-year-old sister, Mary Jean, to whom he had been close, developed lupus erythematosus. Her illness is one of the few subjects that Ted still finds difficult to discuss. "She was sweet as a little button," he says hesitantly. "She worshipped the ground I walked on, and I loved her. A horrible illness." Mary Jean died in 1960, at the age of nineteen. "The only time I ever saw him with tears in his eyes," recalls one of Ted's friends from these early days, "was when he talked about his sister."

Now, only a little more than two years later, it was his father who had abandoned him. In one terrible instant, the world as Ted had known it for twenty-four years had disappeared, and he was cut adrift. "At the end, the bank wouldn't even honor the check for his funeral," he would later recall bitterly. The $2,500 draft was returned marked "insufficient funds." Of his father's death, this latest and most crushing loss, Ted recently told an interviewer, "It was devastating."

CHAPTER 2

In 1938, the Carew Tower was the tallest building in Cincinnati, Ohio, a 48-story skyscraper rising high above the heart of the city, at Fifth and Vine. From its Observation Tower—open to the public for thirty-five cents—visitors had a sweeping, unobstructed view of the city and beyond. To the south lay the Ohio River and Kentucky. To the west, Indiana and Illinois. The steamboat and the river had by 1860 made Cincinnati the sixth largest city in the United States, but after the Civil War it was the railroad and burgeoning industry that were responsible for the city's continued prosperity.

When the Depression hit the country in the early 1930s, it seemed to hurt the South even worse than the North, and small agricultural towns such as Sumner, Mississippi, where Ed Turner had grown up, suffered worst of all. The family cotton farm went broke, and the Turners eked out a living running a bargain-priced, easy-credit hardware store. Every time Ed's father handed him a boxful of rifle shells, he said he expected the boy to bring home one bird for every shell in the box. Robert Edward Turner didn't hold with wasting.

Young Ed saw no future for himself in either farming or hardware. He had been in Tennessee attending Southwestern at Memphis, a small liberal arts college, and he had no intention of going back home. In 1932, however, he had to drop out of school when his money ran out. He stayed on in Memphis for a while, supporting himself by working for the General Outdoor Advertising Company.

Ed was assigned to do traffic counts, toting up the number of vehicles that drove by various billboard locations. He had to jot down not only how many cars went by but how many people were in them—the number of "impressions per month." The better the location, the more impressions, and the more the company could charge its advertisers. Rain or shine, he stood outside with his notepad, a solitary figure watching the world pass him by. It was boring work for an ambitious young man, so when the opportunity came to go to Cincinnati to sell cars for the Queen City Chevrolet Company, he jumped at the chance and headed north.

Because of the diversification of its industry, Cincinnati had recovered more quickly from the economic trauma than many other cities. By 1937, smoke was again rising from factory chimneys. There were over one hundred major industries, including the U.S. Shoe Corporation, Baldwin Piano, and Procter & Gamble. It was a metropolis of nearly half a million.

When Ed arrived in Cincinnati, he was in his early twenties, a tall, slim young man with an easy smile, an outgoing manner, and a thick Southern accent. His boss liked him, the customers liked him, the girls in town liked him. Fond of booze and babes, he was considered a regular "good-time Charlie." It wasn't long before Ed discovered that he could make more money selling advertising than Chevrolets, and he was back in the billboard business.

In the beginning, Ed was living in an apartment hotel owned by the Rooney family. One of its advantages was the presence of the Rooneys' tall, dark-haired daughter, Florence, an innocent young woman who had been brought up in a strict Roman Catholic family and educated at a convent school. Cincinnati at the time had a large Catholic population, and although Ed wasn't Catholic, he was willing to make allowances.

Florence's grandfather, Henry Sicking, had been a well-to-do Cincinnati businessman who had once owned a chain of twenty grocery stores. Despite the Depression, the Rooneys still had money. Ed pursued Florence as if he were selling her billboard advertising and his job depended on it. The fact that she was in love with someone else at the time seemed to make hardly any difference to him. He laid siege to her heart for eight straight months and simply wore her down. "He really was witty," she told an interviewer many years later. "With that thick Southern accent of his, he could be hilarious at times." Her mother, however, had her doubts about Ed Turner. Caught in the middle, Florence—a somewhat passive, obedient young woman—didn't know what to do. When her other suitor proposed, bringing matters to a head, Florence replied weakly: "I told him it looked like I was going to marry Ed. Mother said not without her consent." But in time even Mrs. Rooney was worn down. She reluctantly gave her consent, and on August 17, 1937, Ed and Florence were married in a Catholic ceremony.

Six months later, Florence was pregnant. Ed quickly let it be known that he had no intention of having any child of his brought up a Catholic, and by mutual agreement they became Episcopalians. In this as in so much else in their marriage, Ed would control what they did.

The Turners were living at 918 Dana Street, in a middle-class Cincinnati neighborhood, when, on November 19, 1938, in the early hours of a cold rainy morning, Florence felt the first contractions begin. Dr. Pierce was called, and she was rushed to nearby Christ Hospital. Out in the waiting room, Ed paced, sat down, got up, sat down, fiddled with a copy of the paper.

The headline of the *Cincinnati Enquirer* that morning proclaimed, "GERMAN ENVOY RECALLED; HULL URGES HELP FOR JEWS." In a letter to President Roosevelt, Representative Charles A. Buckley, Democrat of New York, suggested settling Jews in Alaska. At the Albee, Ronald Colman was starring in *If I Were King*. At Keith's, it was Lionel Barrymore in *Young Dr. Kildare*. Among the bestsellers in the Cincinnati bookstores were Daphne du Maurier's *Rebecca* and Ernest Hemingway's *The Fifth Column*.

But for Ed and Florence Turner, there was only one star, and he arrived at 8:50 A.M. Ed was elated. Whiskey and cigars all around! The baby resembled his father more than his mother. A pink-faced, fair-haired, active little thing, he was named Robert Edward Turner III, the continuation of a family dynasty. They called him Teddy.

With a wife and a son to provide for, Ed had no intention of remaining an employee of the Central Outdoor Advertising Company, where he was working. Independent by nature, he wanted his own business. He already had several years' experience in the outdoor-advertising field, and the future looked promising. All over Cincinnati, billboards were going up extolling the virtues of American enterprise. It was obvious to the competitive Ed that any product needed to advertise to beat its competition, and what better way to get your message across than through billboards? Something about their size, their boldness, appealed to him. Most of all, he liked their sales potential in a growing industrial Cincinnati. By the end of 1939, Ed had succeeded in making his dream of a Turner Advertising Company a reality.

There is a photograph taken about this time of Ed and his son. The slim, twenty-nine-year-old Ed crouches down, his white shirtsleeves casually rolled up, his collar open at the neck, his large left hand wrapped around the active two-year-old's body, momentarily anchoring him in place for the camera. A cigarette dangles from Ed's right hand. The knife-edge crease of his slacks and the pen and pencil clipped to his shirt pocket hint at the successful no-nonsense businessman. Hair brushed back, he stares at the photographer with his characteristically rakish crooked smile beneath the mustache he has now grown. Blond Teddy, squinting up at the lens for a second, is fixed in time. He's alert, curious, a good-looking child. His white anklets turned neatly down above his white shoes, he seems the pampered darling of loving parents, a golden boy.

Although things went well for Turner Advertising in the beginning, there was one development Ed hadn't anticipated. On December 7, 1941, the Japanese bombed Pearl Harbor, and America was at war. By one week later, the Clopay Corporation of Cincinnati had already sent off its first shipment of blackout curtains to stores along the Pacific coast. But not all of the city's industries adapted to a wartime economy so well. With the rationing of gasoline and rubber, and the resulting decline in pleasure driving, the billboard business fell on hard times. Before long, even poster adhesive was rationed. And by now the Turners had another mouth to feed.

Teddy's sister, Mary Jean Turner, was born on September 18, 1941. Not surprisingly, there was much fuss over the lovely little dark-haired, dark-eyed girl. Three-year-old Teddy was not the first older sibling to feel neglected. If he had been mischievous before, he became much more so now. Repeatedly, he spread mud over a neighbor's freshly washed white sheets as they hung drying on the line. One Christmas, while his mother chatted with a friend, he pulled all the ornaments off the Christmas tree and stomped on them. "He sure looked good to be so bad," said a family retainer, recalling the escapades of this little handful of a boy. Troubled by business, Ed Turner began to drink more heavily and had little patience for his difficult son. The boy was always getting into things, causing uproars in the family. He had so much energy!

In 1944, Ed Turner volunteered for the Navy. (He would eventually hold a commission in the Naval Reserves.) Assigned to a station on the Gulf Coast, he and Florence decided to take Mary Jean with them and leave the problematical six-year-old Teddy behind in Cincinnati.

Young Ted was miserable. His family had suddenly deserted him, and he was alone. Unable to get along with the other students at school, he became the terror of the hallways, putting stones in his classmates' galoshes and scrambling the clothes in their lockers. Already he was getting the reputation of being a troublemaker. After several of these episodes, he was booted out of the class and then expelled—from *elementary* school! It was this early experience—the first of many in the hard discipline of rejection and abandonment—that Ted Turner would later single out as being especially unhappy for him.

When his family returned to Cincinnati a year after the war ended, Teddy was placed in the public-school system. His experience was none too happy there, either. In second and third grades, he recalls, "I got beaten up all the time, or at least I had to fight all the time. I don't know what it was. Yes, I do know what it was: the other kids thought I was a show-off and a smart-ass." Regardless of what they may have thought, they could plainly see that their classmate was pale, thin, somewhat undersize, and not very well coordinated. What they couldn't see was how unhappy he was.

In the heady, economically expansive years after the war, Ed's billboard business flourished. Eager to enlarge his operation, he began to look toward other advertising markets, canvassing friends and attending conventions and workshops of the Outdoor Advertising Association of America. Prospects in the South appeared especially rosy to him.

Of course, his success hadn't come easily. Central Outdoor Advertising, his old company, had decided to take on the upstart Turner and had been trying to drive him out of business all through the mid-1940s. He was accused of having stolen their leases and snatched away their customers. Ed, never one to back down in a fight, sued the company for restraint of trade, and in a landmark decision walked away with a settlement totaling well over $10,000, a sizable amount at the time.

With a new, large bankroll in hand, Ed was ready to make his move, to branch out. He found what he was looking for in Savannah, Georgia, and in 1947 he bought up the holdings and facilities of two billboard companies, the Savannah Poster Advertising Service and Price & Mapes. He intended to hire a local manager for the new Savannah branch of Turner Advertising, while he himself would remain in the Cincinnati home office. But as is often the case with even well-made plans, that was not the way things turned out.

CHAPTER 3

IT DIDN'T TAKE LONG for Ed Turner to make his presence known in Savannah. In a sleepy, decorous town of Spanish moss and Southern gentlemen, of quaint squares and city greens, red-brick houses and wrought-iron balustrades, Ed's arrival was visible all over town—on billboards that shouted slogans like "Sunshine Cheez-It: GOOD . . . any old time!" and "Krispy Crackers: The Flakier Cracker All Through the Meal!" The outdoor-advertising business was racking up extraordinary sales throughout the country, topping all previous records. "I believe," Turner and a hundred and seventy other delegates were told by a speaker at one billboard convention, "that we are on the threshold of attaining the greatness for which outdoor advertising has always been destined!" Within a few years there would be fifty million cars on the streets of postwar America, and everywhere the passengers looked, another billboard.

The garrulous Turner, now flush with a little cash, cut an attractive figure around town with his handsome face, Panama hat, and ability to knock back a martini or two with the best of them. Ed Turner was generous, expansive. He could spin a tale to tickle the ears. He also had a taste for practical jokes that his friends found hilarious.

Others were not amused. To some in town, Ed seemed too loud, too blunt, too cantankerous. According to Irwin Mazo, who was born in Savannah and whose family has lived there for over a hundred years, "There was no in-between with Ed. People here who were his friends really admired him. But he had so many enemies it was unbelievable."

One of Ed's earliest enemies in Savannah was a man named Shuman—Leiston T. Shuman. Well known and well respected, Shuman had been named "Outstanding Young Man of the Year" by the Savannah Junior Chamber of Commerce in 1944. He was a member of the Kiwanis and of the Savannah Board of Education. Shuman worked for Price & Mapes, and he had always believed that one day he would purchase the company. When the outsider from Cincinnati arrived and snatched the firm away from him, Shuman started his own business and vowed he'd get even. He had been in outdoor advertising for more than twenty years and was a vice president of the powerful Georgia Outdoor Advertising Association. More than that, he knew all the advertisers in town.

Ed, faced with unanticipated competition and fearing for his investment, decided to remain in Savannah, taking charge of his new operation himself. He was certain that only one company could survive in a city barely one-fifth the size of Cincinnati. He was soon joined by Florence and the children, and prepared for what turned out to be a cutthroat battle between the two billboard firms.

Shuman and Turner, according to one local observer, became "absolutely bitter rivals." They took to "jumping" each other's leases, trying to outbid each other for prime rental lots, undercutting each other on the fees they would charge advertisers. Ed Turner's Savannah operation was relatively small, compared with his Cincinnati plant, but Ed simply refused to be beaten by Shuman. John McIntosh, a Savannah real-estate broker and a friend of Ed's, laughs as he recalls their battle. "It was this life-and-death feud. It kept ol' Ed's adrenaline flowing for years. Sometimes on his way to lunch, he would come by my office to tell me something that Les Shuman had done. He'd come in the back, and he'd just be raising hell and waving his arms around." Ed was a litigious man, and whether he was battling Central Outdoor Advertising or Shuman or the IRS (which he did on a number of occasions), he went at his adversaries with all guns blazing. He was not a graceful loser.

The one story Shuman's children will never forget was about the time their father put up a billboard on the old marshy back road leading out of Savannah on the way to Hilton Head, South Carolina. Tickled that he had rented the lot dirt cheap, Shuman sank his posts into the swampland.

A few weeks later, when Shuman and his eldest son were out "riding the plant"—checking on their various billboards—they drove down the road to Hilton Head, turned a corner, and came to a dead stop. There, in the middle of nowhere, Turner had erected another board and placed it squarely in front of theirs. Young Shuman says he'll never forget the sight of his father, up to his ankles in mud, his hands on his hips, staring up at the boards and shaking his head in disbelief.

The Turner children were nine and six when they arrived in Savannah in 1947. Mary Jean was just beginning school, but Teddy already had a record as a disciplinary problem. "He was always mischievous," his mother acknowledged. Convinced that the boy needed to learn to toe the line, his father decided to send him to a military school.

It cannot have been an easy transition for Teddy. He had been forced to leave behind everything that was familiar to him in Cincinnati, and now his father was talking about sending him away from Savannah. Despite Florence's objections that he was too young, Teddy was shipped off to Atlanta and the Georgia Military Academy, which had a reputation for handling problem kids.

Enrolled as a fifth-grade student at GMA, the uprooted young Turner was stripped of his clothes, his family, his peace of mind. "That's being pretty alone," he says now. He was forced to put on the school's Confederate gray uniform. Needless to say, he was not a happy cadet. Every time he opened his mouth, his accent condemned him as the enemy. "It was pretty rough," he remembers. "I was from Ohio, I was a Northerner, and I didn't enroll until about six weeks after the regular term had begun, so I was coming in late."

Rumor quickly had it that the new boy was bad-mouthing General Robert E. Lee, and he was soon running for his life, pursued by a bloodthirsty platoon of about forty cadets brandishing a rope and screaming, "Kill the Yankee bastard!" Turner says, "I hid in my locker for about four hours while they were out there. They were in a rage. If they got me, they would have killed me. I was in terror." He describes what happened next this way: "I became a Confederate to survive." And when he wasn't battling his roommates, he was fighting against the regimentation. Before the school year was out, the lonely, homesick, beleaguered youngster had parted company with GMA by mutual agreement.

If life at school was tough for young Turner, life at home was no picnic either. Ed had dreams for the boy, and his difficult son was not living up to them. "My father," Ted said, "had an idea of what I should do in life and the way I should live, right down to the finest detail." As far as Ed was concerned, the only way to straighten out a young troublemaker was to spank him. Ted said, "He just believed in that old adage, 'Spare the rod and spoil the child.' " Ed had no intention of spoiling his child. He often whipped Teddy, using everything from wire coat hangers to razor strops. The boy's screams made no difference to him. Dr. Victor says, "I think he took out a lot of hostility on Ted." But apparently Ed regarded these sessions as therapy for the wayward Teddy—character-building exercises. His son would later recall his father saying over and over again, "I'm beating you to try and make you do the right thing and grow up to be somebody we're both proud of." Florence claimed that "ninety percent of the arguments I had with Ed were over him beating Ted too hard."

In the summer, the Turners rented a cottage at Savannah Beach, about eighteen miles out of town on Tybee Island. Middle-class Savannah families normally left for the beach in June when school let out and returned to town Labor Day weekend. At the beach, young Ted would wander off by himself, shuffling up and down the shore, staring out across the water. His daydreams were noble, heroic, and sad—fantasies about falling in battle at Gettysburg with the Confederate Army. He still recalls, "When I was a little kid—about nine or ten—I dreamed of dying at Pickett's charge."

Also at the beach during those years was the family of Dr. Paderewski, a Savannah dentist, whose son Jules was exactly the same age as Ted. The two boys sometimes played together, but Jules remembers Teddy as something of a loner: "Even as a kid, he marched to his own drummer. He was pretty independent. I can remember him around jetties at the beach where there would be little butterfish and things. He'd collect them. He always had an interest in fish. He had a net, and he'd scoop them up. He'd spend hours doing it."

In a way, the sea was just big enough to drown Ted's loneliness. And in a seaport like Savannah, you could feel the pull of the water everywhere, in the cobbled streets and boat ramps running down to the river, in the plaques commemorating great moments in naval history—the British battling the French in Savannah harbor, the SS *Savannah*'s 1819 achievement as the first steamship to cross the Atlantic. Savannah is proud of its history, proud of its reputation as the oldest city in the state, above all proud of its links to the sea—a town where a young man could grow up fascinated by boats and boating, thrilled by the intoxicating mix of derring-do and salt water. Besides, Ted's own father seemed a naval hero to. him, having volunteered and served as a naval officer during the war. Small wonder the sea appealed to Ted, evoking some youthful heroic image to which he aspired. Before long, he could be heard reciting at the top of his thin, raspy, high-pitched voice, "Ay, tear her tattered ensign down! Long has it waved on high, / And many an eye has danced to see / That banner in the sky." The following summer, Ted learned how to sail.

When the Savannah Yacht Club reopened after the war at a new location on the Wilmington River at Bradley Point, one of its newest members was Ed Turner. Ed had just bought a 45-foot ketch-rigged yawl and named it after his daughter. Recognizing Ed's skill as a salesman, the yacht club's board of stewards asked him to organize a youth sailing program for the members' children. Ed agreed, with the proviso that Johnny Baker, a young man who happened to be an excellent sailor, handle the instructional part. The deal was struck. Somehow, Ed convinced a number of members that their children might end up in the local pool hall or jail if the parents didn't spend between $200 and $500 each for a sailboat. "He was a damn good salesman," Baker recalls. "To sell those billboards you got to be a good salesman."

Initially, Ed himself had been rather reluctant to buy a boat for his own son. But his business friend John McIntosh, who happened to be one of the stewards of the yacht club and was instrumental in getting the junior sailing program started, helped to convince him. "I think early in the game he thought it was a frivolous present for Ted," McIntosh recalls.

The sailboats that the members bought for their children were compact 13-foot plywood dinghies called Penguins. While the other kids in the Penguin fleet had boats that were silver or white or canary yellow, Ted painted his pitch-black and named it *The Black*

Cat. He enjoyed thinking of himself as the bad boy, the outlaw, the renegade. He was a desperado capable of anything in sailing, and maybe just a wee bit superstitious, too. This was not a boy who saw himself as teacher's pet, or Mr. Congeniality. A few years later, he would get another boat and call it *Pariah*.

Having paid $300 for his son's craft, Ed turned him over to Baker and his young assistant James Hardee, a recent graduate of the University of Georgia who had just landed a job in Ed's advertising company. Up in the sail loft, out on the dock, they drilled the boys on the different features of a sailing boat, the rules and tactics of racing. But the majority of the time, they were out on the water.

Hardee has a strong image of the eleven-year-old Teddy—the same height as the other boys but definitely scrawnier. He wasn't an especially fast learner, but it seemed as if he had more to prove; Teddy Turner was aggressive in anything he did on the water. All he wanted to do was win. He had little patience or desire to learn the mechanics of the boat, its nuts and bolts. According to his instructor, "He really didn't have an interest in that. He wanted somebody else to do that for him . . . to fix the boat. He just wanted to get in and win a race."

Unfortunately for the boy, young dreams of glory are seldom rewarded with real medals. Against the other boys, Teddy proved to be only an average sailor, his competitive fire yielding few results. "My first eight years of sailing," he admits, "I didn't even win my club championship."

During his first summer of sailing, his Penguin was underwater so often that it was more like a submarine than a sailboat. The Penguin is not the easiest boat to handle and can, at times, be downright unforgiving. Unlike self-righting keel boats, which flip back as soon as they turn over, a flipped Penguin simply fills up with water like a cistern. And Ted kept filling his up with water. The yacht club had bought an old navy lifeboat they named the *Penguin Mother,* and Charlie Anderson, the dockmaster, would regularly fish Ted out of the river and tow his sunken Penguin in to shore. There the boy would bail it out, and back he'd go again.

One day, at the very outset of a race, Ted flipped his boat over within yards of the starting line, drenching his sail. Somehow he managed to get the dinghy bailed out in time and went on to win the race. After that, according to John McIntosh, "he developed the habit of turning over, sort of accidentally on purpose, before a race." If turning over before a race somehow gave him an edge, the skipper of *The Black Cat* would do whatever it took to win. "Ever since he was a little boy," his mother recalled, "Ted's always been trying to win. When he was first sailing Penguins the wind would come up, and all the other kids would get their sails down and wait for a tow. Not Ted. He would just keep going until he turned over. . . . He doesn't get it from me. I never cared if I won or not."

Baker remembers how mad Ted would get when he lost. But despite his failures and being labeled by the other kids as "Turnover Teddy" and "The Capsize Kid," there was something about sailboat racing that he loved. Teddy Turner was no ballplayer—he had no skills on the gridiron or up at the plate. In fact, he was almost a klutz. He himself says, "I didn't have the ability to play baseball. Couldn't swim—almost drowned. I tried track—ran the hundred-yard dash in about fifteen seconds." But sailing was something he knew he could do. He was determined to succeed.

That summer he was tenacious. Turner recalls, "I just kept working and working and working." He calls it "the secret of my success." And young Ted needed to be successful if he wanted to win his father's approval. Though his dad provided him with the boat and the lessons, Ed had no sympathy for failure. Any kid of his had to be a winner. He felt that pats on the back and comments such as "Sorry, Son, better luck next time" only nurtured also-rans and pantywaists. One club member recalls that on one of the few occasions when Ed was actually there to watch his son race, he took the drenched loser by the arm and growled, "Now I'm gonna tell ya where you went wrong!"

The Penguin—though it can be handled by one—was usually sailed by two, a skipper and crew, in the yacht club's junior races. From the very beginning, Ted was always the skipper. "I think," says Hardee, laughing, "he was skipper when he got out of the crib." Though he got along with the other boys and was a leader, Ted was not a leader who had many followers. He found his fun in heading the charge, playing Caesar at Alesia or Napoleon at Lodi, but he made no great effort to befriend the troops.

Bunky Helfrich, a neighbor, was one of his few close friends and often crewed for him. He would talk about how Ted, as skipper, was always looking for some "secret" way to beat the competition, some clever strategy to achieve an advantage. It was a habit of mind that he easily applied to his relationships with the adult world. Lacking power, he sought to gain it by manipulation.

Hardee tells the story of a race in Beaufort, South Carolina, that Teddy and the others in his Penguin class wanted desperately to compete in. The young sailing instructor had mixed feelings about taking so many hyperactive, mischievous "little devils" on such a long trip. "But somehow Teddy finagled to get me to say that I'd be glad to take them over there in Teddy's father's boat [if his father agreed]. Well, Teddy, he runs home and says, 'Daddy, Jimmy Hardee wants to know if you'd let him use your boat to take us to Beaufort.'" Hardee shakes his head, "He was always playing the angles." Imagine Hardee's surprise when Teddy's father agreed to lend him both the boat and Jimmy Brown to help him look after the children.

Though seventeen-year-old Jimmy Brown had originally been hired by Ed to take care of his boat, he would eventually develop into a much more important figure for the Turner household. He was to become their cook, chauffeur, crew, companion, nurse-maid—in short, the family factotum. He was dedicated to Ed. And when things began to fall apart for the Turners, Jimmy Brown did what he could to help hold them together. According to Hardee, who knew him well, Brown spent more time with Teddy and his sister than their parents did. He would take them hiking, fishing, and sailing. He was the boy's earliest sailing teacher. "He *raised* those two children," Hardee says. One of the novelists that Ed Turner read and greatly admired was a fellow Mississippian, William Faulkner. Like Faulkner's Dilsey Gibson, the faithful black servant in *The Sound and the Fury* whose fierce loyalty shields the Compsons, Jimmy Brown supplied some of the same glue to help hold this family together.

Ed often talked to friends like John McIntosh about what he expected of his son. Ed had a plan for the boy. He wanted him to go into the family business. He dreamed his son would be one of the best in outdoor advertising. But in order to realize that dream, Ed believed that his son shouldn't expect everything to fall into his privileged lap. In fact, he should expect very little in the way of warmth, or praise, or special treatment. Ed wanted his boy to be ambitious and, according to Ted's first wife, Judy, insecure: "Ed's theory was that you raise a child to be insecure, and from that

comes greatness." Ed said, "You don't praise them all the time.
You don't give them a lot of money. They have to go out and earn
their own. Make their own way." If Ed's notion was, as she implies,
that self-doubt leads to "greatness," he failed to take into account
the human cost of his experiment. McIntosh says, "Ed was always
raising the bar. I really think he expected a level of performance
that's not normal for a kid." Jimmy Brown agrees: "Mr. Ed was
very rough on Ted. I used to feel bad for him, to tell the truth."

A hard, demanding father, Ed was an equally harsh taskmaster
with the people who worked for him. He thought nothing of having
a sales meeting for his staff at five in the morning on a Saturday,
especially when he was taking the offensive against Leiston Shuman.
According to McIntosh, "He would rant at them, and rave, and
preach, and carry on almost like an evangelist." And he was just
as hard on himself. He personally handled all of his firm's contact
work with advertising agencies and would frequently have to go
on the road. He was, of necessity, away from home a great deal.

Florence didn't like his traveling, but what could she do? It was
Ed's business, and Ed was a good provider. Florence didn't care
for his drinking either. What had started out as moderate social
tippling was becoming more and more of a social embarrassment.
He'd have parties aboard his boat for eight or ten couples, and
everybody would be drinking. To make Florence think he was only
having a beer or two, Ed would empty a beer can and fill it with
scotch. It was what John McIntosh calls "his little trick." He'd take
his guests for moonlight boat rides, and before the evening was
over he'd be leading the singing. A great raconteur, he loved sto-
rytelling, and soon poetry would come rolling off his tongue as if
he had dreamed it. It wasn't so bad when he got sloppily senti-
mental, but when he was really plastered, he could become nasty
and belligerent. At the yacht club, his behavior was eventually so
notorious that no sooner did he come in and sit down at the bar
than the older kids began placing bets on how long it would be
before Mr. Turner got into a fistfight.

There were some who agreed with Johnny Baker that when Ed
was sober, he "was the most charming, the nicest guy in the world."
On the other hand, knowing how quarrelsome Ed could become,
Baker said, "I always avoided him when I knew he was drinking."
But as long as Florence wished to remain married to Ed, she didn't
have that option. His womanizing put a further strain on their

marriage. According to one of Florence's friends, Ed Turner always had "an eye for beautiful gals." And since he was a big, good-looking, successful, generous man, the gals often reciprocated. There were arguments, of course, including one particularly ugly one in public at the yacht club, and Florence threatened to leave.

Another bone of contention was Teddy's education. The boy had been enrolled in Charles Ellis, the local public school, which was only a few blocks away from the Turner residence on 46th Street. Charles Ellis was, as the sign in front of it proclaimed, "The Home of the Bobcats." Though more of a wildcat than a bobcat, Ted had calmed down a bit and successfully lasted out the school year. He was clearly glad to be home, and his mother was glad to have him.

But whether Ed didn't trust Florence to raise his son or whether he felt that the boy needed a different environment and more discipline, he determined once again to send Teddy away to boarding school. A number of well-to-do Savannah families sent their children to McCallie, a military academy in Chattanooga, Tennessee. Emblazoned on the school's letterhead were the words "Honor Truth Duty." Ed liked the sound of that. It appealed to his conservative values. By reputation, McCallie had high academic standards, an unusually dedicated staff, and numbered among its students the sons of the best Southern families. Ed wasn't a snob, but along with his great expectations for his son, he had, according to Charles Wessels, "an innate fear that Teddy wouldn't live up to them."

Wessels' father, president of the Southern Bank & Trust Company and the Atlantic Mutual Fire Insurance Company, was one of the wealthiest men in town. He was also a graduate of McCallie. Mr. Wessels' eldest son, Frederick, was already enrolled in the school, and young Charles would be attending in another year. Ed, who bought real estate for his billboards from Wessels, was impressed by what he heard. At $1,400 for room, board, and tuition, McCallie in 1950 was expensive, but nothing that Ed couldn't afford. All he wanted was the best for his boy.

Ed made an appointment with the headmaster of the school and went up to Chattanooga to visit the campus. He liked what he saw but was surprised when Mr. McCallie didn't seem particularly eager to accept his son as a seventh-grade boarding student. They sat in his comfortable, well-appointed office with its leather chairs and wood paneling while the headmaster explained that all of their seventh graders lived in town. An eleven-year-old was too young to be a boarder; it was school policy. But Ed Turner didn't think it was too young at all. Ever the persuasive salesman, he pulled out all of his magic tricks to sell his son, and before Ed left for home, Teddy had been accepted in the fall semester of 1950 as McCallie's only seventh-grade boarding student.

Florence didn't want her boy to go, but eventually she gave in. "There was nothing I could do," she said. "I cried. It did no good. Ed told me he had the purse strings. I had to do what he said."

When Teddy learned that he was being sent away again, he was crushed. He had already had one disastrous experience at a military school. And this new one was in Chattanooga, one hundred miles farther away from home than GMA had been, and in another state. He was furious. He had no intention of going to McCallie. He absolutely refused to go.

CHAPTER 4

IN SEPTEMBER OF 1950, when Teddy Turner arrived in Chattanooga and was driven up Missionary Ridge to the entrance of the McCallie School, his face was as white as bleached bones. What did he know about Braxton Bragg's glorious rout of Union troops at the nearby battle of Chickamauga and how he laid siege to the town from those very heights? Teddy Turner himself felt utterly defeated. But he had already been kicked out of one military school and, with a little luck, he just might be kicked out of another.

Of the forty new seventh-graders that year, thirty-nine were day students who lived in Chattanooga. "I was certainly the only boarding student," Turner says, still remembering how it felt to see the other boys going home to their families every afternoon. He was on his own.

Turner was assigned to room with some older students in a dormitory called Douglas Hall, a three-story brick building that housed in its basement the school laundry and heating plant. From time to time, clouds of steam from the boiler would rise up whimsically and without warning into the first floor. Initially, the closest thing to a family Ted had at McCallie was that of his dorm master, Elliott Tourett Schmidt. Schmidt, now retired, was then head of the history department and also taught public speaking. A heavyset, big-faced, toothy man, the energetic Schmidt has a throaty voice and a hearty manner. He was not insensitive to the fact that there were youngsters in Douglas who felt themselves dispossessed from their homes. But Teddy, according to Schmidt, was never given to self-pity. What he was given to was getting into trouble.

The scrawny kid from Georgia had few friends at the start. Another instructor, recalling those early years, says, "Ted Turner's name was sort of a joke." He was somebody people told stories about rather than took seriously. They nicknamed him "Terrible Ted." He seemed to be constantly into things, seeing how far he could stretch the rubber band before it would snap.

The beds in Douglas Hall were double-deckers and Turner, the new boy, was assigned to a top bunk. It occurred to him that in order not to have to get up in the dark on cold winter mornings, he could weld a brass chain from the lightbulb to the metal frame of his bedpost, so that he could turn on the light from under the covers. Schmidt can still hardly believe what Teddy did. "He blew out just about every circuit in the building." Even more astonishing to his dorm master was that Ted appeared not to think he had done anything at all outrageous. "What's wrong, sir? All I was trying to do was—" Schmidt tried to control himself. "Yeah, but Ted, you could have electrocuted yourself. You could have started a fire in the circuits. You could have burned down the building."

Since McCallie was a military school, there were, of course, inspections. Everything had to be spic-and-span. Nothing out of place or, worse yet, "exotic." One Saturday, the new cadet brought an aluminum pan filled with dirt and grass up to his room. Spotting the offending pan, Mr. Schmidt said, "Ted, it's nice to have grass *outside,* but inside it's not part of the military structure. Is that clear?" Turner replied, "I'll keep it neat, sir." And with that, he produced a pair of cuticle scissors and promised to cut it regularly.

One day not too many weeks later, Schmidt, discovering that the dorm's lone seventh-grade boarder had adopted a stray animal as a pet, decided to look the other way. Before long, the boy's room held a zoo of creatures he had decided to save. "Ted tried to take in dogs, birds with broken wings, gerbils, snakes, anything he found," Schmidt recalls.

For these and countless other infractions of the rules there were demerits, and the punishment for demerits was the hated bullring. A contemporary of Ted at McCallie, Lewis Holland, says that he can't imagine a more disciplined environment. "Every demerit you got," Holland explains, "meant that on Saturday afternoon you walked the bullring in full-dress uniform and carried a weapon."

The boys in the senior school carried regular 12-to-14-pound Army rifles; the seventh and eighth graders in the junior school toted wooden facsimiles. Ted, with his little wooden rifle propped on his shoulder, would march around the oval track—"doing donuts around the bullring"—long after all the other less flagrant offenders had paid their penance and departed. It seemed to his mother as if "I had to buy him new shoes every time he came home. He wore them out walking punishment tours." She summed up her son's first year at the school this way: "Ted hated McCallie. He was a devil there."

Boarding students were permitted to go home only on the major holidays, such as Thanksgiving and Christmas. Ted counted the days. When leaving or returning to school, the boys had to wear their military garb, a midnight-blue McCallie uniform with four gold-colored buttons down the front of the jacket and on each lapel a pin with crossed muskets. The shirt was white, the tie black. The regulation service cap was worn squarely on the head. On some occasions, Jimmy Brown would be sent to pick Ted up and drive him back. Other times, he went by plane with boys from the Savannah area. While the others were eager to return to McCallie following vacation, Ted dreaded the thought. Charles Wessels recalls, "He would always find some excuse not to go."

Typically, at the close of one such holiday, Ted and his mother were given a ride out to the airport with Mrs. Wessels and her sons in their limousine. Just as the plane was ready for boarding, Ted announced that he had lost his ticket. Mrs. Turner became very upset. "Ted, you *are* going," she insisted. Mrs. Wessels, too, began to chide the boy. Young Ted made a big show of looking through his pockets but found nothing. "Well," said his mother, "we'll just have to buy you another one." Seeing that the game was up, Ted suddenly discovered that his ticket had been tucked into the inside of his hat all along. He was hurried aboard.

There is an honor system at McCallie that says, "You do not lie, cheat, or steal." Though Ted as a younger boy at the school was always in trouble, he was, according to his instructors, "as honest as he could be." In fact, he almost seemed to take a perverse pleasure in owning up to his misdeeds, a quality that continues to this day. Whatever his foibles or sins, lying has never been one of them. In this context, the plane-ticket anecdote suggests just how desperate he must have been not to go. If the scene at the airport wasn't so funny, it would be heartbreaking. No matter how tense things might be at the Turners, Ted yearned to stay home.

In recent years, Turner has been both enthusiastic and critical about his education at McCallie, which is no longer a military school. On return visits there, addressing the student body, he has said, "I thought the education I got here was terrific. I learned to think here for the first time." But elsewhere, he has expressed reservations about the "Fundamentalist Christian-type schooling" he received. Any school that could produce two such different graduates as the evangelical president of the Christian Broadcasting Network, Pat Robertson, and former president of MIT James Killian might well give rise to such divided opinions.

Founded in 1905 by Spencer and Park McCallie (sons of Dr. Thomas H. McCallie, a Presbyterian minister), McCallie began as a Christian preparatory school for boys. Forty-five years later, when Ted arrived, the emphasis on Christian principles was to be found in almost every aspect of the school, from faculty to curriculum, from daily chapel to the column "Christian Contemplations" in the student newspaper. Of the four courses a semester that junior-school boys took, one was in the Old and New Testaments.

Ted's bible instructor was John Strang, who was then a new member of the faculty. John Strang likes boys as much as Dickens' Betsey Trotwood disliked them, and in his warmhearted sincerity he puts one in mind of her friend, the amiably eccentric Mr. Dick. Currently in his mid-seventies, Strang continues to teach the bible course in the junior school and always seems to have a pocket full of sourball candies for his students. They still fascinate him, and none did more so than Ted Turner.

He clearly remembers the first day Ted walked into his classroom: "He was frail-looking and his face was very, very white." On the blackboard, Strang had written the McCallie School motto. "I figure they should know the motto first of all," he says. "And McCallie's motto is, 'Man's chief end is to glorify God and to enjoy Him forever.' It's not an actual verse in the bible. It's found in the shorter catechism of the Presbyterian Church." Then beneath it, he had placed the motto of their crosstown archrival, Baylor. "'But seek ye first the kingdom of God and His righteousness, and all these things will be added unto you.'"

"Each year the boys want to know why they have to learn that one, too. 'Well, two reasons,' I tell them. 'First, it's a wonderful verse. And secondly, you'll know something that even the Baylor boys—most of them—don't know.' So that makes it all right." Strang left the two mottos on the board for a week. Friday, he erased them. As Ted's class marched in late Friday afternoon, they stood at attention and Strang told them to recite from memory the McCallie motto. "They gave it perfectly," he reports, his face lighting up as he remembers. "Now give Baylor's," he said. In one voice, they replied, "Baylor school motto. St. Matthew 6:33. 'All we like sheep have gone astray,'" and they all burst out laughing. Strang soon discovered that it was young Turner who had found the verse about sheep and passed it around to his classmates. The bible teacher shakes his head in wonder and beams. "That's just Ted!"

Turner grew fond of his bible teacher, who was known affectionately by the boys as "Yo." Yo is the sort of teacher who has incredible patience and is more likely to be sad than angry if a boy falls short of his expectations. No matter how many tricks Ted played on him, Strang never seemed to take offense. One day Ted raised his hand in class, stood up, and said, "I don't think I'd like to go to Heaven."

The class turned to see if this was yet another of Turner's jokes. But he was serious, so Strang was, too. "Why?" he asked him. "Well, I just can't see myself sittin' on a cloud and playin' a harp day in and day out."

The boys waited expectantly as Yo thought this over. Then he asked, "What's the best day you can imagine?" That was easy for Ted. "Being with my best friend on a golf course with some money in my pocket and no one before us, no one behind us, just ourselves on the greens. Then we play another round. Then we go swimming. Then we go to the clubhouse and get a sandwich."

"Ted, that's wonderful!" Strang said. He then had him turn his bible to 1 Corinthians and read, "Eye has not seen, nor ear heard, nor has it entered the heart of man, the things"—"Plural," Strang emphasized—"that God has prepared for those that love him." Then he looked at the boy and explained, "To me, Ted, that means that either there will be the best golf course that you have ever seen, or something so much better you'll forget all about golf."

Ted nodded his head thoughtfully. Though he didn't smile, he plainly liked what he heard. In those early years, Ted would become, in his own words, "a very religious person." He remembers the series of fiery-eyed evangelical preachers who arrived at chapel with their urgent message that people who weren't born again and hadn't taken Christ as their personal savior were going to burn in Hell forever. And then they'd say, "Now would any of you boys like to come forward and profess your faith?" Several times, Ted stepped forward.

"Well," Turner says, "that's a pretty clear-cut choice, if that's true, so I was thinking about all these people in other parts of the world where they didn't have access to Christianity, they—you know, I'm just a little kid, you know. I thought, Geez, these people are all going to Hell, and I thought I'd go out and try to help as many of them as I can." If the idea of saving little animals and birds appealed to the young boy, saving lost souls took an even stronger grip on his imagination, and in a few years he would seriously entertain the possibility of doing missionary work.

As for the rest of the McCallie junior-school curriculum, when it came to English, math, and history, Turner was not much of a student. "Just average," his math instructor, Houston Patterson, reported. "I don't think he was particularly academically motivated. I'm not sure he ever was." Ted himself says that in the early years in particular, "I was content to get Cs."

Where sports were concerned, he was interested but mediocre. He tried "termite" football and boxing that first year. Not being especially big or fleet of foot or well coordinated, Ted was something of a dud as an athlete. He finally quit boxing when, as Elliott Schmidt put it, he "almost got his head knocked off." Ted says, "I tried and I tried, but I could see it wouldn't work." The fact that he was extraordinarily competitive only heightened the level of his frustration.

At the end of May, Ed Turner arrived to take his boy and his trunks back to Savannah, but discovered that he couldn't. Ted was not permitted to leave, not allowed to receive his final grades until he had finished walking off all the demerits he had accrued during the course of the year. While Ted marched, his father took a room downtown at the Read House and waited. John Strang estimates that "if they had stretched that track out, he'd have walked all the way to Lookout Mountain and back," a trek of about twenty miles.

Ed Turner probably didn't mind waiting. His money was buying exactly what he wanted for his son—an all-white college preparatory school with high academic standards, rigid military discipline, and Southern conservative Christian values. The boy seemed well on the way to becoming as ultraconservative as his father. What Ed didn't care for were his son's grades. That summer he had Ted read a book every two days. The boy had to know more about the great books, had to be able to quote poetry. Ted admits, "I never considered not doing it, because I was instructed with wire coat hangers when I didn't get them read."

Also that summer, twelve-year-old Ted for the first time went to work for his father's advertising company—plucking weeds, creosoting posts, painting the wooden billboards. It was hard, dull work, for which he got paid very little. Eventually, Ed would raise the amount, but, as if he were operating a company store, he then began charging his son $25 a week for room and board. When Ted complained, his father advised him to check around town. If he could find a better deal anywhere else in Savannah, Ed told him to take it and get out. Ed had no intention of mollycoddling his kid. Either he earned his keep, or he sponged off some dumb sucker elsewhere. It was all part of the plan.

Though Ed Turner treated him sternly, Ted loved him and was convinced that his father reciprocated that love. But that summer his parents were often fighting with one another. During these nasty exchanges, the boy hated to watch his mother being browbeaten. Ted's few friends could see him being torn between his parents. Charles Wessels thought that "Ted was *very* close to his mother. I know it because I've seen the way he acted around her. I think he felt like he had to take her side."

Wessels' brother Frederick was a year older than Ted, and though born deaf, he had learned how to speak. Ted had adopted him as if he were another wounded animal he wanted to save. Frederick, to his credit, recognized as much. "In a way, I think he had sympathy for me because I'm deaf. I think he was looking after me the whole time."

There was one visit to the Turners that stands out in Frederick's mind. At that time, they were living at 302 East 46th Street, in a large, comfortable upstairs apartment that they rented in a two-story red-brick building with a Tudor top. Frederick recalls, "It had a living room, a dining room, three bedrooms, two baths. It was really nice."

On this particular occasion, Ted had invited him over for lunch, and Mr. and Mrs. Turner were there, too. "Well, we were sitting at this table," Wessels remembers, "and while we were eating lunch, his parents, I could see, were staring at each other getting ready to start an argument." Every word between them seemed hostile, threatening. They glared at each other across the table. Wessels liked Mr. and Mrs. Turner, but the obvious tension between the two made him feel terribly uncomfortable. And his talkative friend was unnaturally silent. Finally, Ted looked at him and mouthed, "Come on, let's go in the living room." Frederick may have been deaf, but it was clear to him that his unhappy young friend was, in many ways, in worse shape than he was. "I felt sorry for him," Wessels recalls.

Back at McCallie, Ted, according to his teachers, became even more of a "holy hell-raiser" for the next two years, displaying "a general disregard for authority." He'd sit in the back row of his Latin class, where, according to one of his classmates, "all mediocre students prefer to be," and when caught with a "pony" on his lap was not the least repentant. In the third-floor study hall one balmy spring evening, he climbed out the window when the proctor's head was turned, scrambled down the fire escape, went for a quick swim and a hamburger, and then climbed back up and in the window without having been missed.

No matter how outrageous his escapades, McCallie refused to reject Ted, and in time he grew to love the place. Unlike his home, it was utterly predictable. More than forty years later, the McCallie daily schedule is still fixed in the memories of its graduates: "Breakfast was between 7:00 and 7:30. Then we went to chapel. Then we went to classes. Then we had lunch at 12:30. After that, we had two more periods of classes, and then athletics or whatever you wanted to do until 5:30. Dinner was a sitdown meal between 6:00 and 6:30. And then, if you didn't have a B average, you had to go to a supervised study hall." It was a system you could depend on any hour of the day or night, and to this day, according to his aides, Turner still favors the same kind of regimented timetable for his meals.

Ted even became enthusiastic about the military part of the program, which was rather unusual among the students. It was considered "cool" at the time not to like it. "But Ted liked to wear

the uniform," Houston Patterson recalls. "He liked spit and polish—inspections and so forth. His own life tended to be set up that
way. Everything black and white." What he obviously didn't like
were the gut-wrenching surprises at home, the sudden violent outbursts between his parents, the public humiliation of seeing his old
man picking fights in bars. Ted has made no secret about how
important McCallie was for him, and not only in forming his values.
"I've always felt," he said, "that McCallie was kind of like my
second home."

Ted frankly acknowledges what a terrible cadet he was in the
beginning. "But then," he says, savoring his accomplishment, "I
turned it around. I'd been the worst cadet, and I determined to be
the best." Cast in melodramatic language, it's the sort of sudden
theatrical conversion that came to have a particular appeal for him:
Turner battling the odds and converting certain defeat into stunning
victory. "I am the master of my fate: / I am the captain of my
soul." In reality, of course, it wasn't quite that way. Still, over
time, there was a gradual change in his attitude toward McCallie
and vice versa.

Perhaps it began in 1953, Ted's first year in the senior school,
when he got to know Houston Patterson. "Up until then," Patterson says, "he'd just been a kind of funny little kid whom I knew
by name." One day Turner mentioned that he could sail; Patterson
had just learned how to sail the previous summer. They decided
they'd like to do some sailing together. Shortly after the Christmas
break, Patterson returned to Chattanooga with an old sailboat he
had bought. "And Ted and I worked on it and fixed it up and put
it in the water, and just the two of us would go sailing."

An ex-Navy man like Ted's father, Patterson is described in the
student yearbook as "friend, counselor, and a doggone good guy"
with strong Christian values, always ready "to give up most or all
of his spare time" to the boys. The two of them sailed on Chickamauga Lake, the math teacher and the teenage boy, both wearing
T-shirts, both barefoot. Realizing that the boy had a lot more experience than he had, Patterson told him, "You're the skipper. I'll
do what you say."

Ted took over, eager to show what he could do. Patterson says,
"As far as I'm concerned, he was a pretty good teacher. If I did
something wrong, he'd tell me. If he wanted me to do something,

he didn't hesitate to say so." It was obvious that Ted loved teaching the teacher. There was one memorable occasion when, according to Patterson, Turner demonstrated a wisdom beyond his years, and given his wild reputation at the time, "it certainly would have been a surprise to a lot of people at McCallie, had they witnessed it."

The two of them had just arrived at the marina, looking forward to a pleasant Saturday of sailing, when Ted announced, "We can't sail today."

"Why not?" Patterson demanded to know.

"The wind's too strong."

Patterson thought that was the dumbest thing he had ever heard. A sailboat *needs* wind! He brushed the boy's caution aside. "Let's go. It'll be exciting. Come on!"

But on this occasion, the devil-may-care Ted held firm, refusing to budge. "No," he told the teacher. "Too much wind. We can't do it." Patterson reluctantly decided to take his advice. He would in time learn the prudence of young Ted's actions—how tough it is to gauge unsafe wind conditions, how easily little centerboard boats can flip over.

Before the year was over, about six or seven other boys had joined Turner, and under the direction of Houston Patterson, who sacrificed his own family's weekends to the group, McCallie for the first time had a sailing club. By Ted's senior year, club membership had grown to nearly fifty, with a fleet of seven sloops and three catboats, and he had become one of its most skilled and reliable members.

Another event that occurred during 1953 was to have a profound effect on Ted. Back in Savannah, his twelve-year-old sister suddenly took ill. The family doctor was called. After extensive testing, Dr. Fenwick Nichols diagnosed Mary Jean as suffering from systemic lupus erythematosus, a potentially fatal form of the disease. It is an autoimmune disorder in which the body's immune system, for unknown reasons, attacks the connective tissue as though it were foreign. Ed's friend Dr. Irving Victor comforted him as best he could, but he knew that even though the child's pains, weakness, nausea, and other symptoms might temporarily subside, she would probably live only a few more years. Victor remembers this period as "a tough time for everybody. Terrible. It was a tough time for

Florence, and it was a tough time for Ed. It was very tough for Teddy."

Few people ever saw Ted's sister after that, but many around Savannah came to hear of her illness. She was very sick—bedridden at home in a separate room with the door closed. The Turners tended not to speak about her, as if she were the family's guilty secret. But when visitors came to the apartment on 46th Street, they noticed that Ted was always going into Mary Jean's room to talk to her.

Because lupus weakens the immune system, its victims are subject to a host of opportunistic diseases, and Mary Jean, in time, contracted encephalitis. She was taken to see specialists at Johns Hopkins in Baltimore, where for weeks she lay in a coma. Though she recovered, she had suffered brain damage, and when she was brought home to Savannah, she was in considerable pain and given to seizures. People who knew the Turners then could in later years still recall hearing Mary Jean's screams. "God, let me die, let me die!" she would shout, as she thrashed about.

It was impossible for any member of the family to avoid the pain, though Ted would subsequently try to distance himself from it by turning the experience into theater. "She came out of a coma with her brain destroyed" is the way he describes what happened. "It was a horror show of major proportions. Padded rooms. Screaming at night. It was something out of *Dark Shadows*."

Ed couldn't bear it. His friend John McIntosh recalls, "It preyed on him. He did every conceivable thing, spent every conceivable dollar to try to find a solution. It was a horrible downhill ride, and it preyed on him." Ed had taken his daughter to every expert he could think of. Nothing worked. Powerless to change the situation, he decided he wanted the child out of their house. He wanted to put her in an institution, but his wife refused. Florence wasn't going to give up her daughter. She felt her husband was coldhearted, indifferent. He felt she was unreasonable, obsessive. They argued furiously. Ed finally compromised by hiring a full-time nurse and fixing up an apartment for their daughter above the garage in back of the building in which they lived—a separate little compound to keep this family tragedy at arm's length. McIntosh remembers it as "a kind of nursing home / jail. I mean the child was violent." The stresses on any marriage in such a situation are, Dr. Nichols says, enormous. Now after vacations Ted was glad to get back to McCallie.

It was about this time that Ted started to ask his classmates and teachers what he had to do "to come around." He wanted to clean up his act. According to the Wessels brothers, "If you make an effort at McCallie, they'll help you." Elliott Schmidt was one of several who did. Everybody at the school had to work on building vocabulary, and Schmidt made Ted study his list of words again and again, until he had them down. Thanks to Houston Patterson, sailing was becoming increasingly important to him. After doing his homework in study hall, he'd keep busy and stay out of trouble by reading about the sea. "I read C. S. Forester's books, and Nordhoff and Hall about ten times—*Mutiny on the Bounty, Men against the Sea,* and *Pitcairn's Island.* I read about the War of 1812, and about the *Constitution*—you know, the ship." After serving as a lowly private for three undistinguished years, Turner was elevated to the rank of corporal.

When Schmidt moved from Douglas Hall to McClellan as dorm master, Ted went with him. Clearly, he had become attached to his mentor. In one way, at least, this was a curious match, inasmuch as Schmidt was probably the only member of the faculty who was a political liberal. He describes himself with amusement as being McCallie's "bad L" for many many years. "I don't know if I made any impact on Ted," he says smiling, "but he's a liberal now in many ways."

Schmidt, in addition to teaching history and political science, also supervised the Debate Club, which consisted of only a handful of members. Among its newest additions in 1955 was Teddy Turner. Had Schmidt been coaching sumo wrestling, the sixteen-year-old boy might well have put on 400 pounds and a loincloth. He was obviously eager to win Schmidt's respect. Though at the start he was hardly one of the best debaters on the team, he worked hard to improve and that year went to tournaments in Nashville and Atlanta and Knoxville; the club finished third in each of the meets.

By the end of his junior year, Ted, though still rail thin, had grown to be almost six feet tall. He seemed more serious now, more responsible, more determined to succeed. He was promoted to sergeant and named the Neatest Cadet in the Regiment. On one occasion, Schmidt and his wife were even willing to trust him as a baby-sitter for their two young children. The kids were fine when the Schmidts returned later that evening, but with Ted there was always something unexpected, unpredictable, disquieting. Schmidt says, "We had a Shetland sheepdog—Cappy. He was as sweet as sweet could be. And after that one night's visit, he would have nothing more to do with Ted." Schmidt thought that perhaps the boy had "disciplined" the dog, but, almost as if he preferred not to know, he never asked him what happened.

Though Ted wanted to return to Elliott Schmidt's dorm for his senior year, Schmidt privately informed the faculty member in charge of boarding students that after five years he had had enough of Turner. "I thought it would be better for some other teachers to experience him," he says. His colleague replied, "El, you can't do that. If the boy likes you that much and wants to stay in your dorm, you have to consider that as a tribute and be happy."

But Schmidt wasn't happy. "That didn't convince me, and so I wrote a rather lengthy note to the headmaster about my involvements with Turner, and that I thought I'd had enough of him. And the headmaster agreed." The itemized list of Schmidt's problems with Turner over the years covered three pages. Frankly, Schmidt was exhausted. He acknowledges that when the news of the headmaster's decision reached Ted, "he was somewhat offended"; nevertheless, he had to learn to live with it. Ted began his senior year at another dorm.

Turner at seventeen was a hopeless romantic, but it's hard to imagine any young man of that age—with hormones raging—who isn't. He saw himself as a young Alexander, hero of countless battles from Issus to Guagamela, bending the future to his will. "That stuff just knocked me cold," he says. Alternately, on bad days he envisioned himself as the innocent victim of an injustice so profound that he might not make it to eighteen. It's said that on at least one occasion that year he contemplated suicide, going so far as to write a farewell note at the Read House in downtown Chattanooga before climbing out a window of the hotel onto a ledge, a victim of unrequited love.

But at seventeen, Ted discovered that strong emotions were better suited to poetry than suicide notes. McCallie, like many other schools at the time, required students to learn poems by heart, and

Ted, who found that he loved poetry as much as his father did, committed all sorts of work to memory. Keats was a favorite, and he would recite sections from *Endymion* to anyone who would listen. The poem had so much that appealed to him. Ideal beauty! Love! Immortality! Even sailing! "And, as the year / Grows lush in juicy stalks, I'll smoothly steer / My little boat. . . ."

When Ted began to write poems of his own, they were less sophisticated efforts, shorn of all pagan exoticism, and he seemed to favor a simple A,B,C,B rhyme scheme no matter what the subject. His "The Ship That Never Failed" appeared in the February 17, 1956, issue of the school paper, *The Tornado*. It tells of the confrontation at sea of two warships, one English and the other American, the latter "triumphantly" revealed as Old Ironsides in the final quatrain. The influence of Oliver Wendell Holmes is strong, almost suffocating, as young Turner celebrated heroism, patriotism, muscle, and drama on the high seas.

Two months later, the paper published "For Lack of Water," a poem totally different in mood and focus from Turner's earlier effort, but still employing the same bouncy rhyme scheme. Set in the Sahara, it describes the plight of a man wandering aimlessly across the desert and dying alone without a trace. "Awful sin had he committed? / Wrath of God had he incurred? / As he plodded on he wondered, / But his mind and thoughts were blurred." Ted seemed to be stressing the innocence of the man and the pathos of his fate. To be punished unjustly for a crime one didn't commit. To disappear without leaving a trace of achievement. To fail through weakness. ("One could see his strength was failing / Two more steps and he'd be through.") All of these were very much on Turner's mind in his last year of high school.

Ted must have asked himself what sin he had committed that his family was falling apart, what sin Mary Jean had committed to be suffering the way she was. Her illness had shaken even his father's faith. Ed Turner made no secret of his feelings. "If that's the type of God He is," he announced one day, overwhelmed by his daughter's pain, "I want nothing to do with Him." Ted, too, found it difficult to reconcile such suffering with a loving God. In *The Tornado* that fall, the "Roving Reporter" had asked what advice upper classmen would give new students, and one had replied, "I especially urge a new student to think seriously of his Christian responsibility." Of late, Ted had been doing just that. He had once thought that he might like to become a missionary, to save lost souls, but he was now having serious doubts.

Though there were emotional lows such as this during his senior year, Ted, by and large, came into his own. Houston Patterson was relieved to see that he now had a small coterie of friends: "I don't think Ted really became fully accepted by the rest of the student body until maybe his senior year." He was named Captain of Company E, one of the ten captains in the regiment.

The debating team had begun the year with great expectations, because all of its former members were returning. The debate question for that year had to do with federal aid to education. "Resolved: Governmental subsidies should be granted according to need to high-school graduates who qualify for additional training." Ted wanted to debate the same way he sailed. He'd do whatever it took to win. "I want a case that can't be beaten," he told Schmidt. "Well, Ted," the teacher said, "it almost makes it a little unethical if you do that. That's no longer debate." But Ted insisted. "If I can work it out, I want to do it that way."

Schmidt describes what happened next. "So he wrote the proposition out and diagrammed it. Everybody thought that 'need' referred to the students. He diagrammed 'need' to the federal government. Then he got Bob McCallie, who I think was head of the English department at that time, to approve the diagram. He then went down to the university [of Chattanooga] and got some professors down there to agree that need could apply to either the government or to the students." Ted built his case on the government's need. If, for example, the government felt that there was a need for more physicists in the country, it could offer grants to students interested in specializing in physics.

Turner's initiative, self-confidence, and ability to seize an advantage led to his and his teammate Bill Cook's victory in arguing for the affirmative in the East Tennessee division of the Tennessee Interscholastic Literary League. Then came Knoxville and the state finals. Turner remembers it this way: "I beat a girl in the finals. She broke down into tears because I challenged the basic premise being debated." His opponent had simply not prepared herself for such a sly approach. "It was a little below the belt," Schmidt admits, "but he won with a perfectly legal, nondebatable subject." Cook and Turner were declared the Champion Tennessee Affirmative

Team. Ecstatic, Schmidt and the boys hugged one another. It was a highlight of the year for Turner.

Another highlight was Miss Nancy Drake. Although Ted escorted Miss Silvia May, and Miss Nancy Goose, and other young ladies from the nearby Girls Preparatory School to the formal dances held at Davenport Memorial Gymnasium, it was Nancy Drake whom he was most interested in. Ted wasn't a great dancer, but Nancy didn't seem to mind. The captain of Company E looked resplendent in his gold braid, his medals, the huge gold chevrons on the arms of his uniform. Nancy's formal gown billowed around her. They danced together at the Patrons' Dance, and at the Annual Officers' Ball, where they walked arm in arm under an arch of glittering drawn sabers, and then—most important of all—they went to the final dance of the season, on June 4. Beneath the magical multicolored lights strung from the basketball rims, Ted and Nancy glided cheek to perfumed cheek across the hardwood gym floor to the music of Roy Cole's orchestra, the night filled with whispered promises and unspoken dreams, as the chaperones for the evening, Dr. and Mrs. J. P. McCallie and Dr. and Mrs. R. L. McCallie, looked on approvingly. But Ted was about to graduate and would soon be on his way north to Brown University. It lent a special poignant urgency to the evening, with its final tearful kisses and leave-taking.

The decision to attend Brown was not an easy one for Ted. His first choice was the United States Naval Academy, at Annapolis. He liked military discipline, he liked uniforms, he liked the sea. But in this, as in all things that concerned his boy, Ed Turner wanted control. He gave a lot of thought to where his son should continue his education. Jimmy Hardee, who worked for Ed, remembers him talking about the Naval Academy. Ed was thinking that maybe Teddy should go to Annapolis "as long as he was interested in it," but he finally rejected the idea. His son would attend Harvard. Ed was a great believer in marquee value and prestige—in education, as in everything else—and as far as he was concerned Harvard had it.

Though Ted's grades had improved in his last two years, they were still far from scintillating. His IQ was 128, which, while respectable, in no way qualified him for special consideration as a genius. On the other hand, it was nine points better than that of

Harvard graduate John Kennedy. Kennedy's father, however, had not only gone to Harvard himself but was U.S. ambassador to Great Britain. Harvard turned Ted down. The choice of Brown was softened for him by the fact that it had an excellent inter-collegiate sailing team.

At graduation from McCallie, Ted was awarded the Holton Harris Oratorical Medal. Elliott Schmidt presented him with the award, shaking his hand enthusiastically. Ted may not have been a cum laude student or a member of the Monogram Club of varsity athletes or a senior-class officer, but Ed Turner was still proud of what his boy had achieved in the six years he was at McCallie. As a tangible expression of his pleasure, he bought Ted a new Lightning-class sailboat to replace his old one. It was a Lippincott Lightning, the top of the line in 1956, and cost over $2,000—about the same as a new Cadillac. But lest his son get profligate notions, Ed insisted that the boy fork over half the price from the money he had saved while working for him the past five years. Ted wasn't surprised. It was his same old dad, still running the company store.

Before leaving McCallie, Ted gave Houston Patterson his old Lightning as a donation to the sailing club. Other McCallie boys might enjoy it as much as he had. He hoped so anyway. It was important to him to leave something of himself—something he cared about—behind at the school. Turner said, "When I left there, I cried. It was such a perfect place."

THAT SUMMER IN SAVANNAH, Ted once again pulled up weeds by the back braces and postholes of his father's billboard signs. In the sweltering heat of a Georgia July, with the sweat running down his bony frame and gnats and mosquitos swirling around him, Ted pushed the mowing machine up and down one lot after another. Above loomed the mammoth signs: "Taystee Bread for *Tasty* Toast," "Mom—We Need More Ritz!," "Please Fill Our Tummies With Birds Eye Yummies!" And at the bottom of every sign was the oval emblem with the name "Turner."

Knee-deep in weeds, Ted had dreams that had nothing to do with his father's billboards. He dreamed of glory. A few years earlier, he had longed for the return of World War II so that he could be a fighter pilot "and go up there and shoot those guys down. It just looked exciting." Turner says, "I wanted to be a kind of knight in shining armor, and at that time Alexander the Great, Napoleon, and George Washington—the military leaders—were [the heroes] I looked up to the most."

But now, at seventeen, what he longed to do most was sail his new boat, which he named *Greased Lightning,* in a major international competition. The 1956 International Lightning Regatta was scheduled for the end of the summer in Canada, and Ted had made up his mind to compete in it. He dreamed of being the youngest skipper to win his first international race.

A Lightning-class boat is 19 feet long and, in addition to its skipper, has a crew of two. Ted easily convinced his friends Frederick Wessels and Bunky Helfrich that a trip to Canada would be a wonderful adventure. He somehow persuaded his father that Point Abino, Ontario, was on the way to Providence, Rhode Island, and Brown. In September, the three friends loaded *Greased Lightning* on a trailer behind Ted's car and headed north.

Ed knew all about his son's enthusiasms, his aspirations. He also knew that the boy could be wild at times. Worried about his safety, he had placed a governor on the engine of Ted's Plymouth—a device that would prevent the car from being driven over the speed limit. Along with everything else, Ed wanted to control his son's driving, too. Unhappy with this arrangement, Ted went to a gas station and learned how to disable the governor. Under his father's watchful eye, he would drive off at a sedate 35 miles an hour. Once out on the open road, he slammed his foot down on the pedal and rocketed away.

Ted's plan was to get to Canada as quickly and as cheaply as possible. They'd drive nonstop through the night. Somewhere in the hills of West Virginia at three o'clock in the morning, they began to run out of gas. Bumping along the darkened highway, the boys spotted a big two-story farmhouse with a single gas pump out front. The place was pitch black and so quiet it was eerie—like something out of a horror movie. Frederick felt a little spooked. But not Ted. Jumping out of the car, he ran across the road and banged on the door. "Hey, wake up!" he shouted at the top of his lungs. "Wake up! We don't have any gas." A light went on upstairs. Ted's companions half expected to see somebody poke a gun out the window. But the man who came down and opened the door couldn't have been nicer.

Soon they were on their way again, driving past Niagara Falls to the north shore of Lake Erie, and finally checking into a motel near the Buffalo Canoe Club, the race headquarters. Traveling on a budget, the three of them shared a room for $15 a night. Ted couldn't get over it. He thought that the place was ridiculously expensive. Frederick says, "Ted was very cheap at that time. He was watching out for his nickels and dimes. He thought his father never gave him enough money."

When they arrived at the Buffalo Canoe Club, Ted was still

carrying on about the expense. As they signed in, he made such a commotion about the $15 that others at the club turned to stare. Hearing how upset he was, a local family felt sorry for him. Assuming that the boys were short of cash, certainly not the children of millionaires, they invited them to stay in their house.

That night the young Georgians learned a little about Canadian weather. The temperature after midnight dropped to 22° F., and the next morning at breakfast somebody in the family asked casually if they had antifreeze in their car. Ted didn't. Fearing the worst, he cried, "Oh, my daddy's gonna get mad at me. My daddy's gonna kill me!"

The car turned out to be all right, but the racing was another matter altogether. In this, his first international competition, Ted appeared all business, but inside he was wound tight as a hairspring. He was up against some of the best Lightning sailors in the world. He recalls, "I was so impressed just to be there and meet all the real big cheeses." As he watched the other forty-four boats coming up to the starting line, he jockeyed for position, ready to pull out all the stops. *Greased Lightning* shot across the starting line, and Ted was off.

Unfortunately he had jumped the gun, along with several other skippers. Turner and his crew were disqualified, scoring zero for the day. Ed's friend John McIntosh, who was also there racing, tried to intercede with the race committee on Ted's behalf: "My argument was that there were several boats over the line, and the race committee could not have picked him out clearly in the group." The race committee didn't buy it.

On the following day, a memorable one, Ted and the *Greased Lightning* crew finished the race a glorious sixth. "Frederick," he called elated, as they were coming in to the dock, "how do you feel about winning this sixth place?" At the time, it didn't really mean a great deal to Frederick. But he recalls vividly how proud Ted was. He got up on the bow of their boat and shouted at the top of his lungs for everyone to hear, "Look at the best sailors in the world! Here we are! Comin' in sixth place!"

After five days of racing, the 1956 Lightning regatta was won by Bill Cox. Two-time defending champion Tom Allen placed third. And in twenty-third place, halfway down the list of finalists, was Turner. Frederick and Bunky drove *Greased Lightning* back to

Savannah, and Ted, his head still in the clouds after his first major competition, went on to begin his freshman year at Brown.

When he arrived in Providence, Ted was well over six feet tall, still thin, and good-looking, with deep-set, pale blue eyes and a heart-breaker dimple in his chin. He had a funny, down-home accent that was hard to take seriously in Rhode Island and a gap-toothed smile that made you want to laugh. He showed up on campus in his best duds, a salt-and-pepper suit that was just a little old-fashioned and more than a little out of place. Turner, initially, struck some as "quite a country boy!"

If seventeen-year-old Ted was provincial, the same could be said of many other college freshman at the time. But young Turner seemed to feel his discomfort more keenly than most. Earlier, he had been an Ohio schoolboy in a Georgia military academy; here he was a redneck Southerner in a liberal Ivy League college. Once again he was out of step. It must have seemed to him as if all the other members of Brown's entering class of '60 were from fancy Northern prep schools, graduates of well-known institutions like Choate, Andover, Deerfield, and Exeter. Was it possible that they were *all* class presidents who had scored 1600 in their SATs and could swim 50 yards in 22 seconds? Ted could swim, but on one occasion he had almost drowned. He wondered if he was in over his head again. He was afraid that his classmates would be snobs who had never even heard of McCallie, and Ted hated snobs.

Looking back, he would later say that there was, indeed, "a lot of bull at Brown. It mattered who your father was, how much money you had, what your clothes looked like. I was used to a certain directness, and there was very little of that." But if he felt insecure, Ted Turner was far from daunted in his first year at college. In fact, he seemed to take pleasure in going against the grain, to revel in being a majority of one.

Maxcy Hall was a freshman dormitory, and there was general agreement on campus that it was one ugly-looking building. According to legend, it was so hated that when fire broke out in the first-floor lounge and firemen arrived to save the structure, the residents mercilessly pelted them with snowballs. The letter from the university said that Ted had been assigned to 310 Maxcy. He gave the dorm one more look to make sure he had come to the right place, marched in and climbed up to the third floor. He found room 310 at the end of the hall, next to the fire escape. His two roommates—Carl Wattenberg, from Missouri, and Doug Woodring, from Pennsylvania—had gotten there first.

As Ted strode into the room, he stretched out his arms and introduced himself: "I'm Ted Turner from Savannah. I'm the world's best sailor and the world's best lover." It was the first thing out of his mouth and said matter-of-factly, like an open-and-shut case. The startled Wattenberg remembers thinking at the time, "Gee, he's full of it!" It wasn't until much later that he would acknowledge that Ted really believed what he said and, still more astonishing, that it might even be—partially—true.

Maxcy 310 was too small a room for three large young men, and to make matters worse it was cluttered. There were three desks, three beds, three bureaus; the walls were covered with highway road signs that Woodring had collected traveling cross-country. Wattenberg brought with him a large motheaten moose head called "Roscoe," picked up at some local flea market. Ted didn't care for the moose, but he did bring his own contribution to the room— several rifles. He was probably the only freshman to arrive that year at Brown with his own arsenal. If he couldn't hunt coon or boar in Providence as he loved to do back home, he'd settle for partridge.

As soon as he unpacked, Ted made no bones about his plans. His daddy wanted him to have a good liberal education, he announced, and a part of that was to have a good time. Whether or not his father actually equated "a good liberal education" with a good time, Ted plainly intended to behave as if he did. He struck his new roommates as a bright guy with an "attitude." Woodring says, "He was just going to shoot for a C average and have the time of his life, which he pretty much did."

If Ed Turner wanted his son to have a good time at Brown, he certainly did not want alcohol to play any part in it. He had seen the problems it was causing in his own life. Things were getting worse between himself and Florence, and she was again threatening to move out. So Ed made a bargain with his son. If Ted didn't take a drink until he was twenty-one, he'd give him $5,000. It was a lot of money in 1956, but it was simply one more instance of Ed Turner, as father, meaning well and not knowing how to achieve his goal.

Ironically, the college he had chosen for his son had a reputation at the time for heavy drinking.

Wattenberg recalls that there was beer everywhere in the dormitory. Some nights well after midnight, a noisy, reeling Ted would be dropped off back at the room by one of his friends, and handed over to his roommates with the pronouncement "Now he's your problem." Wattenberg says, "He was a loud and boisterous drunk only up to a point, and then I guess he just passed out. Until that point, he was still telling the world that he was the greatest."

Ted's father was a well-known drinker, raconteur, poker player, and womanizer, and in spite of himself, he had taught his son well. Right from the start, Woodring says, "he'd be up playing cards or carousing late every single night of the week." Most mornings his roommates were off to class and Ted was still buried under the covers, sleeping it off. He missed more than his share of lectures. It seemed to Wattenberg as if he really hadn't done that much drinking before Brown, given how it affected him.

Ted was apparently a good poker player. Don Anderson, who lived down the hall, used to play five-card stud with him for nickels and dimes. He thought Turner was a "little tight" with money, but he could play the game. He knew the cards well and had a good poker face. To another freshman who sat in with them, Ted seemed to be "up all the time" while playing—talking a mile a minute, keeping up a constant stream of chatter, so that you never quite knew when he was bluffing: "He was sort of like a snake-oil salesman."

The interest that Ted displayed in women during his last years at McCallie now increased exponentially. He couldn't take his eyes off them. "If it was female, and it got in his sights, it was worth looking at," Carl Wattenberg says. "It didn't matter what school they went to, or if they went to school. Ladies were ladies." And all around, there were plenty of ladies. Only a few blocks away were the women's dorms at Pembroke, Bryant College, and the Rhode Island School of Design.

There were also mixers with women from nearby Wheaton and Wellesley. By now Ted was wearing the Ivy League uniform of the time—conservative narrow-lapel tweed jacket, button-down oxford shirt, loafers, gray flannel pants. McCallie was evident in the squeaky-clean way he groomed himself and the very straight way he stood. "He looked good," Anderson says. "He was a cute guy. He had a very jaunty air."

Ted made no secret about his dates. If he had a good time, a lot of people back at the dorm were likely to hear about it. And whether reporting about a hunting trip, a sailboat race, or an evening out, he loved to spin yarns, to tell long, elaborate tales that featured himself as hero. Even his roommates, who weren't thrilled to be awakened in the middle of the night by an elated Turner, had to admit that he could be very funny, with his Southern drawl and his willingness to tell all. What was important to Ted was that he was on stage and had an audience. "He loved to talk about himself," Wattenberg remembers. "If it was a conquest of a lady or he did well in a race, you were going to hear about it. He'd come back to the room and chat about it."

Woodring, who spent a great deal of time studying in the library, can't recall ever seeing Ted there. Wattenberg, who preferred to study in their dorm room, can't recall ever seeing Ted crack a textbook or, for that matter, any book, "though he did talk about reading the classics." Others he knew claim that he was indeed reading, but nothing that had to do with his courses. He apparently liked popular history.

Often Ted would create noisy scenes in the hallway. He was full of conservative opinions and full of himself. Alcohol only made him more argumentative as he engaged in heated discussions about the Civil War, about Israel and Middle East policy, about the South. He was, his classmates note, "always defending the South." And in doing so, he was always attacking H. L. Mencken. How could Ed Turner's son not loathe someone who spoke of "a raw plutocracy" owning the New South, and of the rest of its population as a peasantry "too stupid to be dangerous"? Perhaps even more galling to Ted was Mencken's claim that of the men currently running the South (men like Ted's father), "not five percent, by any Southern standard, are gentlemen."

The more wound up Ted got, the more outrageous his chatter became. He would argue loudly for the efficiency of war as a way of weeding out the weak in society. With the single exception of his sister, whom he never mentioned, he had no patience for the weak. He defended the ruthless twentieth-century social Darwinism favored by Hitler's National Socialists, and he delighted in the shocked reaction of his Northern liberal classmates. He even seemed to have a soft spot in his heart for the Nazi leader himself. "Hitler," he told a classmate, "was the most powerful figure of all time." And he could make a debater's case for even his most exorbitant claims: "Ted thought fast on his feet," Woodring recalls, "and he was well enough informed in a lot of areas where he was quick and agile." Though Ted may not have participated in the Brown Debate Club's discussion on whether Canada should become part of the United States, he held forth informally on the third floor of Maxcy on just about every other issue, and did so at the top of his voice.

His roommates became less and less amused. They had come to Brown to get an education, and "we were getting disrupted literally almost every night of the week." After a few months of this had gone by, Wattenberg and Woodring, unwilling to play Rosencrantz and Guildenstern to Ted's Prince Hamlet, decided to dump their roommate. And so, before the end of the first semester, they walked out on him, packing up their treasured highway signs and their moose head and leaving Turner in an empty room.

Ted was totally unrepentant. He refused to take his roommates' abrupt departure as a warning. If anything, his behavior became even more outlandish. On several occasions, he cocked his .22 rifle and began firing it out his dorm window, which was only a few hundred yards away from University Hall and the office of President Barnaby Keeney. A lot of people in Maxcy were uncomfortable with guns and, not unreasonably, took a dim view of the Savannah sharpshooter.

Ed Perlberg, who lived across the hall and had come to Brown from New York City by way of Andover, found Turner "a complete wild man." One day while trying to study and frustrated by the noise, Perlberg marched angrily across the hall. There were three very tall guys in the room, all of them over six feet and all three shouting at one another and laughing hilariously. It was Don Anderson, his Chinese roommate Charlie Yuen—a descendant of the last emperor of China—and Ted Turner. "Turner was sort of the ringleader," remembers Perlberg. "Raucous, good-looking, a tremendous salesman, wanting to make everybody his friend and believer. And I think I said something like, 'Would you please shut up?' and then marched back across the hall."

But if Turner was boisterous, he could also be a charmer, and he and Perlberg eventually became friends. One day he invited Perlberg sailing. The Brown Yacht Club had its boathouse on the

Seekonk River, and though the river was a small, undistinguished body of water separating East Providence from the rest of the city, the clubhouse itself was an impressive structure set up on pilings and covered with cedar shakes; the boats and maintenance shop were on the first floor, the second floor was for parties. Perhaps it was merely coincidental that Perlberg had the same first name as Ted's father and also played golf, but he had the distinct impression that Ted wanted "to gain my respect, because all I saw of him was this kind of wild man."

On the water, Ted appeared to be completely at home, calmer, quieter, more serious, more focused. He was a good teacher. In retrospect, it seems to Perlberg that he sailed with a skill comparable to Michael Jordan's on the hardwood: "I don't think I've ever been that close to anyone of that ability in a sport." To Perlberg, the afternoons sailing on the Seekonk were remarkable. As the sail caught the wind and the dinghy carved through the choppy water, Ted would lean back, and talk about his father and the outdoor-advertising business. "He kind of thought of his father as Rhett Butler," Perlberg recalls, noting the hero worship in Ted's account. "He was a very handsome, successful, Southern businessman who believed in the military. That was why Ted was sent to military school." (Or so Ted said, perhaps trying to explain his attendance at McCallie rather than at a more prestigious, New England prep, like Andover.)

One day, out on the river, Ted attempted to show Perlberg one or two subtleties of sailing. He pointed out that the breeze was shifting.

"Wind change?" Perlberg echoed. "What wind change?"

"You can see it," Ted replied.

Perlberg was amazed. "What do you mean you can *see* it?"

Ted told him to "look out there in the atmosphere." Perlberg looked and saw nothing. "Then use your nose," Turner said. "You can smell where the wind is."

That first year, sailing was the one activity—besides carousing—where Ted scored his most notable successes. He became one of the leading members of a very good freshman team, winning many of the races he entered. And, of course, after every win there was the traditional celebration; the Brown Yacht Club was famous for its parties. By the end of the season, Ted Turner had been named number-one freshman sailor in New England.

If his sailing record was brilliant, his academic studies more or less hobbled along. With a certain native intelligence and a better-than-average private school background, Ted had managed to pass his courses with a minimum of effort and not flunk out of school, but it was a tumultuous freshman year. He was not the first McCallie graduate to face a difficult period of adjustment to new-found freedom. The McCallie faculty often declared themselves "very *very* conscious of the upheaval, the trauma involved in moving from the highly structured situation at McCallie—not only highly structured as far as environment is concerned, but highly structured academically—into a more permissive society."

Ted's feelings about Brown were complicated. He reveled in his newfound liberty, but at the same time he seemed to fear it, as if at any moment he might go hurtling off course and self-destruct. He told one of his new friends how oppressed he had been at McCallie. He confided, "It was like being a tiger in a cage." Now, at last, he was Turner unchained. But later, looking back, he would say, "I was happier at military school. I really didn't like it when there were no rules."

If things were complicated for Ted away from home, they were no simpler back in Savannah that summer. His parents were getting a divorce. Although they had talked about it numerous times, threatened one another with the prospect, separated and come back together again as if preferring the thorny relationship they knew to no relationship at all, Ed swore that this time he was going through with it. He wanted to make all the arrangements as quickly as possible, before he changed his mind. His attorney told him that if he went to Reno, he could get a quickie divorce, but first he'd need a financial statement. His attorney recommended the accountants' office of Hancock & Mazo, next door.

Irwin Mazo, CPA, was alone in his office when Ed came in. "Ed and I hit it off immediately," Mazo says. Within six months, Turner Advertising would become Mazo's chief client. Ed, on this occasion, promptly told him what he wanted and why. He had a sick child who belonged in an institution, and his wife refused to allow it. He couldn't live that way any more. He was getting a divorce. Mazo prepared the required financial statement, and Ed left for Reno.

The case of Robert Edward Turner, plaintiff, versus Florence Marie Rooney Turner, defendant, was heard in the Second Judicial District Court of the State of Nevada on the 22nd day of August, 1957, with only the plaintiff present. By agreement, Florence made no effort to contest the action. The court granted the divorce, finding in favor of the defendant. Henceforth the parties agreed to live "separate and apart for the rest of their natural lives."

Ed called Florence to tell her the news, but she had already heard from her attorney. Ed, however, had something else to add. He was having second thoughts. As Florence recalls, "Ed said, 'But I don't want a divorce.' He said he was coming East. He arrived in a few days and handed me the keys to a black Chrysler Imperial I had been wanting. He said we were going to drive to Florida and get remarried. I told him the curtain had come down. He said it could go up again. I told him the players had changed."

All around Savannah, tongues wagged about the divorce. Everyone had known it was coming. Ed and Florence didn't seem to fit together. She was so aloof, so private, so Northern. Ed was so explosive, so driven, so outgoing. And, of course, he kept running around on her. But the affairs, the drinking, were nothing compared to their terrible battles over their dying daughter.

Still, in the end, Ed had done right by Florence. Rumor in town had it that it was one of the largest divorce settlements ever involving anyone from Savannah. The actual amount the court awarded Florence was $15,000 per year for a period of fifteen years, plus ownership of the apartment on 46th Street, which they had bought. But one of the most interesting aspects of the settlement agreement was the custody of the children. Their future was mapped out in the cold legal prose of the court documents: Florence was given exclusive charge of their ailing daughter. But Ed was granted "the sole and absolute care, custody and control of Robert Edward Turner III during his minority, and shall have the right to make all decisions relating to his health, education and welfare." It was a strange settlement—especially at that time—for a mother to take one child, the father the other. But Ed had plans for Ted. He had no intention of letting go of his son. The court had merely made *de jure* what had been well established *de facto*.

The extent of Ed's alienation from his daughter can be seen in

the X-ing out of a standard paragraph dealing with the custody
of the children in the event of either parent's death. Handwritten
into the agreement is the statement that should Ed die before Florence, she would be responsible for their minor son. But if she should
predecease him, their daughter would be placed in the custody of
her maternal grandmother, Mrs. George Rooney. Ed wanted nothing more to do with the long, drawn-out tragedy of his dying child.

The appended trust agreement for the benefit of the children is
noteworthy: "In the happy event that said daughter shall regain her
health and ability to handle her affairs," she shall receive one-half
of the entire trust when she reaches the age of thirty. Although the
typed agreement allocates the other half of the trust to Ted at the
same age, the word "thirty" is crossed out and "thirty-five" written
in and initialed by both parents. Doubtless they knew their son.
Perhaps they had heard reports from Brown and thought that an
additional five years of maturity would help.

Ted felt the emptiness of his house. He was now living alone with
his father. He wanted his mother and father to patch things up, to
get back together again, but it didn't look as if it was going to
happen. As usual, he was working for Turner Advertising that
summer, and when he wasn't working, he was sailing. Sailing
helped to take his mind off the divorce, and his father was out of
town a lot. Ed was away on business, then he was away in Reno,
and then there were those times when he was just away. Irwin
Mazo soon discovered the rhythm: "Every two or three months,
Ed used to go off on a binge. He'd disappear for a week or ten days
and come back and be as good as new. It was a regular pattern."

Ted was no stranger to these disappearances. He sailed obses-
sively, as if his very life depended on forgetting about everything
else. Ted had, in some ways, become the classic child of an alcoholic
parent, sporting a growing drinking problem of his own, subli-
mating his emotions in fierce competitiveness, thriving on crises.

John McIntosh, an excellent sailor, had been keeping an eye on
Teddy at the Savannah Yacht Club. He had seen him develop from
a very aggressive, so-so sailor—with an unfortunate habit of capsizing—into a good one. Though he had "learned hard," he had
learned. "I kept telling Ed that Teddy had a lot of talent. But Ed
had other ideas about what his son ought to be doing with his
time." Ed was proud of the boy's victories when he started to win,

but his plan for Ted didn't include having him fritter away his future playing with sailboats when he should be working.

Still, John McIntosh was sufficiently impressed with the young man's ability that, with Ed's permission, he asked Ted to join him and his wife, Barbara Ann, and crew for them on their Lightning. In effect, they became surrogate parents for Ted on the water. And when Ed wasn't out of town, he'd sometimes watch them race. Afterward, he would invite a group of people aboard his new motor yacht, *The Thistle,* a 54-foot boat manufactured by the Herreshoff Company (considered by some the maker of the classiest line of boats at that time), and throw one of his informal shindigs. Decked out in his fancy Hawaiian shirt and white ducks, Ed was the perfect host, much preferring drinking to racing.

The Kappa Sigma house at Brown was, from the outside, a tidy red-brick structure on a campus quadrangle surrounded by other fraternity houses. Inside, it looked as if it had been lived in—*really* lived in—except for the bar, which had been so "lived in" that it had had to be redecorated. Ted's sophomore room was on the first floor—number 138. It was cluttered like all the rest, with mattresses on the floor, posters on the wall. Upstairs on the top floor was a pool table, where Ted would often go to drink a few beers and shoot pool. At the beginning, he liked his fraternity. It was hardly the wildest Greek house on campus, but one night his pledge class got blind drunk and roared off to Newport. There, in a solemn act of fraternal solidarity, they visited a tattoo parlor and were dutifully inscribed with a one-inch Greek ornament on the shoulder—the initials of Kappa Sigma.

Ted quickly gained a reputation in the house as a gambler. All night long he'd be up playing poker and bridge, and he was good at both. Ted was also known as someone who would do anything to win a bet. One bet the Kappa Sigma pledge class of 1957 will never forget. Ted proclaimed that he could chugalug an entire fifth of whiskey without stopping or falling over and passing out; if he died, needless to say, he lost.

Raising the bottle to his lips, he began to drink. His classmates started to cheer. The more he drank, his Adam's apple bobbing furiously up and down, the louder the yelling became. Finally, the bottle empty, he slammed it down on the table to riotous applause. Then he went into the bathroom and threw up.

It was a prodigious act of drinking, and it won the bet, but for Ted it was less about alcohol than about being "shrewd." Those few who knew him well in the house were aware of his secret: before downing the whiskey, Ted had lined his stomach by swallowing nearly a pint of olive oil. According to one of his frat brothers, "It really had to do with the fact that he had some kind of trick he could use to his advantage." From his earliest days in sailing, Ted had learned to employ whatever tactic gave him an edge.

There were two Jewish students in that Kappa Sig pledge class, and one of them, Lawrence Brenner, lived in the room directly above Ted's. Brenner's impression of his pledge brother was that he was very tweedy, very bright, very charming, and very strange. "Ted said that he had never ever met a Jew before." It wasn't true. In fact, one of Ted's earliest childhood friends in Cincinnati was a Jewish neighbor. Ted got a kick out of saying things to see what effect they might have, but Brenner had no idea of that. "And I thought, Shit, do I have horns?"

Few if any of his classmates at Brown mistook Ted for a serious student. He was considered a goof-off. Some put him down as a cracker, a redneck, or even a racist. Classmate William Kennedy, who would subsequently become assistant director of university relations, told Curry Kirkpatrick of *Sports Illustrated,* "We caroused together. Ted was also a bigot, as maybe all of us were in a sense at that time." He recalled going out in a group with Turner and, after a lot of drinking, singing Nazi songs outside the Jewish fraternity. On another occasion, according to Kennedy, Turner put up signs reading WARNINGS FROM THE KU KLUX KLAN on the doors of the few blacks at Brown. Turner later dismissed these as "childish pranks." Perhaps they were, but it's obvious that, even then, sensitivity was not one of his strengths.

In an era when *in loco parentis* was taken seriously on the nation's college campuses, women—including mothers and grandmothers and great-grandmothers—were not allowed above the first floor in men's dorms, except on major national holidays and then only with the door open. At Kappa Sigma, the fraternity had to have a permit from the university to have a party. Girls were allowed only on the first floor in the lounge or in the party room downstairs. To be caught with a girl in your room meant summary expulsion. It was every college boy's dream to beat the odds, but few were willing to risk the punishment.

Ted, of course, was never one to turn his back on a challenge. He had a girlfriend from Pembroke, the women's college then affiliated with Brown. One evening, he gave her his fraternity pin, and his Kappa Sigma brothers, spruced up in coats and ties and shining like lampposts, stood beneath her window and serenaded her, solemnizing the pinning ceremony in the time-honored Greek tradition. Soon afterward, Ted and his girl were shacked up regularly in his room at the house. In the morning, she would make his bed, adding a nice note of domesticity to their relationship, as if she were prepping for their future married life. "She was a nice girl," one of his Kappa Sig brothers recalls, "but everybody knew it wasn't going to last." Still, Turner—much to everybody's envy and amazement—got away with breaking the rules. In the house, it was suspected that he was paying off one of the campus guards with liquor.

In college sailing, the boats are more or less identical, and they're interchanged between opposing teams during a meet, so it's a question of who can squeeze the most speed out of any one boat. Ted had persuaded a pledge brother named Mal Whittemore—a wiry, athletic undergraduate with a crewcut and a gift for sailing, who wrestled at 123 pounds and had coxed on the freshman eight—to crew for him. Together, they tried to squeeze out every second. Mal remembers Ted's being very aggressive in sailing, always trying to take full advantage of the rules, always pushing for an edge. "He was a real rough, competitive guy," he recalls, "but Ted also had a lot of intuitive feeling about a boat and how to set up the various stays and shrouds to get the utmost out of the rig." Whittemore remembers that he was always looking for the wind puffs and chattering nonstop. "He'd talk continuously: 'Look for wind shifts' and 'Let's get the weight forward' or 'Get the weight aft.'" He also remembers that Ted "paid a lot of attention to detail. That's why he was such a consistent winner."

But at the start of his sophomore year, Ted wasn't winning. It wasn't the hangovers that prevented him from doing well. One of his teammates, Roger Vaughan, noticed that he'd appear in his tweedy three-piece suit on race day, no matter what his condition. "The suit would be rumpled, Turner having slept in it the night before, and the smell of stale drink would be emanating strongly from him. But his mouth would be in gear, the 'ol' buddies' streaming forth at high volume."

On October 12, however, the gloomy results of the Danmark Trophy Regatta at New London didn't augur well for the rest of the team's season. Brown placed seventh, and Turner and Whittemore lost one of their races embarrassingly in the last few yards. The team commodore and spokesman, Charlie Shumway, didn't know how to explain Brown's poor showing.

After that, the team practiced on the Seekonk at least two or three hours every day. By the time Thanksgiving rolled around, Brown had qualified for the Midwestern Championship in Chicago—the Timme Angsten Trophy Regatta. The meet consisted of twelve races held over the holiday weekend; it would prove to be one of the most thrilling series of races that Turner and Whittemore had ever sailed in.

The weather was extraordinarily cold in Belmont Harbor, and there was ice in the water. Though unusual in November, such sailing conditions are not unknown in Chicago. Brown's performance on Friday was as bleak as the weather, and the team at one point was as low as twelfth place. But by Saturday they had surged up to first, with Michigan close behind.

The two final races were scheduled for the next day. Sunday dawned gray and bitterly cold. "I mean it was *cold!*" Shumway remembers. "And it was blowing quite hard." He and his crew, Bud Webster, won their first race, but Brown looked as if it had lost its chance to win the meet when in their next race they were disqualified. The pressure was squarely on Turner and Whittemore.

As the last race began, they got off to a good start. They were going full tilt when Ted decided to risk everything. All the other boats were steering around a patch of ice and losing time. "Ted decided to sail our boat right over the ice," Whittemore recalls, his voice still hinting at the excitement. If the ploy worked, they would save precious seconds. If not . . . it wasn't worth thinking about.

Turner yelled to Whittemore to pull up the centerboard, and he pulled up the rudder. They skidded across the ice, the boat wobbling crazily. Coming off on the other side, they almost capsized. Their boat was half filled with water and their toes were freezing, but they had saved seconds, and they wound up winning the race. Coming from behind, Brown had beaten out Michigan for the Timme Angsten Trophy, the third year in a row that Brown had won. Individually, Turner was the high scorer for the entire meet. Shumway recalls how Ted came in after the race, blustering and tossing off bons mots like "Well, that did it after your boner, Shumway." In victory as in defeat, Turner could often be less than tactful.

It was about this time that two superstar sailors—Bob Bavier and Bill Cox—turned their eyes on the Brown sophomore. They liked what they saw and offered him a job for the coming summer. Would Ted like to be a junior sailing instructor at the prestigious Noroton Yacht Club, in Connecticut? Would he ever! They were offering him $50 a week, more money than he had been making putting up billboards for his father. But money wasn't the primary issue. If the legendary Bavier and Cox wanted him, that was plenty good enough for Ted.

Why Bavier wanted him for the job is interesting: "I met Ted through John McIntosh, who thought he was a fine fellow, and I was impressed," Bavier recalls. "I thought, Well, he's the kind of guy we'd like to have. I knew he was an active young sailor, and I just thought his personality—I thought our kids would eat him up, and they would have. He had enthusiasm and candor—the two things which he still has." As far as Ted's sailing went, Bavier wasn't all that impressed. Turner was hardly what Bavier would call "a whiz kid." "Rather than someone who was hot stuff to start with, he was a self-made sailor. He worked hard at it."

Unfortunately for the kids in the Noroton junior program, they never had a chance to find out how much they would have liked Turner. No sooner did he notify his father that he had a summer job than Ed told him to forget it. He needed him back in Savannah; the boy had no time for such damn nonsense. How was he going to learn the board business if he was off sailing all summer? There was no reasoning with Ed. Once his mind had closed on a subject, it was harder to open than handcuffs. Ted dutifully notified Bavier and Cox that he couldn't take their job after all. If they were disappointed, Ted was crushed.

One of Turner's new friends who helped to console him at times like this was a heavyset, tall boy from New York named Peter Dames. "A nice fellow," according to one of his classmates, "drank a lot, idolized Teddy." They became close friends. "Basically, we both liked to get drunk and chase women" is the way Dames explains their friendship. He, too, had attended a military prep school and, like Ted, was unprepared for the more permissive social life at Brown. "Suddenly there was no lights-out at ten o'clock, no bed checks, no inspection. You could drink and screw all you wanted. And we wanted. If Turner got thrown in jail, I would bail him out. One time I tried to bail him out, and they locked me up, too."

Then there was the night four of them were driving in Don Anderson's car to Wheaton, a women's college in nearby Norton, Massachusetts. They had had a few drinks, and there was an accident. The car was damaged. Everybody was banged around. Dames hit his mouth on the dashboard and knocked out his front teeth. Anderson was arrested.

The next time they headed for Wheaton, they weren't so lucky. On that occasion, they had all done some really serious drinking before starting out. According to Anderson, "Teddy got absolutely shit-faced that night." They arrived at the women's college safe and sound, but late. Though Brown men were free to come and go at any hour of the day or night, college women in those bygone, double-standard days had to sign out after 7:00 P.M., and on weekday nights they had to be back in their dorm and signed in before 11:00 P.M. or else! Though the details of exactly what happened that night are fuzzy, due no doubt to the epic amount of alcohol consumed, it appears that the women were all safely locked away from their amorous visitors, and that the Brown men, frustrated, started to get increasingly bellicose. A major disturbance erupted outside one of the dormitories. Windows were smashed. Property was damaged. The sign-in book at the door was snatched and obscenities scrawled across its pages. In short order, the campus police arrived and seized the lot of them.

For his part in the drunken ruckus, Turner was placed on suspension for one semester. Anderson had his scholarship lifted and quit Brown, finishing his education at U.C. Berkeley. The two condemned vandals drove out of Providence together, with Ted dropping Anderson off at his home in Washington, D.C., before he headed back to Georgia. Under the circumstances, he wasn't looking forward to facing his father.

CHAPTER 6

ED TURNER WAS HAPPIER than he had been in a long time. After his divorce from Florence that summer, he wondered if he had made a mistake, but now he felt certain that he had done the right thing. Bill Dillard, a wealthy and powerful figure in Savannah, had invited him to his house one night to play poker, to substitute in a regular weekly game. Dillard was an important business contact, the president of the Central Georgia Railroad (subsequently merged into the Norfolk & Southern). While the other poker hands were turning up flushes and threes of a kind, Ed turned up Bill's daughter, Jane.

Prettier than Florence, Jane was a tall, slender woman with short brown hair and Southern airs and graces, polished by money. Jane had a thirteen-year-old son, Marshall, from her first marriage, but that made no difference to Ed. Suddenly he was all charm. Jane was taken by his humor, his enthusiasm, his drive, his directness. He loved to tell stories. He told her about how he had learned to fly in Cincinnati and about the time he had flown between the pilings under the Covington Bridge. She found him to be "a very very attractive personality."

Soon Ed was advising her about her son's education. Marshall should go to McCallie just as his own boy, Ted, had. And Marshall went. Ed also told her what to read. "He was appalled that I had never read *War and Peace*," she remembers. "And Dostoyevsky, and things like that. So he made out this list for me to read and improve my mind." In short order, after a four-month whirlwind courtship, Ed took charge of Jane's life. By the time Ted came home from Brown in December, Ed had a big surprise for his son. He was getting married.

And Ted, of course, had his own surprise for Dad. On the long drive home, he had given some thought to exactly how he was going to tell him what had happened. Clearing his throat, he began.

"Suspended!" Ed hit the roof. Ted quickly explained his side of the story. "He would not disclose the names of the other boys who were with him," Jane recalls, and as a consequence the university officials had punished him more severely than he would have been otherwise. Somehow she and his father got the impression that the Wheaton episode involved honor and gallantry as well as women and booze. It was like something out of Faulkner. Ed had once recommended that Ted read the novels of his fellow Mississippian. Perhaps he actually had.

Ed decided that his son should join the Coast Guard. It might end his silly escapades and help him grow up. Make a man of him. There was still a draft on, and Ted had had to register for it. Although undergraduates in good standing were not generally called up by draft boards, Ted was no longer in good standing. If he enlisted in the Coast Guard Reserves, he could serve his six-month active-duty tour and be back in school for the beginning of the fall term. Ted agreed to go, but his father insisted that he wait until after the wedding.

Ed and Jane were married later that month in a small ceremony in the Dillards' house. Irving and Terry Victor were there, plus a few other close friends. Ted got along with his father's new wife, but he wasn't overjoyed about the occasion. To Jane, it seemed as if he wanted his mother and father to get back together. "He always wanted that," she says. "He was always very loyal to his mother." But after Ed's marriage to Bill Dillard's daughter, Savannah people who knew Ed Turner felt that he now had a social status he never had before.

Not every new husband takes his nineteen-year-old son along on his honeymoon, but Ted's father was by no means your average bridegroom. About a week after the wedding, Ed, Jane, and the two stepbrothers went to Miami, where Ed chartered a 60-foot sailboat and set out for Nassau in the Bahamas. No sooner had they started than a terrific storm blew up, and they had to return to port to wait it out. The next day they tried again, and it was a rough crossing. Jane was seasick almost all the way. Ted was none too happy either. He had very little in common with Marshall, who was several years his junior, and, as Jane recalls, "He basically went down and got into his bunk and went to sleep." After about a week, the four of them flew back to Miami, and Ted prepared to go into the service.

On February 12, 1958, Ted enlisted in the Coast Guard in the USCG 7th District. He took the usual battery of placement tests and scored respectably, if not brilliantly, in such areas as arithmetic reasoning and mathematical knowledge. Approximately three weeks later, he was shipped to the Coast Guard Receiving Station at Cape May, New Jersey. He trained there for six months, rocketing up to the rank of fireman's apprentice. He had grown to like the military life at McCallie, but this was far different from his old prep school, and he was no longer the same Teddy Turner. When he was released on September 2, he was happy to be a civilian again.

Ted arrived at Brown in plenty of time to begin his junior year, only to discover that he wasn't a junior. He was listed in the school directory as a sophomore, a member of the class of '61. It was embarrassing. Back once again at Kappa Sigma, he moved into a new room on the second floor, and was glad to see his old fraternity buddies. Soon he had picked up where he left off: gambling, drinking, womanizing, and sailing—and doing it all at the top of his lungs.

With a small group of other Type-A, high-energy personalities from the house, he turned his attention to the sport of kings and began to go regularly to Lincoln Downs racetrack. Less hyper Kappa Sigs marveled at their high-flying pace. "They were out at the track almost every day," one recalls. And at night, they turned their attention to other sports.

One evening, Turner and a few others went to a restaurant in downtown Providence and drank up a huge bill. Before leaving, Ted made his way somewhat unsteadily to the men's room, babbling something about walking out without paying the bill. The alarmed manager called the cops. Later, when they tried to sneak out of the restaurant—the old collegiate "Dine and Dash card"—there was a police car waiting for them at the curb. Once again, Teddy's gang wound up in jail. At the Kappa Sigma house a call came from the police station, and the frat brothers headed down-town to bail them out.

As far as Ted's love life was concerned, he was making up for lost time, going out with women from all the neighboring colleges and, by no means an academic snob, nonmatriculants as well. One woman friend from Pembroke remarked, "Every girl I ever saw him with was pretty." Typically, these relationships had more to do with addition than intimacy. "Intimacy was never his bag," says Lawrence Brenner. And where women were concerned, he had "a sort of trophy mentality." His first wife, Judy Nye, would later recall that Ted made no secret of this. "When he talked to me about the women he had at Brown—I mean, he was counting them." He'd say things like, "I had this gal and that gal. I had ten gals in one weekend. I laid so-and-so twenty times." It seemed to Judy that he was keeping score.

In this, as in other things, he had learned from his father. "I remember Ed bragging one night over dinner," Peter Dames recalls, "that he had screwed every debutante but one in Cincinnati one year, and he would have gotten her if he had had more time. Ted asked him how many women he figured he'd been with, total. The old man said maybe three hundred. Ted was amazed. [His father] intimidated Ted."

During all this time, Turner was also going to school and taking courses. And although he was often absent, he did, by and large, pass most of them. One was in Greek tragedy, another on the influence of classical literature on English literature, taught by Professor John Rowe Workman, of the classics department, one of the few professors at Brown who actually impressed Turner. "His courses," Turner says, "taught me how people think."

Workman was an extremely popular, charismatic teacher, with a special extracurricular interest in athletics. He was advisor to the hockey team, and many athletes took his classes. A member of the faculty who knew him acknowledges that he had a large following among undergraduates but adds dismissively, "He never published anything. He was the sort of guy who tutored the president of the university in his Latin pronunciation before the awarding of honorary degrees." Contrary to what one might have expected, "the classics department at the time had the reputation of being one of the easier departments as far as grades go," according to a Turner classmate. "A lot of athletes majored in classics for that reason."

Despite his father's desire that he prepare himself for a business career, Ted decided to major in classics and wrote home to tell Ed of his intention. Workman remembered Turner, the would-be classics major, as a "colorful" young man: "He was superb in bringing out argumentative tendencies in other students. He had a Southern, postbellum way of looking at things. He would argue until the sun went down. . . . He had great, fiery convictions."

Workman also recalled something else about the young Turner: "I remember he had quite a time with his father." Others on campus also remember that—others who heard Teddy rambling on about his old man, speaking with a strange mixture of fear, frustration, and admiration.

If the father/son relationship was complex—loving and tempestuous—it moved to a whole new operatic scale the morning that Ted went public. On that day, Brown students awoke to find published in the *Daily Herald* a personal letter from Ed Turner to his son. Ed was rebuking Ted for his choice of classics as his major. In yet another act of filial defiance, Ted had raised the stakes and given his father's letter to the student newspaper to print. One classmate recalls thinking, "Oh my God, is he going to catch hell!"

"My dear son," the letter began:

> I am appalled, even horrified, that you have adopted Classics as a Major. As a matter of fact, I almost puked on the way home today. I suppose that I am old-fashioned enough to believe that the purpose of an education is to enable one to develop a community of interest with his fellow men, to learn to know them, and to learn how to get along with them. In order to do this, of course, he must learn what motivates them, and how to impel them to be pleased with his objectives and desires.
>
> I am a practical man, and for the life of me I cannot possibly understand why you should wish to speak Greek. With whom will you communicate in Greek? I have read, in recent years, the deliberations of Plato and Aristotle, and was interested . . . in the kind of civilization that would permit such useless deliberation. Then I got to thinking that it wasn't so amazing after all, they thought like we did, because my Hereford cows today are very similar to those

ten or twenty generations ago. I am amazed that you would adopt Plato and Aristotle as a vocation for several months when it might make pleasant and enjoyable reading to you in your leisure time as relaxation at a later date. For the life of me, I cannot understand why you should be vitally interested in informing yourself about the influence of the Classics on English literature. It is not necessary for you to know how to make a gun in order to know how to use it. It would seem to me that it would be enough to learn English literature without going into what influence this or that ancient mythology might have upon it. As for Greek literature, the history of Roman and Greek churches, and the art of those eras, it would seem to me that you would be much better off learning something about contemporary literature and writings, and things that might have some meaning to you with the people with whom you are to associate.

These subjects might give you a community of interest with an isolated few impractical dreamers, and a select group of college professors. God forbid!

It would seem to me that what you wish to do is to establish a community of interest with as many people as you possibly can. With people who are moving, who are doing things, and who have an interesting, not a decadent, outlook.

I suppose everybody has to be a snob of some sort, and I suppose you will feel that you are distinguishing yourself from the herd by becoming a Classical snob. I can see you drifting into a bar, belting down a few, turning around to the guy on the stool next to you—a contemporary billboard baron from Podunk, Iowa—and saying, "Well, what do you think about old Leonidas?"

. . . There is no question but this type of useless information will distinguish you, set you apart from the doers of the world. If I leave you enough money, you can retire to an ivory tower, and contemplate for the rest of your days the influence that the hieroglyphics of prehistoric man had upon the writings of William Faulkner. Incidentally, he was a contemporary of mine in Mississippi. We speak the same language—whores, sluts, strong words, and strong deeds.

It isn't really important what I think. It's important what
you wish to do with your life. I just wish I could feel that
the influence of those oddball professors and the ivory towers were developing you into the kind of man we can both
be proud of. I am quite sure that we both will be pleased
and delighted when I introduce you to some friend of mine
and say, "This is my son. He speaks Greek."

. . . In my opinion, it won't do much to help you learn
to get along with people in this world. I think you are rapidly
becoming a jackass, and the sooner you get out of that filthy
atmosphere, the better it will suit me.

Oh, I know everybody says that a college education is
a must. Well, I console myself by saying that everybody
said the world was square, except Columbus. You go ahead
and go with the world, and I'll go it alone. . . .

I hope I am right. You are in the hands of the Philistines,
and damnit, I sent you there. I am sorry.

Devotedly,
Dad

As far as Ted was concerned, it was vintage Dad. Ed was a
majority of one, always right, always insisting that he knew best.
Why was *he* the practical one and Ted always criticized for being
the impractical dreamer, the snob, the jackass? Ted wanted to
achieve great things in the world—be a mover and shaker—as much
as his father wanted him to, but he was tired of having his old man
always preaching to him, telling him what to do, making him feel
guilty. And Ted was angry. The publication of the letter shows
just how angry he was.

His father was beside himself when he heard that Ted had published his letter. Though living now with his new wife in nearby
South Carolina, he roared back into Savannah to tell Ted's mother
what had happened. Enraged, Ed was a veritable Vesuvius, fuming
and flailing his arms around. Florence Turner heard him out: "Ed
was simply furious that Ted gave that letter to the newspaper. Our
divorce had gone through, but he came over to rave about it. I
guess Ted was mad at his daddy at the time."

Ted was mad at his daddy and mad at himself. He had more
than enough anger to go around. Earlier, during homecoming
weekend, he had got himself into yet another mess. Part of the

tradition of homecoming at Brown was that each fraternity house set up a large display on its front lawn, usually a papier-mâché construction, urging the school's football team to dismember the visiting eleven. Another part of the tradition was the drinking. Ted liked tradition, and drinking was the part he liked particularly. As usual, he doesn't beat around the bush describing what happened on this occasion. "One homecoming weekend I burned down their display and that was it. Kicked right out [of Kappa Sigma]." The burning of Brown homecoming displays was not unknown. The punishment was usually instant dismissal from the university. Obviously in this instance, Kappa Sigma tempered justice with mercy, moving Turner out of the house but not notifying the campus authorities. Ted still took his punishment hard. As Peter Dames recalls, "It was almost worse getting booted out of your house than the college."

Simmering, Ted found a room in one of the dorms. But ironically, after all the fighting and all the harsh words, he ended up doing exactly what his father wanted. His father, after all, had a plan for him. So Ted reluctantly gave up the idea of majoring in classics and switched to economics, barely passing his courses. There were some obvious reasons for this, but the least likely seemed to be the one that Turner has cited as the problem—politics. "When I got into economics I began running into commie professors who thought everybody ought to work for the government. I was opposed to that and defended the free-enterprise system to the extent I almost flunked the course. To me the capitalist system is still the best way to get things done. . . . What a great system!"

The one patch of blue sky in this otherwise dark period was Ted's election as cocaptain with Bud Webster of the sailing team, a team that went on to win the Schell Trophy, emblematic of the New England championship. Back in September, Ted had begun the 1958–59 school year by driving down to Newport. There he met Mal Whittemore and, boarding a boat belonging to Mal's uncle, sailed out to watch *Columbia* defend the America's Cup against the British challenger, *Sceptre*. The Cup races had been discontinued during World War II, and only now were they being resumed. It was the first time that Ted had ever seen an America's Cup race. Years later he would recall that day: "We were on a sailboat about thirty feet long owned by the family of a friend of mine. We were near Castle Hill, and they towed the boats by—both white, if I remember. The crews were all big, muscular men, with their matching shirts on. I'm sure that at the time I didn't just decide I was going out and win the Cup, but I was pretty impressed." Ted ended the school year by traveling to Boston on May 31, where he fulfilled his Coast Guard obligation of two weeks of reserve training before going home to satisfy his filial obligation of three months at Turner Advertising.

Following their marriage, Ed and Jane Turner had moved to Charleston, South Carolina, taking over the first floor of a vintage Victorian building. It was massive; it was elegant; it had a wonderful front porch that looked out on the water. But as a home it was only temporary. Having always wanted land of his own, Ed decided to buy it on a grand scale. Cotton Hall was a 1000-acre plantation not too far away, in Yemassee. It had belonged to a Savannah friend of his, Harold Allen, and its size alone matched Ed's most extravagant dreams.

Jane Turner recalls how impressed she was when she first saw Cotton Hall. "It was so big. The servants' quarters were as big as a house." The main house was an attractive building, an elegant white structure spread across a cool, tree-shaded lawn. Built during the Depression by a New York architectural firm for Harry Payne Bingham, it had not been well cared for over the years, but it still retained its charm—a stylish, rambling, New England–style wooden building with a pitched roof and lovely red-brick chimneys.

Jane thought the house was beautiful. The living room was enormous; the three large French doors opened onto a bucolic landscape. For all of the wealth she had grown up with, she liked keeping house, cooking, and doing special things for Ed, such as making mayonnaise and cooking quail. She organized large, wonderful breakfasts. Unlike Florence, she enjoyed being out in the country, and so did Ed. He loved the outdoors, the land, and—above all, Jane believes—*having* land: "I think that for Ed, having grown up in hard times and having associated good times with a lot of land, the land might have been the reason that he wanted to have Cotton Hall."

Ed began to raise cattle and hired a manager. He wanted to make good use of the property. "Ed didn't like to see anything that didn't pay." Jane says, "It worried him that he couldn't seem to make it pay. But I don't know anybody who makes a plantation pay."

In order to go to work every day from Cotton Hall, Ed bought a limousine and outfitted the back of it like an office, installing a desk to work on his papers during the sixty-to-ninety-minute commute. Some days he'd operate out of the Charleston office; most days he'd go to Savannah. And, of course, he had a chauffeur, the old reliable Jimmy Brown. Jane says, "Ed worked from the time he left home until he got to his office in town."

That summer Ted worked in the Savannah office of Turner Advertising, on Skidaway Road. He was beginning to learn about the business. And on weekends when he wasn't sailing, he'd drive out to Cotton Hall, and he and his father would go for long walks together around the plantation, or drive around the acreage. Jane recalls how much they enjoyed each other's company. They laughed, they walked, they talked about the future. Ed continued to be vigilant about the subjects his son took in college and how he was doing. He pressured Ted to take more courses relevant to his future in business. Regardless of their many past disagreements, there was nothing like these long outdoor afternoons together tramping across the low-country landscape to remind the Turners of their bond.

Jane describes her stepson on these visits as being full of nervous animation. It seemed to her that "Ted was always buzzing around Ed. He was in constant motion. He either hunted or he fished, or he was just very active." The hyperkinetic young man seemed to be overloaded with adrenaline, she recalls. On these lazy South Carolina weekends, he almost bounced off the walls. He was, she says, "a handful" (which is the same term that John Workman, Ted's classics professor, used when recalling his former student).

Not having made up any of his missing course credits, Ted was reconciled to the fact that he would not graduate with his class at the end of the 1959–1960 school year. Sometimes he wondered if he was going to graduate at all. The primary reason he was glad to be back at Brown was that now he was in sole command of the sailing team. Bud Webster and one of the most brilliant groups of yachtsmen in Brown history had departed. And, as the '59 yearbook said, future success rested squarely "on the shoulders of Commodore and Team Captain-elect Ted Turner." Unfortunately, it was not to be one of the team's most stellar seasons.

Ted was living on the third floor of Goddard House, a dormitory not far from Kappa Sigma, and, with the help of Peter Dames and some other friends, still burning his candle at both ends, still working hard at drinking and skirt chasing. And by inviting different women up to his third-floor room, he continued to court disaster.

It was during Thanksgiving at the Timme Angsten Trophy Regatta in Chicago, where he had had such memorable success as a sophomore, that Turner would score again, but this time his team would not win. There were a few women skippers in the 1959 meet, and by the luck of the draw in his division, Ted was pitted against one of them. Judy Nye, an eye-catching young woman with long straight blond hair and a frank, intelligent gaze, was sailing for Northwestern. Though she was a good sailor, Ted beat her. Afterward there was a party, and Ted asked her to dance. Judy found him funny, attractive, and cute; she decided that she was not a sore loser. But he wasn't much of a dancer. They drank and talked, and Ted discovered that her father was Harry Nye, the famous sail maker and two-time world Star champion, an important figure in sailing. They had other things in common besides sailing; both of them were from the same socioeconomic background. That night Ted turned on all of his Southern charm. He was, Judy says, "charrmmming to the roots."

They made a date for the following evening, Saturday. Together with some of Ted's friends, they went to several bars. Ted seemed to be at loose ends. He was sure that he wasn't going to last out the year at Brown. Judy remembers, "Ted seemed like rather a lost soul." She asked him what had happened. "Ted's not reticent to talk," Judy says. "I don't think he's ever been. Particularly when you ask him something." He told her all about it—"that he was in deep trouble at school, that he'd been having women in his room day and night . . . Now, back then it just wasn't allowed. And he knew it. I don't suppose he much cared."

Judy says that she found his honesty appealing. He came right out and admitted everything. Judy thought that perhaps she could help him, be a positive influence on his life. She felt that if he loved her, he could change for her. And would. So when he invited her to come to Brown before the Christmas holidays, she said yes.

Judy remembers being swept off her feet. Ted had arranged for her to stay with the parents of one of his fraternity brothers. Judy recalls, "He was careful and considerate and treated me as somebody special." He introduced her to his friends, his fraternity brothers, and she felt that whatever women he had known in the past, they no longer counted as far as he was concerned. He was interested only in her. They talked about plans. He told her that he was probably going into his father's business if he didn't do anything stupid. He said he was walking a tightrope between trying to please his father by finishing school and not doing something totally absurd. He talked about all the women he had had, made jokes about it. Judy says, "Maybe I just wasn't listening very carefully." She returned to Evanston glowing. She'd had a wonderful time.

Before the semester was over, Ted had been kicked out of school—caught in *flagrante delicto* with yet another woman in his room. Though not unprepared for the consequences, he was still shaken up. Many years later he would make light of the event, saying, "At Brown I was a rebel ahead of my time. I got thrown out of college for having a girl in my room. Today they have girls and guys living in the same dorm."

News of what happened spread rapidly through the Kappa Sigma house. It was at least the second time that the fraternity had been singed by Turner. By the time he arrived at Peter Dames' door to tell him he was leaving, Ted was already reconciled to his fate. He was full of new ideas. "Look," he told his friend. "It's just a matter of days before you get booted, too. Why don't you come with me?" Ted had bought a car for $125 and wanted Dames to come home with him to the plantation. He hated to be alone on the long drive south. Besides, he figured there was safety in numbers. He wasn't sure how his father was going to take the news.

Turner sold Dames on a bold new adventure. He had a plan for what they would do after he saw his dad. "He had it all worked out," Dames says. "We were going to be the first ever to cross the Atlantic in a Lightning. We would be the toast of Europe. The women would fall at our feet. We couldn't miss." Caught up in Ted's scenario, Dames savored their triumph—the mademoiselles, the press, the champagne, the thrill of being a hero. He jumped at the idea, breaking it down to its essentials. "We were on our way around the world. We would get laid a lot and have a great time." He barely gave a thought to the seasickness, the cold, the dampness, the storms, and the fact that he knew next to nothing about sailing, didn't like sailing, and might very well end up dead. Even then Turner could motivate the troops.

According to Dames, it was only an act of nature that saved their lives.

CHAPTER 7

LIGHTNING SEARED the South Carolina sky, a jagged bolt that ripped through the air, carrying more electricity than all the generators in the United States combined—enough juice to change the course of Ted Turner's life. Flashing and sizzling, the bolt struck the grounds of Cotton Hall like an otherworldly sign, ripping into a giant oak tree and tearing down a limb, a huge hunk of wood that smashed into the appropriately named *Greased Lightning* and more or less totaled the boat. "In retrospect," Dames says, "I'm goddamn glad that live oak tree was lying across the boat. Otherwise we'd be somewhere at the bottom of the Atlantic Ocean." But Ted was disconsolate. Daring transatlantic cruises would now have to wait; his destiny lay in a different direction.

Ed Turner was surprised by the sudden appearance of Ted and his friend. Despite the inglorious circumstances, he seemed glad to see them. Peter Dames says, "I think he was happy Ted had finished with Brown." Now Ed wanted him to buckle down to work. But even the larger-than-life electrical pointer couldn't convince Ted to stay home and enter the family business. Not yet anyhow.

Determined to strike out on his own and eager for adventure, Ted packed up his old jalopy, and he and his college buddy headed farther south. Next stop: Florida, the Sunshine State, where the beaches were lovely, the women lovelier still. What better place to spend the cold months of winter? They'd get jobs as short-order cooks or waiters and lie around on the sand and soak up the rays.

It would be heaven. Once again, young Turner had it all worked out.

But when he and Dames arrived in Miami, it looked as if every other college dropout in the country had had the same idea. Jobs were hard to come by. Moreover, Ed Turner had cut off his son's allowance.

At first, the bohemian, devil-may-care lifestyle was almost thrilling—living above a bar called the White Horse in a run-down, Cuban part of town, lounging around in the warm tropical breezes. But when a cold wave came through and the boys had to huddle under their overcoats at night, when the money ran out and they had to eat peanut-butter sandwiches (or just plain ketchup) day after day, their excellent adventure started to seem merely squalid. There weren't enough dollars to buy restaurant meals or bottles of whiskey. There wasn't even enough cash to purchase toilet paper. Turner and Dames took to tearing out sheets of the Miami phone book on each trip to the bathroom.

It was only a matter of a month or so before Ted Turner's spirit began to ache as much as his backside. He had been through some depressing times before—painful moments when he felt betrayed or abandoned—but here he was at twenty-one with no future that he could see, unless he went back home. He had been booted out of school, his father had cut off his money, and now his grand Florida escapade had turned bleak.

"He had pretty well bottomed out," recalls Judy Nye. As the weeks went by, Turner would write letters to her, sharing his experiences and his unhappiness. "He was depressed, he was upset, he didn't seem to be enjoying his time at all. All of a sudden, he had no status. He was really down to the bottom of the barrel. I can't imagine going from almost complete freedom [at Brown] to nothing. He was really living in abject poverty."

"We were miserable, to tell the truth," Turner recalls. "I said to myself, 'Look . . . I'm unhappy, why don't I go home?' No! I was an independent man. . . . I was determined not to go back to my father."

In this mood, and short of alternatives, he decided his best option was to fulfill his obligation to the Coast Guard—especially when his father forwarded their letter saying he still owed them a tour of duty. On February 12, 1960, he marched into the Coast Guard base in Miami Beach, and by February 15 he had started a

43-day active duty tour on the 125-foot United States Coast Guard cutter *Travis*—an old "buck-and-a-quarter," as it was known in the service, commissioned in 1927 during Prohibition to combat the smuggling of illicit alcohol.

In later years, Turner and his friends would describe the trip as a "dream cruise" to the Yucatán. They'd also speak of how he was eventually singled out as a troublemaker—of the hazing he was put through, and how he toughed it out, cleaning bathrooms and bilges and generally triumphing over dirt and the Queeglike officers by being the hardest working junior seaman aboard. They're good yarns, but in fact, from the moment Turner signed aboard the *Travis,* under the command of Lieutenant H. H. Istock, to the moment he was ordered back to Miami Beach in late March, the boat never left its home berth at Port Everglades for more than a few hours. And in the ship's log, Fireman Turner (number 2020-609) stands out in absolutely no way at all.

But it was during the stint with the *Travis,* scrubbing the head and practicing collision drills, fire drills, and dozens of other drills, that Turner started to make some serious decisions about his life. Judy Nye noticed a change in his letters: "He decided he'd better straighten out. This Coast Guard thing wasn't very much fun anymore. He thought he'd better get himself married, get himself a residence in Savannah, straighten up and fly right, because he sure wasn't going to be in the Coast Guard for the rest of his life, and his father wasn't going to send him back to Brown. There weren't too many alternatives, so he'd better just do this business thing and get on with his life."

His marriage proposal to Judy came by mail in April. Although it had a dash of romance to it (as with all things Turner), it still seemed to her almost a business proposition. Turner had finally come to the conclusion that he should go along with his father's plan, that he needed to settle down and become a young business leader, and an integral part of that was a wife—the right kind of wife. It couldn't be just one of the girls he flirted with or slept with; it had to be someone "presentable," someone from the right social background. "I think if Ted had always married for romantic love, he would have married twenty-five times," says Judy Nye. "But in a way our marriage was sort of a business deal."

Certainly, Turner was attracted to the young blonde with her open smile and her famous sailing family. And Judy liked his good looks and his impetuous charm. Both were clearly drawn to the 1950s notion of a comfortable Ozzie-and-Harriet home: "I think we both had visions of how this would work, how this could be just great. It should have been mutually benefiting. I don't know that either one of us had too many illusions about [a great romance]."

Judy had her engagement ring before she graduated that year, and she flashed it proudly around the Northwestern campus, a pretty diamond that her friends looked on with envy. Early in June, around graduation time, she traveled from Chicago to Savannah for an engagement party. It was the first time she had seen Ted since her trip to Brown in December and—in their whirlwind courtship—only the third extended stay they had ever had together.

On June 22, 1960, on the north side of Chicago, five hundred people packed into the imposing stone structure of St. Chrysostom's for a formal Episcopal wedding. A solid gray church in the English country style, St. Chrysostom's seemed to speak of strength and permanence, with its blunt squared-off tower, its high beamed ceilings, and its heavy wooden pews.

As the crowd started to file in, Ted and Ed were back in their rooms putting the finishing touches on their wedding-day outfits. Both were wearing elegant tuxedos and gleaming top hats. As they checked themselves out in the mirror, it must have been a proud moment for Ed. His son was finally settling down and getting with the program. Now all his plans for the boy might finally amount to something. Ted's stepmother, Jane, remembers poking her head in the door and, seeing the two of them, thinking how alike they were, both decked out in their formal regalia, both turning to look at each other with a gleam in their eyes, and both—suddenly—jumping up onto the sofa. Whooping and hollering, the two of them in their top hats leapt up on the sofa and down on the floor, up and down, up and down, laughing and laughing until their breath ran out.

At the church, the pews were starting to get crowded, and since both Ted and Judy were children of divorced parents, the numbers had spiraled exponentially, even if most of Ted's family was back in Georgia. Judy's friends and relatives were the majority, but Florence was there, too, tall and slender and a little intimidating, decked out in a large stylish hat. The formal ceremony was imposing, and when the Reverend Robert May pronounced Robert Edward Turner III and Julia Gale Nye man and wife, the flustered newlyweds wheeled around and marched straight back up the aisle without remembering to pause for the traditional kiss. As the bells in the carillon tower of St. Chrysostom's rang out in celebration, Ted and his new bride charged nonstop into married life.

Soon the entire affair had moved over to the Saddle and Cycle club for the reception. After the food and the toasts, the couple adjourned to the nearby Convent Hilton, where Judy's stepfather had booked them into the bridal suite. Or so they thought. In fact, as they came in the door, they saw no heart-shaped mattress, no king-size bed, no honeymoon accoutrements at all. Instead they had been given a room with two separate twin beds. "And that," Judy recalls, "was the start of a beautiful friendship."

Renting a small one-bedroom apartment in a complex only two blocks from the Savannah River, Judy and Ted soon settled into married life. Ted was working for his father in the Savannah office, and Ed had made it clear that although his boy had a future in Turner Advertising, he was going to have to earn it. The elder Turner hadn't changed his mind about keeping Ted insecure and hungry. He was starting him off as just another junior member of the billboard sales force, pulling down a skimpy $75 a week. Judy would stay home and cook three meals a day for Ted, shine his shoes, do his laundry. They didn't have an ironing board, so she ironed on a card table—the same card table they dragged out to serve dinner on whenever guests showed up. In the end it got to be too embarrassing for Judy; the dinner table was starting to get scorched from her ironing. But rather than buy a new table, the frugal couple decided that they didn't need to entertain at home any more.

Savannah proved to be only a brief stopover for the newlyweds. Within a few months, Ted and Judy had moved north along Route 16 to Macon. They took up residence first in a small upstairs one-bedroom apartment, then in an only slightly larger basement apartment—a garden flat in a split-level suburban house owned by a family named Everett not far from the center of Macon.

Judy and Ted stocked their apartment with pets—a menagerie

of animals like the ones Ted had collected at McCallie. First came Blackie, a rambunctious black Labrador puppy who was, in both their eyes, "the best dog ever in the whole world." Then there were fish—first goldfish and then flashy tropical beauties in Day-Glo colors. Johnny Baker, Turner's old sailing instructor, visited them and was astounded by what he saw. "By God," he said, "Ted had so many fish tanks down in that basement!" There were birds as well, a parrot and two cockatiels. The parrot was named Homer, because Ted thought that having a bird who could quote the classics would be great. As it turned out, Homer never mastered much more than "Hello."

Perhaps that was because Turner wasn't around the house often enough to teach him. In Savannah, the young couple had lived a more leisurely paced life, with Ted returning home for lunch each day. In Macon, the pace picked up. Ed had bought the Macon plant from an older Georgia businessman named Johnny Jones, and before he retired, Jones had agreed to show young Turner the ropes. Ted put in long days selling billboard advertising space, and by his account his youthful energy paid off handsomely: "I worked fifteen hours a day, six-and-a-half days a week. We doubled our sales in two years. It was phenomenal."

In Macon, Ted the hell-raiser became a director of the local branch of the Red Cross, a Rotarian, a regular Junior Chamber of Commerce kind of guy. He loved advertising, and like his father before him he was now passionate about billboards. When he finally came home and sat down at the dinner table, he was still talking business. Judy recalls his enthusiasm: "He thought billboards were great. They were visible; people would see them again and again. Unlike a magazine or a newspaper, which you throw away, a billboard is in the same place day after day. Then, too, the profit margin was wonderful in billboards. Once a board was constructed, you just had to keep the weeds cut, and paint it every now and then to keep the rim of it nice. And you could get pasters for cheap. It wasn't really a skilled-labor job. So, as advertising, the markup was just wonderful."

These were the things that got young Turner excited. It wasn't as if Ted thought that billboards represented a new art form; he didn't care very much what went on the boards themselves. But he could get really fired up about a solid profit margin. His father, meeting the young man every two weeks or so for business lunches in Savannah or for weekends at the plantation, noticed the change in his son. The newest phase in his education appeared to be successfully under way; Ed's plan was working. "I think he was *so* proud of Ted," says Judy. "He thought the world and all of him. He wanted to create a superman. I really believe that was to be his legacy. His life was going to be through Ted."

But while Ted was settling down in some ways, in others he was as wild as ever. If he was, as he says, out of the house six-and-a-half days a week, it wasn't only billboards that kept him busy. He was working hard, but he was playing hard, too. In the early 1960s, Ted began to get even more serious about competitive sailing; almost every weekend, and parts of some weekdays as well, he'd be off competing at one regatta or another. Judy says, "If there was a sailboat regatta, you can bet that Ted was not on the job."

Often, especially in the early days of their marriage, Judy went along to crew for him, and together they had some success. But Ted was incredibly intense when competing in races, and for Judy "sailing with him wasn't a whole bunch of fun." Not that she wasn't as good a sailor as he was. She had been sailing longer than Ted, and there were even some who claimed that in those years he was learning from her. "Judy was a terrific sailor," says John McIntosh. In fact, it was the love of sailing that had originally brought the two of them together. Harry Nye loved to chat with his daughter and his new son-in-law about sailing. And Judy points out that "the wedding gift from my father was a suit of sails for the Lightning, which I think is indicative of something."

Unfortunately, the twin beds on their wedding night proved to be prophetic. From the start, there were problems with the marriage. As Judy herself would later admit, she was in some ways spoiled and certainly every bit as headstrong as her husband. Even more important, within only five or six months after their marriage, Judy began to find indications that Ted was cheating on her.

It must have been a hard blow for this newlywed of only a few months to find her husband coming home drunk, with lipstick smeared all over him. "It was pretty hurtful," Judy admits. "I was not in control, and I wasn't happy. . . . But when you're with somebody, the hope is always there that you can change things." As time went on, Judy would feel that Ted just couldn't stop himself, that he'd go after anything in a skirt, that his womanizing, like his drinking, was deeply engrained.

On December 15, 1960, one of the most important women in both Ted's and Ed's life died. The long, slow sadness of Mary Jean Turner's life came to its tragic end in Cincinnati at the age of nineteen. Living with her mother, who was nursing her through her last days at 3848 Dakota Street, she finally succumbed to the lupus and encephalitis that had been tearing away at her mind and body for seven years. She was buried at Spring Grove Cemetery in Cincinnati.

In some ways her death was a blessing, but for Ed and Ted, the pain didn't go away with her passing. Both, in their own ways, had distanced themselves from the terrifying illness that they couldn't control—to the point that neither of their wives had even been aware that Mary Jean was still alive when news of her death came. And so, when she finally did die, the suppressed sorrow came back redoubled. In fact, both father and son expressed their reactions in similar ways: God had turned His back on the poor girl, and each of the Turners, finding his faith shaken, felt personally betrayed. "Prior to [her death]," Ted told David Frost in an interview, "I was very religious."

The death of his sister marked a milestone in Ted Turner's life, not so much an ending as the beginning of a new, tumultuous period—a time of upheaval and growth. In July of 1961, his first child, Laura Lee, was born. Ted was too busy to attend the birth. When Judy went into labor and was rushed to the hospital, Ted was out of town sailing. He got the message when he came home that evening. Even then, he didn't race down to the hospital. Actually, Judy recalls, "He didn't come over to see me until the next day. He wasn't terribly interested." She speculates that he probably went out that night, or had somebody over to the house.

According to Judy and their friends, it wasn't that Ted didn't love his wife, and it wasn't that he wasn't proud of his new daughter; it was simply that Ted Turner was wrapped up in himself, his career, his sailing, his needs and desires. He was preoccupied: he was going to become a world-class sailor, a world-class business-man. It was clear that he was already a world-class narcissist.

Even as the head of a family with a young wife and a newborn daughter, Ted couldn't bring himself to stop his philandering. Late at night, as he and Judy lay together in bed, at the end of a long day of advertising sales and diapers, she would try to talk to him about the way he was carrying on, to get him to stop seeing other women, to show him how painful his behavior was. Ted would just roll over and go to sleep. "I think he felt that he was allowed to, since he was bringing home the bacon," says Judy. "He could do anything he wanted to do, because he was the breadwinner. That would have been his father's attitude. The breadwinner calls the shots."

It was part of what Judy came to know as old-fashioned Southern family values. The men made the money, had the careers, pulled the strings. The women were gracious and decorative; they kept the house and the children neat, the husband well fed and happy in bed; they had the babies and stayed quiet. Judy Turner had bought into this life. What she hadn't bought into were all the other women.

Looking back on it, Judy says, "I think it's fair to call his behavior sexually addictive. I think that's always been a problem for him. He's going to sleep with as many women as he can— 'How many times can I do this in one night?' It's like bringing trophies home from the hunt—this great challenge. It wasn't necessarily that he cared for any of these people. It wasn't that he was keeping a mistress. He was going out with scores of different people."

The evidence now became inescapable. Besides the lipstick, Judy would find letters, love notes, earrings in the glove compartment of their car, and on one memorable occasion, even a frilly pair of female underpants that didn't belong to her. Discretion was never one of Ted's strong suits.

By now Ted and Judy had started to fight—screaming bouts that would rage on into the night on those occasions when he was home. Ted, an all-pro verbalizer, could hurl invective with the best of them. He would complain about her in front of friends, make snide remarks; home alone with Judy, he cranked up long rambling diatribes, ridiculing her appearance, faulting the pounds she had put on postpregnancy. But Judy was tough in her own way too, and digging her heels into the kitchen floor, she screamed right back at him. As the arguments got stormier, she would dump ice water on him, fling whatever came to hand. Years later, Ted still recalled the time Judy laid him out with a glass ashtray. "But," he remembers, "I got in a lucky punch and decked her."

Ted and Judy were divorced in the fall of 1962. The only surprise was that their marriage lasted as long as it did. "I just couldn't take the womanizing anymore," says Judy. "I couldn't handle the drinking anymore. I just didn't want to be around. I was not having a good time." Judy packed up her belongings and their toddler, Laura Lee, and, like Ted before her, went off to spend the winter in Florida. Thinking she needed time by herself to heal, she ended up in Fort Lauderdale with no family or friends, and, as the days went by, she hated every minute. It couldn't get any worse, she felt. And then it did. She discovered she was pregnant again.

In one of the rare times Ted had been home, around Labor Day, he and Judy had conceived a second child. When Ted found out, he drove down to Fort Lauderdale and begged her to return with him. "Come back," Judy recalls him saying. "I want you back. We'll have this baby. Everything will be fine. I've really missed you." And he promised her, "As soon as we get back, we'll get married again." Like his father before him, he wanted his wife to return. But unlike Florence, Judy agreed.

They drove back to Georgia and began to put their life together. Almost from the start, however, Ted ducked the issue of marriage. As Judy became more noticeably pregnant, she began to become more persistent. It didn't have to be a large wedding like the first, but she wanted some sort of official ceremony. According to her, Ted replied: "You're big and pregnant. I'm not taking a pregnant lady into City Hall to get married all over again. You can wait until after you've had the baby." And later, when pressed, Ted said, "We've lived together this long; it's not really necessary. I mean, you're my wife, you're living as my wife, why do we need to get married? We've been married before; what's the difference?" In the end, Ted would never make good on his promise.

The best moments for Ted and Judy came when they would drive down for the weekend to visit his father and stepmother across the border in Yemassee, South Carolina—first at the elegant Cotton Hall, then, from mid-1961 on, at the more intimate Binden. Jane was tall and elegant and gracious, the perfect hostess; Ed seemed to be one of the most charming men Judy had ever met, with a twinkle in his eye and a gift for spinning long elaborate tales studded with quotations of poetry and folksy Southern expressions like "the billy blue blazes."

Together, the two couples—on horseback or on foot—wandered about the grounds of Binden. With Ted's Labrador Blackie barking at their heels, the four of them would stroll over to watch the planting of a long row of pines along the winding entrance drive, or to see the cattle herd that Ed was hoping to turn into a profitable venture. And always—between Ed and Ted—the talk would come back to the business, to the father's dream for his firm and his boy.

In September 1962, the Turners pulled off the spectacular deal that would build their empire. Minneapolis retailer Bert Gamble had purchased the General Outdoor Advertising firm, Ed's old nemesis in Cincinnati, and was selling off chunks. For $4 million, the Turners could snap up billboard plants in Richmond, Roanoke, and, most exciting of all, Atlanta. All they needed was 20 percent down. Ed decided to abandon his long-running feud with Les Shuman and sell off the Savannah company to make the down payment. He and his son were stepping up to play on a national level now; their firm was going to be one of the biggest in the country.

Ed assigned Richard McGinnis, who had formerly worked for General Outdoor in Atlanta, to find them new offices in the city. McGinnis says, "He had me out running up and down the main drag looking for great big antebellum homes so we could get one of them as a business headquarters. We had to move fast, because he wanted a high-class operation right away." Ed also wanted a first-class apartment for himself in some fancy neighborhood, with enough room for both Jimmy Brown and his Cadillac.

Ted was ecstatic about the deal. The company's expansion was something his father had been working toward his whole life. "The big time!" Ted would exclaim. And then, when his wife and child finally came back home, he must have felt that all the pieces in his own life were coming together. He took a small apartment in Atlanta, in the upscale Buckhead area, for his regular trips there, and back in Macon he and Judy moved into a stylish modern house.

Built over a stream and set in the midst of trees, their new house was a striking, untraditional structure, designed by an architect and featured in *Better Homes and Gardens*. The grounds were landscaped with gardenias, camellias, and a fragrant Russian tea-olive tree. For the interior, Ted decided that it would be great to have the front hallway framed with birds in cages. So, in a fit of exuberance, he bought finches—dozens of finches—and arrayed them along the front hallway. The only problem was that the bars on the cages were spaced too far apart, and in short order fifteen or twenty finches were flying all over the house, swooping about and warbling fortissimo.

But it was only a few months after his father's purchase that things began to go sour. Burdened by his mounting financial pressures and unable to delegate responsibility, Ed began to agonize over his decision. No longer could he wave off concerns with his usual jokes. By December of 1962, he had checked back into Silver Hill, and was once again trying to cure his alcohol and smoking addictions.

When he returned to South Carolina in early 1963 and Ted and Judy came down to Binden to see him, he wasn't drinking and he wasn't chain-smoking, but he wasn't looking too good. "Ted and I thought maybe he had cancer or some kind of chronic illness that he wasn't mentioning," Judy recalls. "He was losing weight, his eyes didn't have that same sparkle, his color was bad, he just looked awful that winter. Just awful. So we were concerned about him— Jane, Ted, and I. Ed was acting depressed. The man couldn't sleep. And when you can't sleep, your whole life gets out of proportion."

Judy read Ed's prime emotion as fear: "I think he was really scared. I think he was truly frightened. Here's a man who's finally made the big time, and all of a sudden his whole world has collapsed. They'd taken away his crutches at Silver Hill. You can't go to a clinic for a week or two and cut out what you've been living on, then step back into the biggest crisis situation you've ever been in."

Jane saw the fear as well: "He was anxious about what *might* happen, which was something that had never worried him before. He really could see the whole thing going down the tubes. I said to Ed, 'If this is worrying you so much, why don't you sell out, and let's go off somewhere?' My prime thought was that my husband wasn't well. I just wanted him out, at least long enough to recover."

Ted was frankly baffled by his father's fear. Sure, they had a huge debt, and sure, his dad knew the business better than he did, but they had almost five years to pay the debt off, and the cash flow was looking OK. This was their big moment. He was upset that his father wanted to sell. More troubling still, it was hard for a son to understand how his lion of a father—the man who had bullied, raged, and pushed him all his life to excel—had suddenly been reduced to this diminished, almost pitiful state. In conversations at Binden, and in that fateful final telephone conversation, Ted tried to rally his father. He pleaded, he cajoled, and finally he started lashing into him verbally, like a coach with a team down two touchdowns in the fourth quarter, or a general trying to rally his troops. He called him a quitter, a loser—anything to motivate the man, to shake him out of his lethargy. His father's silent acceptance of all this invective must have frightened Ted more than anything else. What had happened to the man who had dominated him all those years? Where was his strength, his drive?

Ted hoped that his verbal prodding had done the trick. "I didn't think his father would sell," Judy recalls. "I thought Ted was pretty persuasive: 'We can do this!' I know Ted was very positive."

But in this affair, as in everything else in Ed Turner's life, the father would not be ruled by the son. And so on that gray Tuesday morning in March, he walked upstairs and pulled the trigger.

For Ted, his father's death was a double shock. First, of course, was the suicide itself, something he could never have imagined happening no matter what difficulties his father was experiencing. And then, shortly thereafter, came the surprise of Ed's will. Seated in the office of Ed's attorney, Tom Adams, the family listened attentively as the lawyer spelled out the terms of Ed's bequest: the books, the paintings, the furniture and silver at Binden were to be given to his loving wife, Jane; Jane was also to receive $10,000 a year from a trust fund Ed had established with the Citizens & Southern National Bank of Savannah; Mrs. Maggie D. Turner, Ed's mother in Mississippi was to get $2,000 a year; Florence, Ed's first wife, was to continue to receive her alimony; and to his son Ted, "all the rest and residue of my property of every nature, kind and description."

When the details of the estate were finally all delineated, Ted realized the terrible truth. He had not won Ed over. He had not been able to convince him to change his mind about selling his new acquisitions to Bob Naegele.

His father had given him the ultimate vote of no confidence. Wanting to make certain that his wife and family were well provided

for, Ed had come to the conclusion that leaving the expanded version of Turner Advertising, with its huge debt, in his son's hands
at this difficult moment would only be frittering away their inheritance. His son had come a long way in a few years. He was much
more responsible than he had been during his college days; he was
a young businessman now. But though Ed loved him, at his core,
in his heart of hearts, at the moment when everything hung in the
balance and he prepared to pull the trigger, he didn't really believe
in him. He didn't think his boy could do the job.

CHAPTER 8

IT WAS ONLY a night or two after the funeral that Ted sat down
to dinner with his father's accountant, Irwin Mazo, at an old Savannah restaurant called the Rex. They had just started in on their
steaks, according to Mazo, when twenty-four-year-old Ted leaned
across the table and announced, "I'm not going through with the
sale." He wolfed down a piece of steak and jabbed the air with his
knife. "I'm not going through with it. My father was out of his
mind when he did it. He gave it away, and it's worth a lot of
money."

Irwin Mazo barely knew Ted, and he was still shaken from
Ed's suicide, but the more he heard, the more the young man's
plan sounded like a bad idea. "Ted," he said, "it wouldn't have
been difficult for your father. Your father had relationships with
banks, he had lines of credit. You—you've got absolutely nothing."
In the back of his mind, Mazo had even graver reasons for doubt:
"Ted knew nothing about the operation of his father's business. He
had no experience. His father had given him no authority what-
soever. He had been up in Macon, working in a small plant—selling
billboards, primarily. It was almost a menial task. It was not a
position of authority. If you were going out to look for an executive
to run your company, he would have been a long way down your
list."

But young Turner would not be swayed. "My father didn't

leave me all this money just to go sailin','" he said. "He left it to me to run the company."

Almost overnight, Ted Turner seemed to have changed. Now he had a purpose, a focus that he had never had before. Perhaps his father's death had made him realize it was time to grow up. Perhaps his father's final slap in the face—selling off a large part of the company—had shocked him into action. Either way, as he sat across the table from Mazo, he now seemed entirely serious, committed.

What stood in Ted's way was a sheet of paper from a legal pad, drafted in Ed's hand. Though informally handwritten, this document was a binding contract, which agreed to sell the main plants of Turner Advertising (Atlanta, Roanoke, Richmond) to his friend Bob Naegele and associates for the price Ed had originally paid. In addition, there was the $50,000 profit that Naegele had promised them. Ed Turner's idea had been to leave his son a handful of money plus a smaller Turner Advertising, with several of the lesser plants still intact. In other words, Ed *had* left him a business to run—a diminished business, but a viable business all the same. Ted, however, wanted the whole company—the dream, not the leftovers.

Judy recalls Ted's emotions at the time: "He was going to hold it all together. I mean, this was the big opportunity. . . . It might have been the only opportunity he was going to have. His dad had been waiting for this for years. So why would Ted want to let it go? No way. I don't think there was ever a doubt in his mind that he was just going to [take the money]. I don't think there was ever any question. . . . I mean, he was groomed for this. There was no way Ted was going to let that go."

Determined to void the contract with Naegele, Ted made it clear that he was prepared to go to court to block the sale. Mazo discussed the matter with the Turner family attorney, Tom Adams, and they decided to send an old family friend—Ted's sailing mentor, John McIntosh—up to Atlanta to have a long talk with the boy and explain the facts of life.

According to McIntosh, Adams and Mazo had asked him to go up there and "see if I could reason with Ted." As two of Ed's closest friends and associates, the lawyer and the accountant wanted to honor his dying wishes. They felt it would be in everybody's best interest if Ted would, too. After all, under the contract, even Ted would make out well. He'd wipe out his debt on potentially shaky companies, he'd wind up with the lion's share of the remaining

Turner enterprises, and he'd come away with a small profit as well.

But most important, McIntosh remembers, "they wanted to make him understand that . . . if the sale was set aside, and he should fail, everybody could suffer. They wanted to be sure that he understood that it wasn't just *his* inheritance he was gambling with, but his mother's and stepmother's as well."

Turner and McIntosh met at a restaurant in downtown Atlanta. If the responsibility of safeguarding the inheritance of loved ones might have caused most young men to pause, it didn't stop Turner. He had been marshaling his facts for days. Looking across the table at McIntosh—a savvy Savannah businessman—Ted gave him, one by one, a detailed financial history of each of the acquired properties, Atlanta and Richmond and Roanoke. McIntosh recalls: "Teddy sat there and laid out chapter and verse what the businesses were worth, what kind of profits they were capable of, and that with good management they would pay off the debt, and everybody's interests would be secure. His point was that each of these properties had been self-sufficient, had been profitable—that his father had not paid too much, that it actually was a shrewd purchase, that the price was right, that the cash flows would handle the debt service without any problem."

They talked animatedly throughout dinner, the experienced businessman probing young Turner's argument with pointed queries. Around 10:30 P.M., they went back to McIntosh's room at the Dinkler Plaza Hotel to continue their discussion. As young Turner paced about the hotel room, his energy and enthusiasm seemed infectious to McIntosh. "Dad just got anxious," Ted said. "He got emotional. . . . John, he was scared. I kept telling him we could pull it off, that it wasn't a problem, and he kept saying, 'Hell no, I'm gonna get rid of it.' He sold it for a pittance. I mean, he sold it for what he paid, plus a little, but not what it was worth."

By midnight, McIntosh was convinced that, barring some kind of national economic disaster, Ted would have the company well in hand and the sale paid off long before the terms of the arrangement. Around that time, the conversation shifted into its critical phase. "OK," McIntosh said, "maybe you can do it. But your father has entered a contract for selling. What are you going to do about *that?*"

"Well, John," said Turner, "I'm gonna take these people to court to break this contract if they insist—if they won't settle."

What ammunition did he have to void a perfectly valid and legal sale? Only his dead father's reputation, which Ted was perfectly willing to trash in order to hang on to the entire company. In *his* eyes, it was, after all, his proper legacy.

Inside John McIntosh's hotel room, in the early hours of the morning, Ted mapped out his whole plan: "I'll go to court and testify that my father was emotionally unstable, that he had been on an alcoholic binge, that his thought processes were unclear— that he wasn't capable of making this kind of decision. . . .

"I am perfectly prepared to get on the stand and say, from my last meeting with him, that my father was very depressed, that his chronic alcoholism was very much in evidence, that he was not in his right mind, and that he was not capable of making a lucid or intelligent business decision."

Ironically, alcoholism had been the least of Ed Turner's problems at the time, since he was in Silver Hill going cold turkey when he made his decision. But anything Ted could use was grist for his mill as he prepared to battle for control of the entire Turner organization.

It was close to three in the morning when their conversation finally wound down, and almost in spite of himself John McIntosh was impressed, both by the sheer force of Ted's determination, and, as a bottom-line person himself, by Ted's grasp of the family business.

Exhausted, McIntosh saw Ted out the door, and fell into bed. As he drifted off to sleep, he was thinking about how much Ted Turner had grown up: "I had known him a long time. I mean, he was a little bit of a skinny kid when we first got to fooling with him, you know? He grew up with us. I saw him through his sailing career, had seen him get more and more competent. . . . And you know, I didn't give a damn how old he was, he sounded like somebody who knew what he was talking about. And I came back and told the attorney and the accountant that I didn't think there was a prayer of changing his mind."

As Turner tells the story, he shifted into high gear at this point, a whirlwind of activity, stealing away his old employees, his old customers—even threatening to build new billboards in front of the old ones. Then, he says, he informed Naegele's people of what he had done: "I told them I'd already hired the entire lease department of the company they're supposed to be buying, and I'd already jumped the leases—that's when you go out and get a new lease, for another company. It's sabotage. If you ever want to steal a franchise, that's how you do it. 'Furthermore,' I said, 'I can delay another two weeks, and then I'm going to burn all the records. You're going to have nothing but a disaster.'" Turner would later suggest that it was this ruckus, this threatened scorched-earth policy, that scared the new owners and forced them to agree to relinquish their claim on his father's company. As he tells it, he single-handedly drove off the big guys, like a David of the billboard world. In fact, the real story was somewhat different.

Ted Turner flew out to Palm Springs, California, to meet with the vacationing Bob Naegele. At the time, Naegele's advertising empire spread across twenty-one major metropolitan markets in ten states, from Ohio to Nevada. Naegele had just spent $3.5 million to renovate his billboards in Detroit alone, in what was being termed "the biggest facelift in outdoor [advertising] history." And he had recently paid $10 million to acquire Walker & Company. Twenty-four-year-old Ted Turner was hardly going to intimidate a man of such wealth and power in the industry.

When Ted arrived in the California desert resort community, intimidation had nothing to do with the events that unfolded. He sat down across from Naegele and simply explained that his father had been sick when he sold his company, and that Ted now wanted to back off from the deal. According to those familiar with the discussions, Naegele said, "You know, Ted, your mother and father were always good friends of ours, and we don't want to do anything to hurt you, so we'll consider it."

Not too long afterward, the two met in Atlanta, with Ted's lawyer Tench Coxe in attendance, and struck an agreement. Naegele and his associates would give Ted a break—for a price. Turner had to pony up approximately $200,000 extra; more important, if, at any point, he couldn't make the principal and interest payments on his debt, he gave Naegele the right to step in, cure the default, and take over the entire business. According to observers at the time, Naegele was being kind but also self-interested. Along with everyone else, he doubted that Ted could succeed. So if he bided his time, he might end up acquiring all of Turner Advertising for a pittance.

Ted returned from these talks elated. He had gone toe-to-toe with the big boys, and they had blinked. That, at least, was how he saw it. "He really thought all his maneuvering had caused them [to change their minds]," recalls Mazo. "But in my opinion, the number-one reason was, here was a twenty-four-year-old guy whose father had just committed suicide. They were friends of his father. They didn't want to get into any litigation with him. If they had wanted to, they could have ripped him apart. But they didn't."

Almost immediately, Ted was on the phone to Mazo. "Look," he said, "we're going to have to come up with this [$200,000]." Mazo sighed, "I wish you'd asked me about that before you made the deal." Turner had plenty of enthusiasm, but cash—liquid cash—was in short supply. "In fact," Mazo says, "I had to personally loan Ted thirty thousand dollars once to make a payroll." The company had no way at all to pay off the debt.

Turner had been trying to scrape together money wherever he could. He sold the plantations in Georgia and South Carolina to settle the various real-estate taxes he owed. And only a few days after the funeral he went through his father's books, looking to collect any debts he could. He found that Ed's close friend Dr. Irving Victor owed him $2,000 for a second mortgage on Victor's house: "Ed had given me the house, and the second mortgage [on it]. He had helped support me," Victor recalls. "It was a 'pay-it-when-you-can' kind of deal. He was a great friend." But Ted wasn't letting friendship stand in the way of business. He needed the cash.

"Look," Ted said, "I know that maybe you and my dad had agreements. But you know, times have changed. Things are different. He's gone . . . and I'm trying to clear up things, and this is it." Victor believes that if Ed had lived, the debt "wouldn't have been called." The doctor's impression was that Ted "was taking charge." He'd later say that Ted was "a hard-nosed businessman, all right. It always amazed me that in all the turmoil he had time to take care of that minuscule item."

And so with a minimal amount of cash in hand, Mazo and Turner flew up to Naegele's headquarters in Minneapolis. In theory, they were going to work out the fine points of the payments. Actually, they were going "to stall, stall, stall."

A gold Cadillac picked them up at the airport, and Ted was really impressed. He couldn't stop talking about it. It seemed to symbolize everything to which he was then aspiring. He loved this Caddy—he went on and on about it. But when they sat down with

Naegele, Curt Carlson, owner of the Radisson Hotel chain, and their associates, Ted kept quiet.

"We're all pretty wealthy guys," Naegele said. "We don't really need this money individually. We'd like to give this money to our children—we'd like to cast the transaction in our children's names." Clearly, they were trying to avoid capital-gains taxes.

This gave Mazo and Turner an opening. "That's a different ball game," said Mazo. "OK. We'll pay it to your children, but then we want to change the transaction, too. Give us some more time to pay it." And so, rather than being forced to pay cash he didn't have, Ted was given five years to meet the commitment.

Surprised and delighted, Turner flew home on golden wings. He had wriggled out of impending financial catastrophe. He had survived the crisis. But the Naegele affair was important for another reason, too. It marked the beginning of a pattern that would dog him for much of his business life: "After that, in the company," Mazo recalls, "it was always a scramble to come up with the money. . . . We borrowed from Peter to pay Paul. We were always scrounging for a period of six or seven years."

Now that Ted had the company, he sprang into action at a pace that must have amazed even him. He was on the phone; he was talking to advertisers; he shuttled from meeting to meeting. It was a crisis, and Ted Turner was at his best in a crisis.

According to psychoanalysts, there is a pattern that seems to play consistently across the generations for certain offspring of the bottle. While some adult children of alcoholics become jokers, or loners, or losers, a certain percentage are destined to be what are called "superachievers." And for these individuals, when the stakes are high, when the odds are stacked against them, the pressure rarely appears to oppress them; rather they seem to bloom under it. Perhaps it's because they have had so much practice living in pressure-charged situations. Whatever the reason, they are at their best when life is a series of battles at long odds against almost insurmountable opposition.

But the other half of the pattern is that in the gray humdrum of everyday life, the children of alcoholics often seem to fare less well, often attempting to stir up new crises in order to make the world regain its edge and clarity. In Ted Turner's case, for example, routine life at Brown was stifling. Again and again, with compulsive regularity, he would act out, break the rules, explode into yet another scandal; and in later life the day-to-day grind and details of running a business would always seem to weary him. But here in this moment of crisis, with his father's company—now *his* company—hanging in the balance, he single-mindedly poured himself into saving the firm.

Of course, he had some extra motivation as well. For Ted, saving things had become something of a passion. From his early dreams of missionary work to his frequent rescues of small animals at McCallie, the act of "saving" held a special attraction for him, a special nobility. But the rescuing of Turner Advertising had to be exquisitely poignant. Just imagine the psychic baggage that came with inheriting a company in trouble from a father who had committed suicide, a father who doubted your ability to run the firm in the first place.

In Ed's moment of pain, Ted had not been able to save him. He may have even inadvertently pushed him closer to the edge with his hectoring. If Ted experienced the feelings of guilt typical of the family of a suicide victim, he also was carrying some extra emotional freight: not only his father but everyone else thought he was immature, not up to the task of running a major company.

These were the forces driving Ted forward like a gale-force wind at his back. Ted couldn't save his father, but he was sure as hell going to save his company. He was smart enough; he was energetic enough. He was a bulldozer of determination. He could not be stopped.

Ed Turner had been in the ground no more than a few days when Ted was back at work, running the company full throttle. He had no time for grief, or at least no time to let it interfere with business. Even his tribute to his father in the Outdoor Advertising Association of America newsletter was primarily a plug for new business:

> I want to thank you for your kindness to my father while he was alive. He loved this business and spent his whole life trying to make it better in his own way. He may have been independent in some ways, but he was honest, fair, and never spared himself.
>
> We have amicably concluded recent negotiations with

the Naegele Companies to the satisfaction of all concerned. We will retain Atlanta, Richmond and Roanoke.

I want everyone associated with this business to know that the Turner Advertising Companies fully intend to continue forging ahead, providing the finest in both plant and services.

The real subtext of the letter was buried in the final paragraph—a clear signal to the industry that the new Turner, while every bit the go-getter that his father had been, had no intention of being a loner like his old man: "We intend to cooperate on all reasonable joint efforts with the dedicated men in this industry in the future growth of outdoor advertising, and we pledge to our advertisers our intent to make dollars that are spent with us reap a bounteous harvest in the form of increased sales."

Although no longer writing poetry, young Turner was now turning his poetic impulse to reaping "a bounteous harvest." Suddenly, his name began popping up everywhere. Right from the start, he seemed to have decided that the only way to run the company was to establish a high profile. His father had worked for decades in the billboard trade and earned only a handful of mentions in the industry newsletters, the longest of which was his obituary. Ted, on the other hand, was turning up in print during this period nearly every other month. He took out a large booth at the Georgian Exposition of Commerce and Industry for his "TriVision" billboards. ("The Turner booth was one of the largest displays in the exposition and was probably the most colorful," enthused the industry newsletter.) He soon announced that all eight of his major plants—Atlanta, Augusta, Macon, Columbus, Charleston, Richmond, Roanoke, and Covington—would become members of the industry group Outdoor Advertising Incorporated.

Appearing at the 1963 Outdoor Advertising Association of America Convention, he delivered a speech on "Leasing—A Key Factor in Operations." After discussing the fine points of single-panel requirements and perpetual leasing, Ted managed to work in a plug: "I'd like to point out that we're tenacious, we don't accept marginal locations. We pick the location that we want and then keep going back until we get it. In fact, we literally hound the people to death in a nice way. Personally, I've had people throw up their hands after forty hours with them on about thirty calls and say, 'OK, OK.'"

There is a photograph of Ted at that convention, sitting with the other speakers on the leasing panel. Presidents and vice presidents all, they are men of substance—solid, square-shouldered executives in their fifties, with wrinkles and glasses and large beefy hands. In the middle of the group sits young Ted, their junior by several decades, fresh-faced, angular, and just a little cheeky, staring out at the camera with a cool look that says, "Hey, I may be young, but I'm going to show these old farts a thing or two."

It was an exciting time, the early 1960s—a time of new frontiers in the White House and across the country. "In this soaring Space Age," said billboard magnate Harry O'Melia Jr. "all business, all media are feeling the Challenge of Change. . . . Growth is seen and felt everywhere. . . . This age is no time for the faint and the faltering."

With the turn of the decade, the billboard trade was trumpeting the story of "Supersaturation": "Think BIG and you'll sell BIG," screamed the *OAAA News,* and it sounded good to Ted. "Bigness itself gets attention and supersaturation is big. . . . supersaturation creates excitement. . . . Every word must be full of excitement. This is an exciting idea!"

Caught up in the ferment, Ted *was* thinking big—literally. Along the newly constructed Atlanta Freeway Connector, he built a giant billboard sign, seven stories straight up. It was, reportedly, the largest painted billboard anywhere in the South, and one of the largest in the nation. While advertising his clients, Ted was also advertising himself—carving a piece out of the Atlanta skyline to announce his arrival as a major player. Three huge thirty-six-inch steel beams held up the sign. Eighty-five feet from end to end, it read "Miller High Life—sparkling . . . flavorful . . . distinctive!"—even the small "i" stood larger than a human being. In the evenings, the entire board was lit up with 15,000-watt lamps and transformed into a dramatic, electric, nighttime spectacular.

Turner Advertising was getting big, too. Hardly more than a year after his father's death, Ted Turner was being featured on the cover of the *OAAA News* in a glowing article examining the company's spectacular sales. And the way they had raised those sales would become vintage Ted. He and Dick McGinnis, who was now company president, had sent a letter to the wife of each Turner salesman, marked "PERSONAL" in large red letters:

> On April 1st, some Turner salesman's wife is going to be driving this new Chevelle. It could be you. It can be you if your husband wins the "Local New Business Contest" that ends March 31. . . . I'll bet your guy can do it—don't you agree?

Turner had taken the toughest, least remunerative section of the business—sales of small, local business accounts, as opposed to national accounts like Coke or Miller—and had managed to squeeze money out of it. The results were $147,814 in brand-new contracts from local firms such as Belk's Beauty Salons, Gibson Tire Service, and Veteran Cab. Ted was wringing out every dollar he could and trumpeting his success at each opportunity along the line.

Amazingly enough, throughout the period of his father's death, the company upheaval, and the new, high-powered start, Ted never stopped sailing. He often sailed the way he ran his business, risking everything, his spinnaker up and flapping when no one else would take the chance, and daring the weather to prove him wrong. In "heavy" air, with the wind blowing like crazy, others might be more conservative rounding a mark, but not Ted. He tried to force his way in, to slip past the other boats. And in calm, still air, he had a way of squeezing out every second: "Ted Turner of Savannah, Georgia, turned out to be a surprise contender," read a newspaper account of one regatta. "He earned the respect and admiration of the entire fleet for his ability to get all he could out of very light air."

In August of 1963, just five months after his father's death, with his company still in turmoil, Ted entered the national championship in the two-man Y-Flyer class being raced that year out of his old hometown Savannah Yacht Club. He was sailing with Judy, and together they made a good team—good enough to win the national crown. It was Turner's first major victory in what would become a long string of sailing championships. The next day the two of them appeared in a big photograph in the *Savannah Morning News,* proudly reclining side by side next to their boat—a former local couple making good.

In fact, Ted and Judy were only barely getting along. "Ted can be a nasty person on a boat," says Judy, "screaming and yelling and jumping up and down. He'd yell, 'You stupid ass! Why didn't you do that faster? Why'd you do this, why'd you do that?' It was abusive most of the time—verbally . . . and physically. He'd haul off and hit you, or elbow you in the ribs. Bounce you off the head, or whatever. Then again, he wasn't selective. I don't think it stopped or started with me."

Judy and Ted's relationship was going from bad to worse. Only a few months after her return to Georgia, with Ted's protestations of love still ringing in her ears, Judy found that he was again seeing other women. "I was really upset," she recalls. "I just assumed that if I was living in the same town that he was, at least he'd be really discreet."

Judy will never forget one night while visiting in Atlanta, when she was almost eight months pregnant and waiting for the baby to be born. As the evening wore on and Ted still hadn't returned, she was feeling miserable and alone. "Maybe he's working late," she told herself. She dragged herself downtown to look for him. Arriving at his office, she poked her head through the door into the darkened interior, and there was Ted—naked on the sofa with a woman. He sprang up guiltily, covering himself with his shirt. Judy never knew who the woman was; she walked home in a daze. "I thought, By God, I've taken enough from him," she recalls. "I really wanted out big time. I was really upset. I just lost it. I mean, absolutely lost it."

Judy had had her doubts about returning to Ted from the start: "If my baby hadn't been [on the way], I probably would not have come back to Ted. I had cut my losses, only to go back and gamble again." Their son was born in May of 1963. In a way, Judy recalls, "he was the child nobody wanted." Or perhaps more accurately, he was born into a family that nobody wanted. For Ted, in fact, was excited to have a male child, an heir. There was no question about the boy's name: Robert Edward Turner IV. They called him Teddy, and while Ted spent no more time raising him than he had with Laura Lee, he was proud to have a son. The satisfaction was mixed with sadness—sadness that his father had died just months before seeing his grandson. "I wish my daddy were here to see him," Ted said, looking down at his newborn child.

Ted and Judy and their two children moved back to the house in Macon, but they were really living separate lives. Ted stayed at the apartment in Atlanta during the week, then came down to Macon on the weekend. He would walk in the door, strike a pose by the mantle, cigar in hand, and thoughtfully quote lines from *The Peloponnesian War* or some other classic. He played a little with the kids; then he and Judy would go off sailing, leaving the children with Jimmy Brown. As Jimmy had taken care of Ted and his sister, he now looked after Ted's kids. He was wonderful, Judy says. If he could have nursed the babies, he would have.

Though sailing held the couple together for a time—with victories like the Y-Flyer championship—it was also sailing that eventually drove them apart. Early in 1964, Ted and Judy went up to Lake Allatoona to enter a regatta that stretched over several weekends. It was merely a series of one-man dinghy races—hardly a national championship—but as the last race started, Judy was winning the series and headed for a clear victory in the overall event. She was feeling pretty good about herself. Leading Ted and the pack into the final leg of the final race, she rounded the last mark and set her course for the finish line. That's when Ted decided that he was going to play hardball.

With Judy's boat just slightly ahead of him, Ted brought his dinghy up into the wind, forcing her upwind as well. She was being driven farther and farther off the optimum course, and Ted, by virtue of his position, was technically allowed to do it. Judy couldn't believe he intended to ride his own wife out of the race. Assuming Ted would soon let her by, she tried to stick to her course, but her assumption was wrong. Their boats collided, Ted ramming into Judy, and the foul was called against her—not just disqualifying her from the race, but costing her the entire meet as well. "I thought he was kidding," Judy recalls. "I didn't think he was serious. Here I was, I was finally sailing in my own boat, I was really doing great, I couldn't believe it. I didn't really think he'd go through with it. . . . I guess he was jealous because I was winning the series and he wasn't, and he wasn't going to let this happen." When Judy got back to the shore, she was fuming. Ted, who saw it as all in a day's competition, laughed away the whole thing.

"That was the last straw," Judy says. "It was really devastating. It was just finally taking away any pride I had. So I just lost it. I had him take me down to the bus station; I came back to Macon, I packed up my stuff, and I told him, 'This is it. I'm going. Goodbye.'" Judy's mother, Bettye Herb Sollitt, and sisters came to take her back to Chicago with the children. A few weeks later, in front of an Atlanta judge, Ted and Judy were divorced for the second time, although they had been officially married only once: "I really felt a deep sense of shame that we hadn't gotten married again, officially," Judy recalls. "On the other hand, my attorney said, 'Well, you've been filing income taxes together and you have an address together, and he said to you "I do" and you entered into this thing.' I had become his common-law wife." Judy sighs. "These things, they're real easy to enter into, but really hard to get out of."

Ted didn't find it hard at all. He shrugged off the marriage with a simple "we weren't really compatible," and turned his sights on Atlanta. There were so many women available for a charming young millionaire of twenty-five like himself. This was 1964, after all, and, as Bob Dylan was singing on his new album, "The Times They Are a-Changin'." Everywhere you looked, there were the new *Cosmo* career girls. Natalie Wood was starring as Helen Gurley Brown in *Sex and the Single Girl*. And Atlanta was hopping.

Atlanta was an exciting city on many levels. The rising new metropolis of the South, it glistened like Oz for ambitious young Southern men and women. "Atlanta was a young town full of opportunities," recalls one businesswoman who arrived there in the early sixties from North Carolina. "It was where the young people who cared about where they were going and what they were going to do went." Gleaming new office buildings were going up. Across the South, a new attitude was being born—an attitude of Southern sophistication—and Atlanta was its hub. These young people, in their tailored navy suits from Rich's and Muse's department stores downtown, weren't good ol' boys or gentleman farmers or redneck hicks—they were young professionals. And they packed the sidewalks of the metropolis.

Just a few years earlier, on Saturday, October 10, 1959, at 9:00 P.M., the population of metropolitan Atlanta had reached the one-million mark, and the city wasn't looking back. One quarter of all the people in the state of Georgia now lived in metropolitan Atlanta, and a new person kept showing up every eighteen minutes, eager for what the city promised. Atlanta in 1964 was the heart of the South.

Ted, newly single, was on the make in every way. He quickly gravitated to young, middle-class Atlanta's hot spot for socializing and networking—the Fulton County Young Republicans Club. "It was the place to be on Friday night," recalls Billy Roe, a friend from that time and president of the club. "We had these 'Thank God It's Friday' parties and everybody came, even if they weren't Republican. It was the young set in Atlanta that was going to do something. They were the movers and the shakers."

Each Friday night at 6:30, nearly a hundred young professionals would gather at downtown clubs, or in members' houses or apartments. Some were drawn by politics—a fiscally conservative, racially moderate approach to government (as distinguished from that of the governor, Marvin Griffin, who was one of the last of the Georgia segregationist Democrats). Many who would later run for office or play a role in state and national politics could be found there on those Fridays, among them Paul Coverdell, who would head the Peace Corps from 1987 to 1992, and the powerful congressman-to-be Newt Gingrich.

Ted Turner was part of that group, although no one believed he was there for the politics. (One friend, much later, would describe Ted as "fundamentally conservative and opinionated, although almost a virgin in politics.") He would come in, chomp a cigar, and stand in the back, drinking and eyeballing the girls, while the speakers droned on. Roe says, "We had the best-looking women in town." More often than not, Turner was accompanied by a small group of friends who worked for him—a handful of sharply dressed young men looking to blow off steam at the end of the work week. "It was a team, like a basketball team," Roe recalls. "They were four or five players and they basically did everything together. And Ted was the head of the team."

Billy Roe's wife, Nancy, recalls Ted at the time as a dashing young Southern gentleman, a Rhett Butler type, which was exactly the kind of image he was trying to cultivate, and in time he'd even add the mustache. "I think every single gal in that room would have been glad to have been the one that he picked out to date," she says. "He was *very* attractive. He had sex appeal or whatever you call it."

One Friday night in the spring of 1964, as Ted was talking to

Billy and Nancy, he told them, "I'm lookin' for a cute gal. Who do you know? Who do you think is nice? Who's a great gal?"

They both thought at once of Jane Smith, a modest, attractive blonde from Birmingham, Alabama, who was a stewardess for Delta. She was rooming with a close friend of theirs in Colonial Homes, an Atlanta development of rental units with the feel of a college dormitory. Jane helped out at the Young Republicans, making phone calls, typing name tags. She wasn't really there for any passionate political belief. "She was basically a conservative Southern girl," says Nancy Roe. "She was this gal that came to Atlanta to meet someone and have a job and get married. Back then, young ladies didn't go bouncing out to the bars, but Young Republicans was like a fraternity or sorority party. This was the set she felt comfortable in. She was a nice girl. Nice as could be. She fit the bill, really. She was pretty, she was quiet, she was a good listener—you'd have to be with Ted."

Billy and Nancy arranged a double date for dinner with Ted and Jane. Ted talked, Jane listened, and Billy and Nancy agreed they were a good-looking couple, with his lean, lanky magnetism and her blond, cheerleader looks and delicate charm. Soon they were dating regularly, and in almost no time, they were setting a date for the wedding.

"I think she was the right woman for him," Nancy Roe says. "She went along with his lifestyle. She did not hold him back in any way. She wasn't going to match wills with him. She was the perfect wife—very attractive, and a gracious hostess."

There was one other thing, too. As several of the couples who knew them most intimately at the time were aware, Jane was pregnant. Ted had knocked her up.

No one was going to put a shotgun to Ted Turner's head. Still there was enough of the Southern gentleman about him to know the proper thing to do. Then, too, Jane really did seem a good match for him—blond and pretty and, above all, deferential, unlike that troublesome Northerner, Judy. Though Ted's emotions were mixed, it was far from the worst hand he'd ever been dealt.

They were married on June 2, 1964, "the same day," he would later say, that he married Judy. (This might have been a sign of trouble to come, for in fact his memory was a good twenty days off; he'd married Judy on June 22.) In Atlanta, a dozen of their friends gathered for a small reception. After a few drinks, Ted took Irwin Mazo and his wife aside and announced to the startled pair, "I didn't really want to get married. I didn't want to marry Janie. I said I'd marry her because she was pregnant—but don't expect me to be a good husband."

Good husband or not, they set up a comfortable home in a red-brick colonial on a cul-de-sac called Carriage Drive and turned out three children in quick succession over the next five years. "Ted named them," says Jane Turner. "Rhett is named for Rhett Butler; Beauregard—or Beau, as we call him—is named for General Beauregard; and he wanted to name Jennie Scarlett, after Scarlett O'Hara. But I wouldn't let him. I thought that might be a little too much for her to live up to. Then he decided to name her Jeannie, after Stephen Foster's song "Jeannie with the Light Brown Hair." But I changed it to Jennie."

Right in the middle of all this birthing, his two older children, six-year-old Laura and four-year-old Teddy, suddenly turned up too. Judy had gotten remarried to a young man who, she claims, "was really more of a father to the kids than Ted had ever been. He took more time with them, he loved them." But, according to Judy, he was also "really into drugs." As a weight lifter, he started taking anabolic steroids and other drugs in the mid-1960s and, Judy says, it changed him. It was sometime between 1965 and 1966, according to her, that he started being physically abusive to the children.

"I didn't know what to do," Judy says. "There was no one to talk to, no one to see. He couldn't control it. And it just got worse and worse. . . . Teddy particularly was black and blue, and it was very bad." That year, with a heavy heart, Judy sent Laura Lee and Teddy down to their father for their Christmas vacation, her emotions an unhappy tangle of shame, guilt, and frustration at her powerlessness.

When the children arrived in Georgia, Turner was shocked. He called Judy on the phone.

"This looks *real* bad," he said.

"It is real bad," said Judy.

"Teddy's been beaten," he charged. "What's going on?"

"It's bad. I don't know what to do."

"You don't want me to send the children up, do you?" Turner said. "I can't send the children back up."

She paused for a long moment.

"I know," she said. "I knew when I sent them down they weren't coming back."

When she got off the phone, Judy went into the bathroom, sat down in the empty bathtub, and began to cry, tears streaming down her face. "It broke my heart," she recalls. "I sat in that bathtub for twelve hours, just going berserk. I don't suppose I really wanted to kill myself, because I had another little one. But I just had to close the door of the kids' rooms. I had to go on, I guess."

In the spring of 1967, shortly before her divorce from the weight lifter, she began to think about regaining custody of the children. One day she impulsively decided to fly to Atlanta and try to get them back. It wasn't a well-thought-out plan; she hadn't even told anyone that she was coming. When she took a cab to the Turner house and walked up to their front door, she didn't know what to expect.

Ted and Jane wouldn't even let her see the children. Looking back after nearly thirty years, she admits, "They were right. It would have been disturbing. You don't just grab up two kids that have started another life. And Ted had money, he had position, he had power, he had a wife. They seemed to be happy together." It was, she's convinced, the right decision for Laura and Teddy, but the pain of that decision drove her to some hard times.

Judy saw her children on only a handful of occasions after that. And the cards, letters, and Christmas gifts she sent were for many years never given to them. Her one unexpected ally was Ted's mother, Florence. When Teddy and Laura went to visit their grandmother, Judy would sometimes get a letter from her daughter, or a phone call, or a photograph of the pair.

Whether there was one child in the house or five, Ted's life didn't change very much. His infatuation was sailing, and following his first national championship in 1963, it turned into a full-blown obsession. By 1964, he was spending $100,000 a year on sailboat racing—a substantial sum in those days. He was also spending almost every free moment he had on the water. And though he would win more championships over the next two years in relatively small boats like the Flying Dutchman and the 5.5-meter class in closed-circuit competition, he had begun to thirst for higher stakes and even bigger action—the 50- and 60- and 70-footers of ocean racing.

In 1964, his old friend Johnny Baker called and asked if he wanted to join a Savannah syndicate that was buying an ocean racer. "Nope," Ted said. "I'm gonna have my own boat by the end of the year." With no large-boat experience at all, Ted chartered a 40-footer called the *Scylla* in New England, and with Irwin Mazo, Jimmy Brown, and a few other inexperienced ocean sailors, set off on a southerly course for Florida, planning to arrive in time for the start of the Southern Ocean Racing Circuit. (The famed SORC consists of a series of six winter races, usually held from January through March in the waters between Florida and the Bahamas.) "It was the damnedest four days of my life," Mazo recalls. As storms racked and roiled the waves and pitched the boat around, they soon discovered that their radio and much of the gear were broken. A galley fire started, almost flaming out of control. And still Turner plowed on through the night. Even he admits that it was "hairy." At one point, he ran the boat aground just south of Cape Hatteras. "Give me a million bucks and I wouldn't go to sea with Turner again," Mazo swears.

By 1966, Turner had not only entered the SORC in a new boat, the *Vamp-X*—his own ocean racer—but he won, and did so by the biggest margin ever. He triumphed by pushing his crew to use every minute of every day, racing through the night with an almost maniacal drive. While most crews were used to sleeping at night in these long-distance races, Turner always worked with a twenty-four-hour racing plan. "You can't win ocean races," Turner told a *New York Times* reporter, "without working harder than the other guys."

Ted had fallen in love with ocean racing—the excitement of the competition and the sheer beauty of it: "Some races are so beautiful that you're sorry to finish," he would later say. "The Montego Bay race in '66 was one. It was a full moon, clear nights, warm, and there was a good wind. The world was *beautiful*. In our sport, you're out there with nature—you're as close to nature as you can be—with gulls, flying fish, whales, the dawn, the sunset, the stars.

"You take a deep breath and you feel alive, really alive. The brilliance of the stars is hard to describe. You think you can reach out and grab a handful of them. It's as if they're ten feet away, and there are millions of them."

His one other passion through this period was work—perhaps not precisely the billboard business (although he would spend hours talking about panel placement and plant depreciation to anyone who would listen) as much as the simple allure of creating an empire, striking it big. Billboards were only a means to an end. Ted was dreaming the big dream. "The dream," Judy Nye says, "was just to build on the dream, till you can't go any further."

Ted's colleagues at Turner Advertising and his friends from the Young Republicans Club circuit can all remember how Turner would get on a roll and start to speechify about the future, rambling on and on like some half-cracked visionary. "Stick with me," he'd proclaim. "If I make it, you're gonna make millions. Stick with me. We're goin' places!"

When he'd get really stoked up, he'd take on all comers, all the real and imagined critics who ever said he wouldn't make it: "Those dumb bastards! We're gonna make something out of this thing. We're gonna make this thing work. If we make it, you're all gonna be rich. And if we don't—well, what the hell! What've you lost? You'll be young enough, you can do something else—go down to Florida and sell advertising in the Yellow Pages. What the hell!"

Turner was in the empire-building mode. Although his company was still on shaky financial ground, Ted was plunging ahead on all fronts. Where many would have paused, at least momentarily, to consolidate, especially when they were chronically short of cash, Ted, a mere fifteen months after assuming the helm of a company that his father considered so tenuous it had driven him to suicide, had no interest in playing it safe.

He was aggressively buying new plants. On July 1, 1964, he took over the Tennessee Valley Advertising Company, with control of forty towns in Tennessee and Georgia, including the plum of Chattanooga. In early 1965, he acquired the assets of the Knoxville Poster Company, which contained the most prominent plant in the city of Knoxville. His company billings shot up over the $4-million mark. By the mid-1960s, the Turner Advertising Company had become the South's largest outdoor-advertising firm, with ten key Southern markets spread over eight states. In terms of its total number of billboards, Turner Advertising ranked as the fifth largest outdoor-advertising company in the nation.

How was Turner paying for these new acquisitions? In part, his enthusiasm and drive were motivating his sales force to new heights of success. Showing a deft understanding of the technique of the carrot and the stick, he was proving to be an inspirational leader. Sometimes he would explode at employees, chewing them out the way he chewed out his sailing crews. But often he was generous with them, too, and it paid off. One six-month sales contest for small local billboards alone generated over $2 million. "When the original idea of a sales contest for our sales force was thought of," Turner recalls in *OAAA News,* "we decided to put up some pretty extravagant prizes by other companies' standards, and it has paid off for us." For the first time, an outdoor-advertising firm was offering a piece of the business—company stock—as an incentive.

But if sales were improving, they were still not generating enough money for everything Turner wanted to accomplish. And so what he did, mostly, was borrow. He became a master at emptying bankers' pockets. "What you do is you get a bank, and you borrow all you can borrow," he'd later declare in speeches to business groups. "You borrow so much that they can't foreclose on you." That, according to friends, was "the secret of his success." Turner himself would be quite open about it. "I don't have any money," he told one startled friend. "I'm just gonna keep borrowing. You know, it's like a Ponzi scheme. The whole thing is a Ponzi scheme."

Of course, not all bankers were taken with Turner. Many were positively put off by the brash young man whose mouth kept flapping away. Irwin Mazo remembers taking Ted to meet Dick Kattel, then the head of Savannah's Citizens & Southern National Bank. The two came together like pit bulls. "Ted told Dick he didn't know what the shit he was talking about," Mazo recalls, shaking his head. "Dick called me aside and said, 'If you ever bring that son of a bitch in to see me again, our relationship is through.'"

"I think you had to take Teddy for what he was," Billy Roe says. "There wasn't anything hidden. No hidden agenda. Here I am. Fuck you if you don't like it." In a lot of ways, Ted had become like his father—*Esse Quam Videri*. Most people didn't know *what* to make of him.

Billy Roe, who was an investment broker, remembers going to visit Ted in the mid-1960s to talk about trading his stock. The meeting was slated for half an hour, but Turner pulled out a big cigar, leaned back in his chair, and started to talk. He rambled on for what seemed like hours about depreciation, taxes, how leveraging was going to be the answer for him. He laid out an entire plan for his business career to the startled Billy Roe. "He just went on and on," recalls Roe. "And I'm sitting there thinking this guy is really . . . you know, this is very unusual. This is a pretty unique guy. I mean, here I was in my mid-twenties listening to a guy talking about how he was going to take over the industry. He had bought Chattanooga, and he was going to keep parlaying that on and on.

"And then . . . he just sat back and said, 'And what the hell! If it doesn't work, I can't do any worse than my father did. I'll blow my goddamn brains out.'"

Billy Roe went home amazed and shaken. Was the guy a lunatic? A genius? There was no way he was going to touch Turner's stock; it was too thin. But what a tale he had to tell his wife: "Hey, Nancy," he said, "I just heard the wildest story I've ever heard in my life! Teddy's really off the wall!" And in that moment, their perspective on Turner changed. "We used to think, Now, there's a businessman on the go!" Nancy Roe says. "Then Billy goes to this meeting, and all of a sudden you realize that Teddy's going to be front-page headlines—one way or the other. Either he's going to succeed, or he's going to fail . . . spectacularly."

Throughout the 1960s, as his business started to grow and flourish, it became clear to those who knew him well that Turner had, in the words of one friend, "a dark side." When he'd had a couple of drinks, he'd wax philosophical and talk about his fascination with Howard Hughes. He had read and reread *The Carpetbaggers* several times, and he felt sure his life would parallel the multimillionaire's. He, too, dreamed of a huge business success; he, too, wanted to be an innovator in diverse fields; he, too, was transfixed with the movies; he, too, was a womanizer; he, too, would dominate an industry. And in the late night hours, over the last cocktail, when the talk turned as soggy as the martini olive, he worried that he, too, might end up like Hughes or William Randolph Hearst or one of those other American titans—alone and unloved.

"In this life, you meet a lot of people," says Billy Roe, "and some of them you envy their success. But Teddy's different. Teddy's in a world of his own as to achievements and lifestyle. You have to respect what he's done. But I wouldn't want to be in his place."

Though Ted Turner was building his father's company into a greater success than it had ever been before, his father's specter was always with him—prodding him forward, dragging him down. Nancy Roe remembers, "He was always saying, 'Come with me, we'll make lots of money—and if we don't, what the hell!' Teddy held an ace, a trump card. He always knew that if he messed up everything, if everything fell flat, he had the ultimate out, which was suicide."

His father had done it, and he could, too. Often he would talk cavalierly about killing himself just to shock people, or to give himself an edge in a business venture. He'd even use it as a way to get bankers to back off on their demands for repaying loans: "Keep pressing me," he'd warn them, "and I'll just kill myself, and then you won't get anything." But the sheer number of times his conversation came back to suicide indicates how much it preyed on his mind. "He always talked about that," Nancy Roe says. "He always knew that he would do that if that was the way he needed to go."

But it certainly wasn't the way he needed to go just yet. His business was looking good, and he was eagerly looking forward to making it grow still larger. Skating toward the future, Turner sped over thin ice, his momentum carrying him safely along.

And just up ahead lay a whole new world.

CHAPTER 9

JIM RODDEY has a sly, dry, tight-lipped sense of humor. Ted, not especially given to jokes himself, didn't hold that against him. He had met Roddey many times at advertising conventions and been impressed. Here was a bright young guy who in the mid-1960s had already risen to become the Atlanta-based head of Rollins Outdoor Advertising.

One weekend, Ted invited the Rollins executive to crew for him on Lake Allatoona. It was a local event for Y-Flyers. Despite the unimportance of the race and the modest level of the competition, Ted was, as usual, all business. But they were still only in second place as they began their final leg going into the wind. Ahead of them in first place was some Sunday sailor and his wife. "You would have thought it was Dennis Conner," Roddey says ironically, recalling that afternoon. An intense tacking duel began. Ted was growing more and more furious. He felt that everything Roddey was doing was wrong and told him so in no uncertain terms. "Goddamn it, Roddey," he screamed, as they sailed across the finish line behind the winners, "it's all *your* fault!" Roddey stood up, his fists clenched, ready to deck the loudmouthed son of a bitch. Suddenly, Ted smiled and stuck out his hand. "Good job, Roddey," he congratulated him. "Great race!"

It was no surprise that when Ted asked him to join the Turner organization in early 1968, Roddey didn't exactly jump at the opportunity. Rollins at that time was the fourth-largest billboard company in the country. "In fact," Roddey points out, "it was quite a bit larger than Turner." And Rollins had also developed a successful broadcasting division, with Roddey in charge. He didn't exactly need Turner. But Turner needed him. Roddey was just the guy Turner wanted for what he had in mind.

Ted had grown bored with the billboard business. Unless two companies are locked in some peculiar grudge fight (like the war between his father and Les Shuman in Savannah), outdoor advertising tends to be more monopolistic than competitive. And by the late 1960s, according to Irwin Mazo, "We had really got the company on a pretty sound footing." Turner, having run out of challenges, was getting impatient. Here he was, on the brink of turning thirty, stuck doing business as usual. So he had begun to think of new businesses, new industries to conquer—like broadcasting.

Roddey says, "Ted knew he wanted to diversify, but more important, he was trying to get money. And the banks had simply told him that he didn't have the kind of management team they would require to provide financing." His pursuit of Roddey was part of an overall strategy to enhance his credibility to bankers.

Ted's father used to tell him that if he really wanted something, if he *had* to have it, then it didn't matter what he paid to get it. Ted had no intention of being turned off by Roddey's lukewarm response to his offer. He had found his ideal man, a guy who combined solid management skills in both billboards *and* broadcasting. "Typical of Ted," says Roddey, "he put on a sort of full-court press." In his attempt to convince the Rollins executive, Turner met with him repeatedly, asked him out to dinner, went over to his house, and in turn invited Roddey to his own.

Hour after hour, Ted told Roddey just what he planned to do with the company. Step by step, he outlined his vision, and what needed to be done to accomplish it. He painted the future in glowing terms, promising Roddey that he would be the president and chief operating officer of the company; Turner himself would be the CEO. And together they would get very rich. After about two weeks of unrelenting pursuit, Roddey took the job.

Almost immediately he discovered that for all Turner's talk, there were some gaping holes in the company. A few years earlier, Ted had tried to diversify Turner Advertising by investing in a Fort Worth firm called Plastrend, which made fiberglass racing sailboats.

It was run by a sailing acquaintance of his named Andy Green. Though Green may have been an engineering genius, he was not a very good businessman, and when Roddey joined the firm he discovered that the Texas company was "a disaster." It was losing $100,000 a year. In addition, Ted owned an entire roomful of computers, sold to him by a friend at IBM, for which he had absolutely no use. For all his penny-pinching ways, Ted could throw away money on a grand scale. "So my job," Roddey explains, "was to eventually get rid of Plastrend, to hire some decent management, obtain long-term financing, and get us into the radio business."

In late 1968, Ted decided to purchase radio station WAPO in Chattanooga. WAPO had been in existence when he was a student at McCallie, and it would be his first station. He and Roddey and Dick McGinnis, president of the billboard division, drove up to Chattanooga to look the station over. Playing it cagey, Ted told the strapped station owners that he'd think about it and left. But before he could actually make an offer, another buyer had bought it. Ted hit the ceiling. He called WAPO's new owner and asked what he wanted for it. As McGinnis later told *Washington Post* reporter Christian Williams, "We wound up having to come up with another $300,000, and keep some people on the payroll, and there was a stock deal, too. It really cost us two or three extra ways. Ted learned a lesson from that. Now he just decides on the spot."

On January 1, 1969, Turner changed the station's call letters to WGOW. It was Roddey's idea. "The letters stood for Go-Go radio," he says. "Sort of a takeoff on the go-go-girl theme." It was, after all, the sixties, the decade of the Frug, the Pony, and the Swim, and everywhere discotheques were popping up. So were Turner radio stations. Having acquired Chattanooga, Ted went on to buy stations in Jacksonville and Charleston. "Guess what," he casually reported to a friend one day while they were out sailing. "Now I got three radio stations."

Although Ted liked the idea of owning the radio stations, he took little interest in their operation. He was, according to Roddey, primarily concerned with sailing and making money and, above all, building an empire. As before, he was constantly trying to acquire more financing, to borrow money to build on—and money was still in short supply. One of the places he went to was Teachers Insurance & Annuity Association, a pension fund in New York.

Irwin Mazo went with him. At one of their meetings with TIAA, as Mazo started to explain the firm's future plans to the gathered TIAA executives, Ted suddenly exploded. "Goddamn it, Mazo," he screamed at his accountant. "How stupid can you be! That ain't what I want to do. This is *my* company, goddamn it!" Though they got their money, Mazo says, "He embarrassed people. He did that to everybody. It was his way. The next day we were friendly."

Another place Turner went to seek financing was the brokerage house of Robinson Humphrey in Atlanta. He thought that going public might be the answer to his problem of dependable long-term financing. If he could sell stock publicly, who knows how much money he could raise? Ted turned to Lee McClurkin, an executive at Robinson Humphrey whom he had known from their Savannah days.

"When Ted came to see me," McClurkin recalls, "he was trying to figure out how to raise more capital. . . . I told him it was going to be difficult to do." But, he added, "if you want to get into TV, I've got an idea."

Remarkably, it was in this accidental, offhand fashion that Ted Turner, the founder of seven television networks, was first introduced to the TV business. He had thought fleetingly about television before, but never seriously, because, according to Roddey, "we always assumed we didn't have the muscle, the borrowing power—our financial situation was too precarious."

McClurkin had a problem of his own at the time, a major one. A year or so earlier, he had arranged a public stock offering to help a wealthy Atlanta coal dealer, Jack Rice, raise capital to get into television. Rice, something of a playboy, spent much of his time down in Florida at the Palm Bay Club, and WJRJ/Channel 17, his UHF station, was soon struggling just to survive. "Quite frankly," McClurkin says, "the station was a bunch of old crap that wasn't working very well." With the station failing, Robinson Humphrey was about to become a serious loser. It was at this point that thirty-one-year-old Ted Turner walked in the door, looking like the answer to McClurkin's prayer.

"My idea," McClurkin says, "was that Turner would be acquired by WJRJ for about 80 percent of the stock in the TV station. And that way he would go public through the back door." Ted liked the idea immediately. "I had never watched the station, because I couldn't even get it on my set," he recalls. "I never watched any television in those days. I had no idea what UHF stood for."

But none of this made any difference to him. Turner had been in the advertising business all his life, and to him television was just another way to sell advertising. He already owned outdoor billboards and audio billboards (his radio stations). Now he'd own video billboards. Whether his signs were strung along America's highways or floated out on its airwaves, when it came to the dollars-and-cents part of the business, he'd be flacking the same toothpaste and soda. And at that time the broadcasting business was experiencing a meteoric rise. Then, too, it was glamorous. "Broadcasting," McClurkin points out, "was a lot sexier than the outdoor business."

Knowing nothing about television, Ted consulted people he trusted for information. One of them was Don Lachowski. Lachowski had at one time bought advertising from Turner and was now moonlighting on special projects for him. Ted explained that he was thinking of buying WJRJ. Lachowski says, "I told him that he was going to lose in the vicinity of seven hundred thousand dollars a year for the first two years."

Lachowski wasn't the only one who had his doubts. Mazo discovered that WJRJ had lost $900,000 the year before. He and Roddey and Ted's lawyer, Tench Coxe, were all dead set against the deal, which would cost the company about $2.5 million in stock. "We tried everything we could do to keep Ted from buying that station," Mazo remembers. Independent television stations were a dime a dozen then, and UHFs had the life span of a grasshopper. "We didn't have any money to lose in a television station." Mazo feared that their billboard profits would be gobbled up by this TV turkey, but when they tried to find Ted and stop him, they couldn't. "They hid him," says an incredulous Mazo. "Robinson Humphrey literally hid him until the deal went through."

When reminded of these events, McClurkin explodes in laughter. "Clearly, as far as I was concerned, it was a marriage made in heaven. I had a helluva problem." If it wasn't exactly a win–win transaction, it wasn't a win–lose one, either. McClurkin dwelled on the upside for Turner. He'd be able to expand, and with luck he might even make something of the station. Ted signed the agreement to buy Channel 17 in 1970, eventually changing its call letters to WTCG, for Turner Communications Group, or, as he later claimed, "Watch This Channel Go."

"When I bought Channel 17, everybody just hooted at me," Turner says, relishing the memory. "The station was really at death's door. I didn't bullshit anybody. I told them I didn't know anything about TV." McClurkin marveled at what he calls Ted's chutzpah. "The guy's really got balls," he says. "If you were given the opportunity to mortgage your house with a 25-percent second mortgage to make a payroll offer on a TV station you didn't need, would you do it? Ted didn't blink an eye."

Located in downtown Atlanta on West Peachtree Street near the expressway, Channel 17 was *literally* at death's door, situated as it was only a stone's throw from a funeral parlor. It was a small, shabby complex painted a dingy, peeling white; a low-slung two-story cinder-block structure housed the studio, and a somewhat taller brick-faced appendage was used for offices. At the rear of the building, there was a small parking lot for employees, a rusty generator, and—the one solitary sign of prosperity—a huge, 1093-foot, pyramidal steel transmission tower that rose straight up into the clouds and was said to be the tallest freestanding tower in the country. One night on a dare, after a few drinks, Ted began to climb it, disappearing into the darkness and coming down two-and-a-half hours later. He claimed that he had made it all the way to the top. Roddey, who was with him, says that the only reason he didn't go up too was that he had better sense. Climbing the tower would become a running gag at Channel 17.

In the winter, ice accumulated on the top of the tower because the station had no de-icer, and large frozen chunks would come crashing down on the cars in the parking lot. Employees often had to park several blocks away. For those brave or foolish enough to use the lot, there were hard hats hanging by the rear door. "We'd put on the helmets and sprint to our cars," one early employee recalls with a laugh. The roof of the studio was covered with old tires and wire mesh to keep all the falling ice from smashing through the ceiling. Channel 17 was far from a blue-chip operation. To furnish the station, Turner arranged what they called "trade deals," bartering ad time with local shopkeepers for furniture and framed prints and posters. A sign on the front door said, "NO BARE FEET ALLOWED."

In the beginning, hardly anyone in Atlanta—with the possible exception of its creditors—knew that Channel 17 existed. Turner called it an underground station. The entire operation was run by two to three dozen people, and initially many of them didn't stay long enough to collect a second paycheck. "Mainly hippies," Turner told a writer for *Television/Radio Age,* "and inexperienced, college kid–type engineers. And the whole operation was terribly under-capitalized."

One who did stay on was Gene Wright, the chief engineer. Wright says, "We had absolutely no equipment at Channel 17 and no real engineering or maintenance being done. We had all kinds of technical problems and kept going off the air." Every time the wind blew and the tower swayed, the transmitter cut off, and Turner either refused or couldn't pay the $75,000 for a new line. By the time he finally did, the transmission line had blown completely, and it was about a week before they were back on the air. Wright was working so many hours a day that he bought a cot and essentially moved into the station. He remembers those early years with a mixture of alarm and amusement. Not only was his boss crazy, but nobody seemed to notice. "The operators would sit around playing banjos and smoking pot while the movie would run out."

For Ted, it was a great time. "I just love it when people say I can't do something. There's nothing that makes me feel better, because all my life people have said I wasn't going to make it." All at once, his life took on a new focus. Unlike his handling of the billboard business, which he had lately left for others to run, he was now once again intensely involved in his company. Television excited him. Every day was a crisis, and there was nothing he enjoyed more. "He would just explode into work in the morning," says Gerry Hogan, who, at twenty-five and with no experience, came to sell advertising for him. "By the end of the day, he'd be exhausted. I think the level of energy was the thing that impressed me most about him, and just his enthusiasm for doing things, for taking chances." Turner seemed to thrive on stressful situations. "Television turned him on. It just sparked him," Jim Roddey claims. "I know that after we bought Channel 17, he could sit down and recite by fifteen-minute segments the entire demographic break-down of the Atlanta market for every television station in the city."

When motivated, Ted is a quick study, and he had given himself a crash course in television. He read every broadcast trade journal and rating book he could lay his hands on. They shared the space in his office at the back of the building with the sailing trophies he had won.

"To me," Ted says, "the whole damn thing was a challenge." He quickly discovered how bad his situation was—and it would soon become even worse. Out of the blue, a company called U.S. Communications opened a competing independent UHF station in Atlanta. This brought the number of television stations in the city to five. When the new ratings came out, Channel 17 was buried at the bottom of the heap, in fifth place.

"One of the first things I learned," Turner says, "is that outside of New York and Los Angeles, there's no way two indies in the same market can make money. Especially two [UHFs]. I knew when they came on that one of us would be off the air before too long. And I was going to be damn certain it wasn't us." Turner began to spend less time at home and more and more at the station. Hogan remembers, "He would roam the halls and try to pump people up and get them excited about things." Though he often overreached in wanting his sales staff to approach big companies like Sears for advertising—companies they had no hope of attracting—rather than small local automobile dealerships or fast-food franchises, he was a great motivator. He didn't fault his troops for failing to win the big ones. "Never look back," Ted would say about the accounts that got away. "Don't worry about it. Just keep moving forward."

He set up a list of priorities he called his "five Ps": programming, personnel, promotion, penetration, and profits. He needed movies for programming, but he couldn't afford anything but oldies in black-and-white. "So I started spending a lot of time with the film people—MCA, Paramount, United Artists, Viacom. All of them," Turner says. "And they helped me whenever they could."

He needed a staff he could depend on, and gradually he put one together. Sid Pike became the manager of WTCG. Working with him were Gene Wright, Gerry Hogan, R. T. Williams, and a young college boy, Terry McGuirk, who in time would handle special projects. Having persuaded a number of disaffected salesmen from the local ABC affiliate (WQXI) to jump ship, Ted went after its sales manager, Jim Trahey. He called him every day for two weeks. "You gotta come, you gotta come," he told him. And Trahey eventually did. Who could resist a boss who promised you the moon only a few years after Neil Armstrong had planted the flag on it?

"I don't know where I'm goin', Jeem," Ted said, "but hang on to my coattails, and we'll get to the stars and the moon."

Although Ted had inherited Irwin Mazo from his father, he now needed a full-time financial officer. He had known Price Waterhouse's Will Sanders from college days, when they competed against each other in regattas, with Sanders sailing for Yale. Shortly after his father died, Ted had asked Sanders to join the company, but Sanders turned him down, fearing that Turner was too highly leveraged and financially shaky. Now it was Roddey who offered him the job of chief financial officer. Sanders thought the company's new president had gotten a good handle on things since taking over. The firm's revenue and cash flow had grown, and Sanders was intrigued by Ted's new television venture. Although he was still anxious about how much red ink the station might bleed, he decided to join the Turner team.

All during that first year, Channel 17 continued to suffer from chronic breakdowns. Even the furniture was broken down. Gerry Hogan remembers that Ted for a long time had a chair with a broken leg in his office: "I kept telling him I could trade [advertising time] for a new chair for him, but he said, 'No, no. I don't wanna do that. When people come in to sell me stuff, I don't wanna look like we're doing well. I want 'em to feel sorry for us, have them feel they're helping me out by giving me a good price.'"

But it was the station's electronic equipment, which was either too old, too unreliable, or simply lacking, that was the real problem. Everything came to a halt, according to Hogan, when they tried to make their own commercials: "There was a piece of equipment called a switcher, and we only had one switcher, and you couldn't both make a commercial and stay on the air at the same time." As a consequence, most of their commercials had to be made in the early hours of the morning.

About six months after Turner acquired the station, Lee McClurkin, in an effort to help him out, told him about a station in Charlotte, North Carolina, which was about to go out of business and might be good for secondhand parts. As McClurkin puts it, "We went up there to cannibalize the station." It was Channel 36, and it was owned by a dentist named Twysdale. The three of them had lunch together, and McClurkin remembers, "It was sometime during lunch that it occurred to Ted that maybe Charlotte would be a good place to have a TV station. Literally, by the end of that afternoon, he had agreed on the outline of a transaction. I tried to talk him out of it all the way back to Atlanta. I thought he was nuts."

If Irwin Mazo considered Ted's purchase of Channel 17 dumb, he was even less sanguine about the Charlotte station. Ted knew his opinion: "Irwin and the others thought I was crazy, and in all fairness it looked pretty dark there for a while." To the accountant, it was obvious that the billboard business was going to be sucked totally dry to pay for the television losses. "This company ain't going to make it," Mazo figured, and decided to bail out. He had also had enough of being embarrassed by Turner's rudeness. "Ted didn't listen to anybody," he says. Besides, Will Sanders was now working full-time for Turner, and Mazo felt more strongly than ever that his advice was no longer wanted.

Jim Roddey was also against buying the Charlotte station, even though Ted had been able to negotiate a very modest price for it because it was in receivership. Sanders recalls Turner presenting his plan to the board to get them involved. "I think," Sanders says, "the board felt we had sufficient red ink at Channel 17, and we simply could not take on another bleeder of that magnitude."

Ignoring their advice, Turner decided to go through with the transaction by personally acquiring Channel 36 with his own money. "I think he paid hardly anything for it," Roddey says. "I think he assumed some notes on some leases and equipment. I doubt if he had more than seventy-five thousand cash in the whole deal." Like many entrepreneurs, Ted had a habit of mixing his own money up with that of the company. But what Sanders and others feared was that the Charlotte station's losses would ultimately be a serious drain not just on Ted but on the parent company as well. "It could have sunk our ass, that TV business," Sanders says. "We were really in it. Our purchase contract included all the debts. There was no way to cut it loose if it didn't work. We all said he was crazy."

Turner was now moving much too fast for those around him. It seemed to Roddey and McClurkin that he had a death wish, eager to bet much too much without any adequate backup. At least once a week, Ted would trot out his old threat, "Hell, if it don't work, I'll just do like my daddy did." Suicide was simply a normal part of his conversation. But no matter what he said, the company had enough of a problem just trying to arrange the necessary cash flow to pay for programming for Channel 17 without taking on another risk. At the same time, one of Ted's bankers, First National, was beginning to get very uncomfortable about its relationship with him.

As a result, Roddey and Ted started to have heated arguments. Ted seemed to want to put all of the resources of the company—not to mention his own—into television, which at that time was only a small part of the overall organization. "Ted always accused me of being too conservative," Roddey says wryly. "His definition of my conservatism was that I wanted to keep at least enough money to make two months' payroll. He thought that was recklessly conservative."

After three years, the parting of the two was inevitable, and Roddey returned to Rollins Advertising. Their relationship, however, remained amicable. Yet it was not without its competitive edge. "We're both competitive guys with strong personalities, I guess," Roddey acknowledges.

Win, lose, or draw, Turner loves competition. In victory he proved that he could be totally obnoxious. (After winning one series of races in 1977 by beating Lowell North's *Enterprise,* Turner crowed to the press, "I didn't just eliminate Lowell North, I destroyed him. He left here a broken, bitter man. I'm three times the man he is.") Turner has been generous in defeat, but there are those who say that even in defeat he can, on occasion, be insufferable. Perhaps this is because he hates so bitterly to lose. Roddey remembers the time he beat Turner at Ping-Pong. Infuriated, Ted not only crushed the Ping-Pong balls but broke the paddles. Years later, on his fortieth birthday, a group of his old friends put together a video satire of him entitled "The Brat Who Ate Atlanta." Roddey, in the lead role, describes how he played the part: "You just try to eliminate all sense of humor, of which he has none. You try to be an egomaniac, which he is. And you try to dominate everyone and every conversation. That's just typical Ted. Or, at least it was at that time."

Turner laughed at Roddey's depiction of him, but it was plain that he didn't think it was very funny, and some ten years later he would have the last laugh. He phoned and said, "Hey, Roddey, do you know I'm now worth five hundred million dollars?" Roddey asked if that was why he had called. "No," said Ted, "I called to say that I made four hundred million of it doing things *you* advised me not to do!"

Like others at the start of the seventies, Turner had foreseen the growth potential of television. Even more important, he had grasped the way he might synergistically use whatever empty billboards he owned to advertise his stations, as well as the opportunity he had as owner of more than one station to share programming and equipment. When he bought Channel 36, however, he could hardly have had any illusions about it. The UHF station, which he named WRET using his own initials, was located on the outskirts of Charlotte, with woods all around and a muddy cow pasture out front. Its highest rating was a 2—about as low as anyone could get in that era. But even with modest expectations, Ted hadn't fully appreciated how much it would cost him to keep the station going. Will Sanders says, "Ted's problem was feeding the operating losses out of his own pocket. At the time, it was several hundred thousand dollars a year, and he was having to do it out of his personal funds." Add to that the more than half a million dollars the company lost that first year keeping WTCG on the air, and Turner was in big trouble.

Though Ted never showed any lack of confidence, he would later admit that there were moments during this period when even he had his doubts about whether his company, Turner Communications, would survive financially. UHF reception was not good; you had to keep fiddling with the tiny dial on your TV to bring in anything at all. And the small number of sets able to receive UHF was chilling. Faced with such obstacles and heavy financial losses, Ted worked long hours and lived on the edge. He would go into an advertiser's office to sell time on Channel 17 and, after being informed that they didn't buy UHF time, he would boast about how smart his audience was. "And they would say, 'How do you know that? How come?' And I would tell them, 'Because you have got to be smart to figure out how to tune in a UHF antenna in the first place. Dumb guys can't do it. Can you get Channel 17? No? Well, neither can I. We aren't smart enough. But my viewers are.'"

Turner began to acquire a reputation for outrageous sales tactics—whatever it took to get advertisers. According to Gerry

Hogan, who would soon become his general sales manager, Ted would do anything to make a sale, and often the more bizarre the better, "like stand on a table, or take a guy by the throat, or kiss his feet . . . *anything!*"

But nothing seemed to help. Turner's two stations continued to hemorrhage dollars. At Channel 36, the financial picture became so bleak that he OK'd a "Begathon." The idea was to plead with the viewers to send in the price of a couple of movie tickets in order to keep WRET on the air. According to Turner's logic, it was a good investment, guaranteeing that they could watch the hundreds of movies the station planned to show in the coming year. But there was no getting around the fact that he was literally asking his viewers for a handout. He promised to return the money *plus interest* if the station ever became profitable—and he actually did. The most amazing thing, however, was that people responded at all. Sanders recalls, "We collected something in excess of thirty thousand dollars." It was a temporary reprieve, but it was enough to give Turner a little breathing room.

In March of 1971, Channel 17 was also barely limping along, well on its way to losing even more than it had lost the first year, when suddenly, without any prior indication, the other Atlanta independent station ran out of money and folded. To Ted, it seemed the handiwork of some benevolent deus ex machina. The general manager of the other station, Bill McGee, says, "Turner was making no headway at all. He was in a sorry situation. Had we not shut down, he would have been in serious trouble."

When the next ratings book came out, Channel 17 had crawled up a notch—now fourth place in a four-station town. Though still dismally last, WTCG was now the sole independent station in Atlanta. Turner was jubilant. He insisted that he had "whupped" the opposition, and to celebrate the event he held a victory party on the air. R. T. Williams produced Turner's two-hour "Thank you, Atlanta" show, filling the studio with balloons and hiring a band. But he didn't like the idea at all. "It was really embarrassing," he says. "Here another station had gone out of business, and Ted goes on the air to thank Atlanta. You got to have brass to do that." Turner made the entire staff come. Playing the role of host, he interviewed each of them.

"Well, Gerry, how long have you been working for WTCG?" he asked Hogan, and Hogan dutifully told him. "Now tell me about your family." Then Turner wanted to know about the advertisers he had sold. Hogan still remembers how embarrassing it was. "Incredibly. It was incredibly embarrassing. None of us wanted to be there. There was a certain amount of controlled substance smoked before anybody would go on the air." From upstairs on the balcony, other staffers watched and toked nervously on their joints as they waited their turns.

Though Turner now owned the only independent TV station in town, he still managed to lose another half a million on WTCG by the end of the year. It wasn't until he decided to counterprogram that he succeeded in turning his station around.

Counterprogramming is the counterpuncher's way of picking shows—making selections in opposition to your competitors' schedules. The tactic assumes an adversarial relationship, and as such it was a natural extension of the Turner personality. From childhood on, Ted had always seen himself as the outsider, the loner apart from the pack, and here was a perfect opportunity to exploit this as a strategy, a way to do business. The three network affiliates in Atlanta all programmed in what he considered to be a similar fashion. Turner says, "I felt the people of Atlanta were entitled to something different from a whole lot of police and crime shows with murders and rapes going on all over the place. I believed, and I still believe, that people are tired of violence and psychological problems and all the negative things they see on TV every night. I wanted to put on something different and give them a choice. Also, I wanted to prove that a small, *locally owned*, independent UHF station could make it in a big market."

When the ABC affiliate in Atlanta was forced by the network to run its evening news at 6:00 P.M., Ted saw his chance and ran reruns of *Star Trek* against it. The results were encouraging for Channel 17. Ted presented his own news in the early hours of the morning, and as little as possible. "As far as our news is concerned," he admitted, "we run the FCC minimum of forty minutes a day." His principal fare was old situation comedies, including *Gilligan's Island*, *Leave It to Beaver*, *The Beverly Hillbillies*, *Petticoat Junction*, *Gomer Pyle*, and *The Andy Griffith Show*, a massive back-to-back dose of mind numbing, lightweight comedies and canned laughter. "We're essentially an 'escapist' station," Turner said without apology.

Though he was outspokenly against violence, Turner was more than willing to do violence to reality and good sense in his battle for ratings. While calling his offerings family entertainment, he seemed to have an instinctive sense that appealing to the lowest common denominator of taste wasn't going to hurt his pocketbook. And he didn't hesitate to show physical violence when it upped his audience share.

With the help of a former girlfriend who had married a wrestling promoter, he managed to acquire television rights to professional wrestling from the weak ABC affiliate. Sweaty overweight titans in spandex pounding each other's head into the mat—you couldn't beat it for ratings, his sales people told him. Later, he would add *Roller Derby* to his offerings. Explaining the program to some foreign visitors, he described it as contestants beating each other up on skates. He added proudly, "It's a big hit." Contradictions that might trouble others passed over Turner like a cloudless sky. To fill the gaps in his scheduling, he gathered a large library of feature films and was soon claiming to have more programming than any other station in the country.

The leading television outlet in Atlanta was the NBC affiliate, WSB-TV, owned by the Cox family, which also owned the *Atlanta Constitution* and *Journal*. Ted took them on when he bought five NBC shows that WSB had turned down and then announced all over town on his billboards: THE NBC NETWORK MOVES TO CHANNEL 17. It caused quite a fuss. Ted says, "We got a whole bunch of phone calls and every other damn thing." Most of the calls were from angry Cox lawyers and people responding to the articles that appeared in the Cox papers. Turner was ecstatic. He didn't expect to knock the Cox people out of first place, but he had succeeded in rattling their cage. As a consequence of the publicity, he knew that curious viewers would now be looking for Channel 17 on their screens.

Turner in 1973 also attacked WSB in another area—baseball. When he heard that the Atlanta Braves weren't thrilled with their twenty-game WSB contract, Turner gave them an offer they couldn't refuse, committing himself to televise three times that number of games. It cost him $1.3 million, but Turner was more than willing to take the risk. He had a plan to make baseball pay off.

Ted's notion was that no matter how much he had to pay for

the three-year package, he was going to wring as much value out of it as he could. First, he got all his salespeople together, yelling and screaming at them about what a great opportunity this was, even though they all knew that the Braves team was a disaster. His pitch was that his salesmen were up against only network reruns during the spring and summer, and here was fresh programming. And women love baseball, so they could sell the female audience. But most important of all, as far as Turner's thinking was concerned, was *where* they'd sell it.

By 1972, Turner had started to think big again, and just as in the old billboard days, he figured that supersaturation might be the ticket. He had come to know Andy Goldman, a tall, outgoing Northerner with a fondness for sailing and suspenders, who also happened to be a marketing director in Alabama for the largest cable operation in the country at the time, Teleprompter.

Back then, cable was basically just another way to get the same old network programming for those who couldn't get a clear signal, a static-free picture. In Alabama in 1972, Teleprompter had about two hundred thousand subscribers, an incredibly large number for a cable operation at the time, and Goldman liked the idea of offering them a bonus, something extra and different in the way of programs. He went to Turner and said, "Ted, we'd like to get your signal in Huntsville, Muscle Shoals, Tuscaloosa." (The signal was carried by microwave land lines, an expensive series of amplifying stations located on towers. Turner's broadcasts and Teleprompter's Alabama subscribers would eventually be linked by a common-carrier microwave highway owned by a man named Frank Spain.)

With the independent WTCG, Goldman could bring his viewers major-league baseball on a regular basis, old movies, old sitcoms, and cartoons. At the time, cable subscribers were paying $4.95 per month, and now Goldman had what he calls "a product worth somebody putting up $4.95 for."

Ted wasn't getting a share of that $4.95, but the deal had plenty of allure for him nonetheless. Suddenly WTCG was being seen by another quarter of a million viewers, another quarter of a million people who might buy advertised products. Like an early form of the home shopping networks, WTCG carried low-cost direct advertising—giving subscribers an 800 phone number and inviting them to call in to buy their very own "incomparable recordings" of country music's greatest hits. Turner's station was flacking a whole bucketful of dubious products: mood rings and genuine family coats of arms and Popiel's Pocket Fisherman—basically a fancy ball of string. Then there were the Ginzu knives—they sliced, they diced, they were dirt cheap and so sharp that you could cut a can with them, though why anybody would want to cut a soda can with a knife was never explained. Goldman says, "The whole concept of the Ginzu knife was unique. And Ted sold the hell out of Ginzu knives."

"Jesus!" Turner said, when he saw the calls coming in. There's a helluva business here." Who needed local ratings? Now all of a sudden Turner's shabby rust-streaked bunker on West Peachtree had grown into a regional station. It was soon carried by microwave to cable stations all over Alabama and down into Florida. One day, Turner took his sports announcer Bob Neal down to a small room in the basement of the station. "I thought he was taking me into a closet," Neal says, "but he opened the door, and there was a roomful of women opening mail. He started picking up the letters, which had postmarks from places like Valdosta, Tallahassee, and Opelika, and saying, 'Look at this! Look at this! Look where they're from. Our audience is not in Atlanta. I've found a way to make money with this television station!'"

Baseball and movies became two of Turner's programming staples. "He had quite a library of movies," Andy Goldman says, and quite a way to use them, too. Turner would pay modest rates to broadcast movies in the local Atlanta area; but because of his new reach, he could turn around and show them throughout the Southeast. Cable companies were then allowed by the FCC to carry one distant independent station.

Equally shrewd was the way he used baseball. Turner could make baseball pay for him, because he now had a large regional network and was able to syndicate the games in Georgia, Alabama, North and South Carolina, Florida, and Tennessee. Jim Trahey recalls that the Braves were carried in approximately thirty-three markets in the Southeast. The Braves games became the top-rated locally produced program in Atlanta. Signing the Braves, Gerry Hogan says, was crucial in changing the station's image. "Atlanta went from a three-station market plus WTCG to a four-station market. We were in it after the Braves signed with us." By the end of 1973, Channel 17 had turned its first real profit—over a million dollars. And only one year later, Turner was trumpeting his claim that based upon the size of its audience, Channel 17 was the number-one UHF independent station in the entire country.

Ted Turner once said that for him first comes sailing, second comes business, and third—bringing up the rear—family. Though big on family values for television, he was often too busy to devote much time to his own wife and children. When he was in Atlanta, he focused most of his attention and energy on saving his two TV stations and keeping a step ahead of his creditors. He was definitely a hands-on boss, working eighteen- and nineteen-hour days. With energy to burn, he seemed to be in perpetual motion, sparks shooting from his heels. "I programmed the whole station myself in those days," he says with pride, recalling his frenetic pace. "I sold ads, I signed all the payroll checks, I went to parties, I met people, I asked a million stupid questions, and I educated myself. I would wander around in a daze all day thinking, 'What am I gonna put on at four-thirty?' No committees, no studies, no bull. I would ask Janie what she thought of a certain movie, or I'd ask one of my friends, or the girls at the office."

Janie Turner liked to watch movies, but she often had to watch them alone. "He's almost never here," she told an *Atlanta Constitution* reporter doing a story about her life with her whirlwind husband. She now had five kids to look after, his two from his first wife, Judy; and their own three. In 1968, when her second child, Beau, was four months old, Jane had discovered that she was pregnant again. "I cried a lot when I found out," she says. In their staunchly conservative, upper-middle-class circle in Cobb County—an Atlanta suburb where they had moved—such flagrant fecundity was not the norm. "My friends thought it was vulgar. I love babies, but I nearly died. I thought my back was broken. I stayed home for five years. It would have made anyone crazy."

From the mid-1960s on, when Ted wasn't working, he began to devote more and more time to sailing. He was often away from Atlanta for weeks or even months at a time. He told a foreign visitor as he showed him around the city, "It's great in the spring and fall. In the summer I'm off sailing, the winter is broken up nicely with SORC and Montego Bay, and then I come back in the spring. It works out nicely."

For Janie, it worked out less well. She had few friends of her own. Her socal life was limited to her husband's business and sailing associates. Though Ted's friends liked her and found her sweet and attractive, a gracious hostess, and a terrific mother, they couldn't figure out the marriage. She and Ted were such different people— the quiet, withdrawn, proper Southern lady and her loud, crude, outrageous, womanizing husband. "Janie's a lovely girl," one of his sailing pals said, "but she's sort of dumb. She's not quick like Ted."

Lee McClurkin, who with his wife used to get together socially with the Turners, recalls, as do many who knew Ted and Janie well during this period, that Janie "had to put up with a lot." Trained by his father to be careful about money, Ted could be incredibly penny-pinching with his wife. He could also, like his father, be thoughtlessly blunt, even cruel. After dinner at the Turners one night, Jim Roddey complimented Janie on the meal. "Yeah," Ted said, "she's gettin' to be a good cook. Almost as good as my first wife." Demanding that everything be shipshape at home, Ted made his displeasure known in no uncertain terms if he found the refrigerator empty or his socks still in the dryer. He lashed out at his wife, the way he would berate any member of his crew, or his business, who screwed up.

"He could be very impatient with her if she didn't meet his expectations," says Mike Gearon, a Turner business associate and friend of the family, who believes that over the years Ted has undergone a change for the better. But back then it seemed to him that Janie and Ted had a marriage in name only. "He'd treat her horribly. There was no giving on his part to speak of. He wanted her to do the kinds of things that his woman was supposed to do, and not object to his own lifestyle." Ted would be sitting at the dinner table one minute and then suddenly, without a word, he was gone. Janie would have no idea where he was going or when he'd be back. Often the only way she'd find out was to check with his secretary, Dee Woods.

It may be, as someone who knows her well claims, that Janie's marital fantasy was as uncomplicated as a white picket fence, a rose garden, and a house in the suburbs. If so, she got much much more than she bargained for in Turner. Overwhelmed by her repeated pregnancies and his sudden departures, she felt increasingly lonely and sorry for herself.

One night, Billy and Nancy Roe, who hadn't seen the two of them for some time, had dinner with the Turners. It surprised them to see that Janie was drinking now. During the course of the dinner, Ted apparently became so annoyed with his tippling wife and her failure to answer questions that he began to mock her in front of their guests. Finally he said dismissively, "Oh, just give her another drink." Nancy Roe says, "I just wasn't comfortable with that."

Finding sympathetic ears in Irwin Mazo and his wife, Janie began to complain about Ted. "She once told me," Mazo remembers, "that it couldn't get any worse." She couldn't understand why he was so mean. "I was miserable a lot of the time," Janie admitted. "Every chance I got I would load up the car with cribs and diapers and stuff and drive home to Birmingham. Three times, Ted was away from home sailing over Christmas. One Christmas I was so pregnant I couldn't even go home, and my parents couldn't come to Atlanta for some reason, and so I just stayed in Atlanta. I cried and cried."

When he was gone, Turner left a little something of himself behind for the family. Every Sunday morning, they could see him on videotape as host of WTCG's *Academy Award Theater*. Seated in a large wing chair, he would introduce the morning's movie in a few words. The show was another of his counterprogramming efforts, scheduled to run opposite the Sunday religious programs of his Atlanta competition.

For *Academy Award Theater* (as for almost all of Channel 17's programming), Turner not only selected each movie but decided when it would run. Most important of all, he put himself in front of the camera. It was, after all, his station, and he could do what he damn well pleased. "It was kind of a joke," R. T. Williams, the producer of the show, said of Ted's role, "but he took it quite seriously." Billy Roe remembers on one occasion watching the program together with Ted. They rolled on the floor and howled about his performance, but there was no doubt in Roe's mind who Turner thought the star of the show was. "I enjoyed it," Ted said, "and it was great during those winters I was off sailing in Australia. The kids could turn on the set and see their daddy at Christmas." Turner, in his innocent, solipsistic fashion, genuinely thought that it would cheer them up in his absence. Though really too busy to do the program, he was willing to make any sacrifice for his family but stay home.

In 1964, Ted had attempted for the first time to qualify for the Olympics. Teamed with Andy Green, his sailing friend from Plastrend, Ted came a cropper at the Flying Dutchman trials in Acapulco. "We did everything bad you could," Green says. "We hit other boats, we hit a mark, we screwed up the start."

Four years later, in 1968, Turner was sailing a 5.5-meter and was considered a shoo-in for one of the three spots on the Olympic team when the trials began in May. His boat was designed for him by Britton Chance Jr., a young naval architect, and Turner, out for revenge after his earlier fiasco, named it *Nemesis*.

One of the members of the *Nemesis* crew was Charlie Shumway, the former commodore of the Brown sailing team. "We were seeing a lot of one another that summer," he remembers. One evening, Shumway and his wife invited Ted and Janie and another old sailing friend from Brown, Roger Vaughan, and his wife to dinner.

Not having seen Turner in ten years, Vaughan was astonished at the sea change in the callow young redneck who at school had so often been involved in drunken escapades and had been ingloriously kicked out. That night, Vaughan says, "he presented a commanding image in his double-breasted blazer with the ornate crest of a foreign sailing society sewn over the left breast pocket. He was tanned and healthy from his latest ocean venture, and a touch of premature gray around the temples completed the Hollywood image of a world ocean sailor." To Vaughan, the twenty-nine-year-old Turner, seen across the dinner table in candlelight and through rose-colored glasses, appeared to be "a true gentleman," America's ambassador to the world from Atlanta. It must have been difficult for him to reconcile that image with Janie's obvious unhappiness, her bitterness about Ted as a husband and father.

Two months later, an ecstatic Turner and his two-man *Nemesis* crew were 3,000 miles away in Newport Beach, California, having qualified for the 5.5-meter Olympic final eliminations. *Nemesis*, however, fared poorly, and even after a strong start in the last race, she came in second to Lowell North's boat. Overall, Turner and his crew placed seventh and were eliminated. When Ted returned to Atlanta, Jim Roddey remembers how depressed he was. The last time Roddey had seen Ted so low was after his father's death. But Ted was never down very long. "He had," according to Roddey, "a typical salesman's mentality. After every defeat he was down, and after every victory he was sky-high. Sort of a roller coaster."

Failing in one area, Turner typically explored two or three other possibilities. He couldn't stand to lose, and he couldn't stand still. The romantic idea of being in command of a sleek 12-meter competing for the America's Cup, which first flamed in his imagination during undergraduate days, now took hold stronger than ever. He talked of forming a syndicate for the 1970 Cup defense and approached Britton Chance again to design the boat. Chance describes what happened next: "Ted said he had the money at his fingertips. We drew up a contract and got ready to go." But a year went by, and prospects began to look bleak. When Ted finally called the designer to say that he couldn't raise the money, Chance was furious. He felt that Turner had oversold his ability to deliver, and the delay had effectively eliminated Chance from consideration by other syndicates. If Chance was disappointed, so was Turner. But Turner is no brooder, and he had other irons in the fire.

One was *American Eagle,* a boat Ted really wanted to own. The glistening white 12-meter yacht was a beauty that had been designed for the 1964 Cup defense but had failed in the trials and lost in the '67 preliminaries as well. The prospect of turning a proven failure into a success always appealed to Turner. In addition, the *Eagle* had been converted for ocean racing by its owner, Herb Wall, and Ted loved the challenge of racing a 12-meter boat in the SORC, when most knowledgeable yachtsmen thought it was nuts. He made Wall an offer of $70,000, but it was refused. Later Wall changed his mind, and on December 15, 1968, Turner flew back from Australia, where he had been sailing on someone else's boat in the Sydney-Hobart race, to take command.

At first, Turner's critics seemed to be right about the unsuitability of a 12-meter boat for ocean racing. It tended to go *through* steep seas rather than over them. Then, on March 3, in the Miami-Nassau race, the *Eagle*'s mast hurtled over the side. "It happens all the time," said Turner, trying to put a brave face on the debacle. Within three days, he had replaced the mast and had new sails made, and *Eagle* was once again ready to race.

That year, 1969, Turner sailed his boat in the World Ocean Racing Championship, which consists of eighteen races all over the world. (To qualify, a boat must sail in seven of these events within a three-year period.) Turner at one point was flying between Denmark and England in order to enter major races, and trying to squeeze in a 5.5-meter championship in Sweden as well. In his own inimitable unbuttoned fashion, he described what he was doing this way: "Sailing is like screwing—you can never get enough."

Later on, Turner would attribute his manic behavior and inability to stay put at this time to insecurity. "Particularly at the beginning of my life, when the family first appeared," he says. "I was still very insecure. And I was trying to be the world's greatest yachtsman." By 1970 he had already competed in both transatlantic and transpacific events, racing in Ireland, Jamaica, Bermuda, Denmark, Norway, Australia, and Hawaii. The big thrill for Turner had always been in winning, in looking over his shoulder and seeing the rest of the fleet behind him, and with *American Eagle* he began to win often.

The first SORC race of the 1970 season began on a January afternoon in choppy seas and low temperatures, with a fleet of eighty-seven boats charging down the Gulf of Mexico in a bone-chilling nor'wester, their colorful spinnakers bellied out in winds that sometimes gusted to as much as 35 knots. Turner was in his element. Commands shot from his lips in broadsides. He shouted, he cursed, his mouth seemed to be constantly in motion. Bob Bavier, who often sailed with him, says, "You have to be a little thick-skinned to sail with Ted, because in the heat of battle he comes out with outrageous things." Under pressure, Turner has been known to become Captain Bligh, even allegedly smashing a crewman in the face when he thinks he is not doing his job right. "Usually just in the back," he says defensively.

As the 105-mile race from St. Petersburg to Venice, Florida, progressed, many sailors became too seasick and numbed by the cold to work. Eventually, twenty-one of the boats that started dropped out, but Turner's converted 12-meter sloop, according to one participant, "sailed erect as a New England church spire." Overtaking the favored *Windward Passage* in the final 28-mile leg, Captain Turner and his gung-ho crew drove *Eagle* to an elapsed-time record in the event.

Then, taking a couple of days off, Turner packed his toothbrush and flew sixteen hours to Australia, where he won the 5.5-meter Gold Cup. Immediately after that, he caught a plane back to Florida and, aboard *Eagle,* picked up where he had left off. Turner estimates that during these years he was spending about $20,000 a year on plane tickets alone and logging more flight time than some commercial pilots. By the end of March, his boat was first in four of the five offshore races in which it participated, and was declared the SORC champion. One reporter labeled *American Eagle* "the brightest star on the current ocean-racing scene." Nine months later, Turner, after sailing more miles with more success than anybody else in 1970, was named Yachtsman of the Year and presented with the elegant Tiffany-designed Martini & Rossi trophy.

It was a close vote. Ted had beaten out the skipper of *Intrepid,* Bill Ficker, that year's successful defender of the America's Cup, by only a narrow margin. Earlier, Turner had brought *American Eagle* to Newport—self-declared yachting capital of the world and site of the Cup races—and, flaunting a Confederate flag, succeeded in irritating the establishment by sailing her as a trial horse for the Australian challenger, *Gretel II.* Though he had behaved in the best international tradition of good sportsmanship, Turner was condemned by some members of the New York Yacht Club for assisting the challenger in her tune-up preparations, or as Cup historian Doug Riggs puts it, "giving aid and comfort to the enemy."

In the decade of the seventies, the New York Yacht Club was the center of the yachting establishment, and its landmark building on West 44th Street in Manhattan was home to the celebrated America's Cup. It had been won for the first time in 1851 from the British Royal Yacht Squadron and had remained in the yacht club's possession ever since. The 27-inch-high, silver 100-guinea pitcher—breathtakingly beautiful in its symbolism of unbroken sailing triumphs but actually a lumpy, potbellied eyesore—was bolted to a pedestal in the club's trophy room. Nearby, in the huge model room, with its three bay windows, are scale models of every winner and loser in the illustrious history of the America's Cup. There are about three thousand members in the club. Although Turner knew many of them, he was not one himself. Frankly, he didn't really give a damn. It was recognition that he seemed to crave, not approval.

By and large, Turner wasn't a big joiner, and the New York Yacht Club had a starchy reputation that hardly seemed to suit his style. He was, to be sure, a member of the Capital City Club in Atlanta, which had a well-known antiblack, anti-Jewish policy, but he claimed he joined only to have a place to eat lunch and so that Janie and the kids could swim in the pool.

In the seventies, in order to participate in the America's Cup races, you had to be a member of the New York Yacht Club, and in order to become a member of the New York Yacht Club, you had to be nominated by one member and seconded by another. The names of the nominees were then circulated among the membership for approval. One naysayer can, hypothetically, blackball a candidate. Defenders of the Cup had always been members of the club, and it looked as if the 1970 Yachtsman of the Year was serious about wanting to play a more pivotal role in the next Cup defense. But when Turner's name was put forward for membership, he was blackballed. To his critics, the potential that he posed for embarrassing the club was obvious. One prominent member wrote him off by saying, "I think he's a jerk."

But Turner refused to go away. If anything, his frantic pace in sailing only increased in the next few years, as if he were trying to make his critics look foolish by his repeated victories. In 1971, *American Eagle* set a course record in the Fastnet Race, off the coast of Great Britain, and then won the World Ocean Racing Cup in Australia. The next year, Turner was first in the Sydney-Hobart regatta, and in 1973 he again won the SORC, but this time at the helm of a new boat, *Lightnin'*. Turner's success in ocean racing is attributable in no small part to a skillful crew, many of whom, such as Bunky Helfrich, Marty O'Meara, Billy Adams, and Richie Boyd, sailed regularly with him. And at a time when yachts in ocean competition were being handled by gentleman skippers using cruise control, Turner, with fierce concentration and furious intensity, sailed every nautical mile as if he were in a dinghy race with his life hanging in the balance. In 1973, Turner for the second time was named Yachtsman of the Year.

But the America's Cup still eluded him, and his big shot at defending it came in the person of George Hinman, a former commodore of the New York Yacht Club. A white-haired man in his late sixties with a distinguished air, Hinman had the weathered look and sharp eyes of someone who has spent more than a few years on the water. In '73, he headed the Kings Point Fund, one of the syndicates that was putting together a team to defend the America's Cup. Not surprisingly, one of the possible skippers he was considering for his boat was Ted Turner.

Hinman checked the thirty-four-year-old Turner out by talking to almost everyone who had ever sailed with or against him. The reports were satisfactory. The only problem with Turner seemed to be that he was "a bad boy" ashore. "I asked him if he would like to be considered," Hinman says, "and he said he would. I also asked him not to tell a soul. Three days later I read it in the newspaper." Turner had told a friend of his, and the news leaked out. Hinman was angry. He considered dropping Turner but was finally won over by Ted's enthusiasm.

On the second floor of the New York Yacht Club is a room that few nonmembers have ever seen. The Commodore's Room is covered with a crimson carpet that deadens sound, and it contains a large oval conference table. From the walls, portraits of the bearded early commodores look sternly down upon committee members. It was Turner's old friend Bob Bavier, a former commodore himself, who presented his name the second time to the membership committee. Bavier knew that it might be a hard sell.

"Look," he told them frankly, "you want to know the faults about this guy? Ask me. I can give you a bunch as long as my arm. But you won't find one that you haven't already heard about. All of his faults he wears right on his sleeve. But I know this fellow pretty well, and he's a straight shooter. I like him."

In his typical candid fashion, Bavier told them what he thought, and there was no doubt that it made a difference. This time, according to Bavier, Turner also had more letters of support "than almost anyone in the history of the New York Yacht Club." In December of that year, six months before the 1974 Cup trials were slated to begin, Ted Turner was elected a member of the club.

Though he had behaved as if he didn't care one way or the other, Ted was delighted. He knew how much his membership in the club would have impressed his father. Ted said, "I remember him telling me about the New York Yacht Club once. How it was swank and ritzy and all. My father never would have dreamed that I'd be a member." There were some longtime members of the club who felt the same way. They were certain he was a ticking time bomb ready to go off.

CHAPTER 10

NEWPORT, RHODE ISLAND. The sign in the Visitors Bureau on America's Cup Avenue reads, "The Birthplace of American Yachting." Shipbuilding began in Newport as early as 1646, and by the end of the century it was a major seaport. Situated on an island in Narragansett Bay, it has been described as "the sailing capital of the world," and Ted Turner had no doubt that he was in sailing's red-hot center when he blew into town in June of 1974, eager to begin the trial races that would select the defender for the America's Cup.

With ship chandlers and carriage-lamp streetlights, Newport is small and quaint, but it also has palatial nineteenth-century mansions, the luxurious summer "cottages" that line Cliff Walk and Ocean Drive and were built for the Astors, Belmonts, and Vanderbilts. And because it is a seaport, there is even an occasional touch of the exotic. Standing on the cobblestones of Thames Street in front of a laundromat, four black crewmen in matching nautical uniforms of sky blue T-shirts and white ducks chat quietly among themselves in French as they wait for their wash. A block away, on the waterfront crowded with pleasure boats, the sleekly elegant 12-meter racing yachts are tethered to the wharves, their busy crews readying them for competition.

That year there were four American contenders: *Courageous, Intrepid, Mariner,* and *Valiant.* Bob Bavier, winner of the Cup ten years earlier and one of Turner's youthful sailing idols, was the skipper of the favorite, *Courageous*. Turner's boat, *Mariner,* was a new aluminum yacht with a radically innovative "fastback"—an unconventionally stubby rather than streamlined rear, designed by Britton Chance. Turner had done well with his Chance-designed 5.5, *Nemesis,* winning the Australian Gold Cup in 1970. Although later they had had a falling-out, Turner still had a high regard for Chance's ability. So, too, did George Hinman, who had chosen him as *Mariner*'s designer for the Kings Point Fund syndicate.

With Chance as designer, Bob Derecktor as builder, and Turner as skipper, Hinman had selected a talented trio, a "supergroup," according to Roger Vaughan, who in his book *The Grand Gesture* gives a detailed account of their attempt to create a "perfect" 12-meter racer. A 12-meter derives its name not from its length, which is approximately 65 feet, but from a complex formula involving such dimensions as sail area, ballast, and beam that must always equal exactly 12 meters. Great things were expected of *Mariner*. If not the favorite, she was certain to be a strong contender. It was not Ted Turner's preferred position.

Turner was always most comfortable in the role of underdog. With a kind of theatrical perversity, he seemed to relish being the dark horse who, against all odds, somehow triumphs. Coles Phinizy, writing in *Sports Illustrated,* says that Turner revels in adversity. "He loves to battle with his back against a wall. If there is no wall close by, he will go miles out of his way to find one."

Like the underdogs he identified with, Turner often presented himself as the disadvantaged outsider; his have-not attitude, by and large, had nothing to do with his actual material circumstances—growing up, as he did, a millionaire's son—but perhaps it captured his emotional state as a product of a dysfunctional family. The terrible injustice found in teenage Ted's poem "For Lack of Water" was, ironically, still haunting the life of the wealthy, successful thirty-five-year old businessman and two-time Yachtsman of the Year.

On May 10, Ted Turner had been at Kings Point, Long Island, for the commissioning of *Mariner*. The Kings Point Band played "Million Dollar Baby," cannons were fired, speeches were made. Ritual and pageantry are very much a part of the America's Cup, and Turner had been fond of both since his military-school days at McCallie. He considered his being named an America's Cup captain a very "big deal" indeed and used a reporter for the *Atlanta*

Constitution to publicize that fact. In his gleaming red blazer with its breast-pocket *Mariner* crest, Turner stood with his crew, ramrod straight and confidently smiling. They watched as the admiral's wife broke the traditional bottle of champagne on the gleaming red boat and the band played. The future seemed rosy with promise.

When the trials opened early the next month, the roseate glow that had bathed *Mariner* and its crew only a few weeks earlier had begun to fade. Despite his reputation in sailboat racing for being quick off the mark, Turner was losing most of the starts against Bob Bavier on *Courageous*. Then came their fateful encounter on June 26. Turner not only lost the start of that race, but *Mariner* crossed the finish line nine minutes and forty-six seconds behind Bavier's boat. "An equivalent score in football," wrote a reporter for the *New York Times,* "would be 72-0." Everyone knew that radical surgery would have to be performed on *Mariner,* even though that meant its crew would miss almost a month of training. The three months of trials to select the Cup defender are like the baseball playoffs before the World Series. No one goes to the finals without winning the playoffs.

At the end of June, *Mariner* was sent back to Derecktor's shipyard in Mamaroneck, New York, to have its square back modified. Turner was furious. He held Brit Chance personally responsible. "Really, I would like to rearrange his face for him," he swore angrily to crew members, "but that probably wouldn't do any good, would it? Probably not. But my God, we all knew the boat was a dog down in Long Island when we first went sailing."

The differences in temperament and style between the flamboyant Turner and the tall, patrician Chance, which had led to friction early in the campaign, would soon break out into open warfare. While acknowledging Turner's skill as an ocean racer, Chance didn't think much of his record on a closed course. And the America's Cup matches were sailed on a closed triangular course of 24.3 nautical miles, some 10 miles south of Newport. He also faulted Turner's choice of crew; with the exception of his tactician—the rising young star from San Diego, Dennis Conner—they were all Turner regulars, such as Bunky Helfrich, Marty O'Meara, and Legaré Van Ness, who had little or no Cup experience. Lacking confidence in Turner, Chance wanted him replaced. He needled him at every opportunity and made no secret of his criticism. "Turner isn't steering the boat right," he insisted. "In tacks, he is spinning it too fast. He isn't using the right sails. He never was that good, you know."

But Turner could, as usual, give as good as he got. He let the designer know that the reason there were no square-tailed fish was that all the pointy-tailed ones had caught and eaten them. On the day the overhaul of *Mariner*'s chopped-off stern was completed, Turner and Chance's bickering reached a new level of nastiness. "Damn it, Brit," Turner told him at one point, "even shit is tapered at both ends."

Although it was already August and the last month of trials, Turner still felt he had a fighting chance with the new, remodeled *Mariner*. He became almost cheerful. "It's going to be the old rabbit-in-the-hat trick," he promised his crew, "the eleventh-hour comeback, practically impossible, but we can do it. If the boat is at all competitive this time, we can do it."

But then George Hinman assigned the young hotshot Dennis Conner to take over as helmsman of *Valiant,* the syndicate's trial horse, and Turner sensed that his job was in jeopardy. According to Roger Vaughan, who was there, he became a nervous wreck. In the crucial races between the two boats that followed, Turner was sky-high, talking a blue streak from start to finish, and it plainly interfered with his concentration. His performance was erratic. In one race, he fouled his rival three times. When asked by a friend who had just arrived from Atlanta how things were going, Turner made no excuses. Everything was up in the air. Then throwing an arm around his visitor's shoulders, he confided, "This has been worse than when Dad passed away."

On the 14th of August, syndicate manager Hinman informed Turner that Dennis Conner would be the new skipper of *Mariner,* and that he was now assigned to *Valiant,* the junior-varsity boat. Though Turner tried to accept the switch with good grace, it was a terrible disappointment for him. "Sure it hurt," he admitted. "It hurt bad, I mean I was in tears." Money aside, he had invested a great deal of himself in the campaign. And there was, of course, the public humiliation of his demotion. Turner, who never saw an audience he didn't love, had brought family, friends, and business associates to Newport from Atlanta and hired the luxury yacht *Kanaloa* for them to witness his triumph close up. Even his sixty-eight-year-old mother, Florence, who hadn't seen Ted race a boat since his childhood days in Savannah, was there. Putting on a brave face, he told his new crew, "Tomorrow after we beat *Mariner* and they start begging me to go back, I'm not budging. This is it."

But Turner did not beat *Mariner* the next day, or the day after. Nor did he beat anyone else. After one particularly decisive defeat, he acknowledged Conner's skill with a curious gift, a photograph of himself, his mouth gaping like a yawning hippo, and an inscription that read, "To Dennis, a good friend and a great helmsman." Although Conner often sailed brilliantly that week, he was clobbered when *Mariner* went up against *Intrepid* and *Courageous*. So on August 20, after both *Mariner* and *Valiant* had failed to impress the New York Yacht Club Selection Committee, the Kings Point syndicate's double entry was eliminated from Cup competition.

Dressed in *Valiant* pale green for the final ritual, Ted Turner and his crew waited on the dock like condemned men for Commodore Henry Sturgis Morgan (grandson of the banker J. P. Morgan) and the rest of the Selection Committee to formally pass judgment on them. Commodore Morgan, a short man whose emotions were hidden behind dark glasses and beneath the downturned brim of his straw hat, approached the tanned and tired-looking Commodore Hinman. Hinman held himself stiffly, as if awaiting a blow. Placing his hand on Hinman's shoulder, Morgan shook his head and said, "Sorry, old man." Then all the members of the committee went down a receiving line of the eleven crew members from each boat and shook hands. After the melancholy ceremony was over, someone offered Turner a Coke. But a Coke was not what he had in mind at a time like this.

"We're gonna drink until we pass out," Turner announced to his friends, and hurried aboard *Kanaloa* to party. With a few minor exceptions, he had held himself in check all summer, and it was time to cut loose. He had known very well the sacrifices that were involved in Cup competition. "You're not sailing your own boat," he noted later, when trying to explain some of the pressures he felt to readers of *The Racing Edge*, the book about sailing he cowrote. "You're working for a committee, the New York Yacht Club Selection Committee, and you're working for your syndicate. So you have to be a little bit of a politician, in the better sense of the word. You have to be a gentleman and you have to do what is expected of you on the water as well as off. I mean, going around and writing 'turkey' with a grease pencil on another guy's boat, like you do in the Finn class, doesn't make sense in a twelve. If they found out who did it, you'd be taking a walk down the dock the next day."

Turner was no longer worried about the next day. No one had to write "turkey" on *Mariner* to reveal the truth about the boat he had skippered that summer, which Doug Riggs has judged "the single greatest design disaster in the modern Cup era." In order to sail *Mariner,* Turner had made a bargain with Hinman, exchanging his usual freewheeling ways for the New York Yacht Club's chains of decorum; he called them "chains of gold." And he was, by and large, as good as his word. There were, of course, a few times that summer when he had rattled his golden chains by drinking and talking too much. On one high-spirited occasion, he declared provocatively that had the South won the Civil War, he would have been a Confederate challenger rather than a United States defender. And then there was the time when he slipped his fetters completely, with the arrival of the strikingly attractive Martine (Frédérique) Darragon.

Darragon, a wealthy French heiress with a love of polo, travel, and sailing, was very fond of Ted. They were old friends and had done a good deal of sailing together. Earlier that year, in Florida, she had been the cook aboard his boat *Lightnin'* when it placed second in the SORC. Before the races, Martine was in the habit of casually sunbathing in the buff on the boat's deck, as if she were at St. Tropez. She had a model's figure and didn't mind showing it off. "My tits are all right," she acknowledged matter-of-factly, in her throaty, strongly accented English. "But my ass is better." Outspoken and independent, she was no more likely to be intimidated than Turner himself. A free spirit, she came and went as she pleased, and was the sort of woman who, on a whim, was entirely capable of invading the masculine sanctity of the New York Yacht Club in a see-through blouse. "She ain't just pretty," an FBI agent said some years afterward, having been assigned to follow her as part of an investigation. "She looks like a goddamn blond movie star, and she acts like she owns the world." Martine had come to Newport to be with Ted and was there just long enough to create syndicate storm clouds before she was gone. Like Ted himself, Martine had a way of generating a great deal of electricity wherever she went.

Following the defeat of the syndicate, Turner left immediately to race his Flying Dutchman in Canada for a week, but he did poorly and was unable to get the taste of failure out of his mouth. He returned to Newport in September to watch the Cup finals, which pitted *Courageous* against the Australian challenger *Southern Cross*. Like a nautical Eve Harrington, the quiet Dennis Conner— last seen aboard the ill-fated *Mariner*—had once again ended up on his feet, being named *Courageous*'s starting helmsman. And shortly thereafter, Bob Bavier was replaced as skipper by Ted Hood. *Courageous* ran away from the Aussie challenger in the Cup finals, winning four straight races. It was no contest.

Aboard the *Kanaloa* for one last party before leaving Newport, Turner and a group of friends sat around on deck drinking heavily and commiserating with one another. Turner admitted that he was "broken up and a little crocked." B. J. Beach, a cute boatgirl whom he knew and liked, was there to comfort him. Earlier that summer, after a bad *Mariner* defeat, Turner had said, "I'm like the grass. I get trampled down one day and spring right back up the next. I've been beaten so much that one more loss doesn't make any difference. Losing is simply learning how to win." But this time his spring seemed to have broken, and even B. J. couldn't cheer him up. He began ordering her around, talking wildly about sailing the world and never coming back. "I've got to go," he told her. "I don't want to spend time ashore anymore. I'm a man without a country. I want to switch ships on the way into port and live like the Arctic tern."

In the background, Frank Sinatra was singing "My Way." It was a favorite Turner tape, so he let Frank sing it again. It didn't help. The fun-loving captain was clearly in pain. Though it hurt to have to return to Atlanta a loser, he had no choice. Turner was soon back in his cramped, second-floor WTCG-TV office, nervously pacing up and down in front of his wall of honor, its shelves heavy with silver cups, bowls, trophies, photos, and other sailing memorabilia that swore he still was a champ.

Janie and the children had spent the summer in Newport, too, but they might as well have been on another planet. Though they were living in Conley Hall, the empty Tudor dormitory of Salve Regina University, where Ted and the *Mariner* crew were also billeted, they saw little of him. Fortunately for Janie, Jimmy Brown was there to help with the kids. He would take the boys out in a boat to nearby Goat Island, instructing them in the basics of sailing as he had their father.

In November of the previous year, when George Hinman and the *Mariner* supergroup had first got together at Bob Derecktor's boatyard in Mamaroneck to lay plans for their Cup campaign, Turner had said, "I'm so busy I don't know where to turn. And here I am in Mamaroneck when I should be home watching my sons play football. It's gotten so I can't even remember my kids' names." And the demands on his time got progressively greater during the 1974 Cup summer. While *Mariner* was being modified, Turner did on one occasion take time out to show his kids the nearby campus at Brown where he had gone to school, but that sort of family outing was a rarity.

Ted had always known exactly what his priorities were. Though already named Yachtsman of the Year twice, he was no Father of the Year. Teddy, his eldest son, felt that his father regarded them all as a necessary evil. "In Dad's defense," Teddy adds resignedly, in an attempt to come to terms with his father's behavior, "he had no way of appreciating what it means to have a family. Winning was all he understood. When his family came along, all we did was get in the way."

Turner didn't want his children to have any special treatment, and in that he was very much like his own father. If anything, he wanted it to be harder for them than for other kids. Above all, he demanded that they toe the line. In addition to his commandments against skateboards in the driveway, lateness for meals, and rudeness to adults, there were his special taboos. If he happened to come home and catch any of them watching a television program he didn't approve of, he hit the ceiling. But crying was even more frowned upon. "If he caught you crying," says Teddy, "that was the worst thing you could do. You never expressed your feelings at our house. I was a fairly disturbed child. Dad didn't have time for me."

Even when he did find time for the boy, it wasn't necessarily a pleasant experience. Teddy recalls a canoe trip that he and his two younger half brothers took with their father. Turner, he says, "yelled and screamed the whole time. It was a nightmare. So when we had finished and we were just going down the Chattahoochee River and Dad said, 'Well, did everybody have a great time?' I said no. And, boy, he smacked me hard."

Although Teddy was no better at baseball than his old man had been, he joined a Little League team. He seemed to like playing and was well liked by the other boys. Rating him "subpar" as a ballplayer, his coach thought he was "just a nice young kid." He remembers that unlike the other players who came to the games with their parents, Teddy was usually brought there by Jimmy Brown or, on occasion, by his mother. He could recall seeing the boy's father at only one game. "In fact," he says, smiling, "I had a little problem with him in the stands."

The stands were very close to the sideline, and Ted, rather than cheering for his son's team, was loudly engaged in trying to get people to bet on whether some kid or other was about to strike out, or drop a fly ball. Finally the coach became fed up. A big guy, he approached Turner and said, "Look, please stop that, or I've got to ask you to leave. We just can't have that here. It's disruptive to the kids." Expecting an argument, the coach was surprised when Turner just quit without saying a word.

Some years later, when Turner was advocating brotherhood and thinking about saving the world, he noted that true happiness comes "not from obtaining personal wealth or possessions or materialism but from helping others. That means coaching a Little League team, working with Big Brother programs. You can do it all kinds of ways." Turner has always been better at seeing the big picture than the little people in his life. Once when Janie told him that one of his younger sons was feeling badly neglected, Ted said to the boy, "Well, son, I've only been paying attention to those who talk the loudest, and I guess I didn't hear you. You've got to speak up, son. Make yourself heard."

Laura Lee, little Teddy's older sister, was about thirteen at the time and staying with her grandmother in Cincinnati while going to school there. She, too, was already having problems, and her father would soon describe the rebellious teenager as being "on the threshold of delinquentville."

That Janie, even with the help of Jimmy Brown, had her hands full with five active young children is obvious. That she resented her husband's absences is equally clear. "He never had to worry about our children," she says proudly, but her bitterness is visible just beneath the surface. "I did that for both of us." Janie had been reared by "a mother who was a mother," and she thought her children deserved the same. She prided herself on being just that sort of mother. Her marriage, however, was quite another matter. In 1976, she told a reporter for the hometown newspaper, "Sometimes it's boring that Ted isn't here." Trying to put a good face on things for the neighbors, she added, "Then it's exciting when he is. I like to go places with him—usually because the people I meet are fun and interesting."

But it wasn't always very much fun at all. Things were getting much worse for Janie. On one occasion, at a large cable-convention dinner party and much to the discomfort of everyone within earshot, he shouted at her, "I'm taking you home. You're fat, you're ugly, and you're drunk." She apparently sought refuge from Ted's dismissive treatment and womanizing in alcohol and self-deception. Of course, she knew what was going on with her husband. "I've seen them, the groupies, when we go sailing," she was quoted as saying in *Newsweek*. "They're always there. I think it's kind of pathetic, frankly, because they're young and yet that's all they can get." Faulting the naughty young girls rather than her wayward husband, she then declared with astonishing guilelessness, "That's why it really doesn't bother me." Although they were hardly impregnable, Janie did have her defenses.

She needed them. Ted was competitive about everything, and even playing backgammon with him—one of the few hobbies they shared—was a put-down. Ted let her know bluntly that he didn't think much of her ability. "Janie can't play too well, and I win," he told a visitor, completely indifferent to her presence. Then rubbing it in, he said, "But even that's just winning my own money."

Playing backgammon with Martine Darragon was a much more sporting proposition. She was a crackerjack player, had money of her own, and more tricks than a circus. Martine, for example, liked to play topless, which could distract an opponent even more determined and focused than Turner.

During these years, Ted was hardly home. And when he was, it was only long enough to cause an uproar before hurrying back to the office. Business was on his mind now. With the end of the Vietnam War, Turner had expected 1973 to be a good year for him, but then came the OPEC oil embargo, and the energy crisis. He had complained to his Newport sailing buddies about the declining state of business in Atlanta. There was no gasoline, no cotton, no chlorine, and pretty soon, he joked, there would be no boat building. He had had to turn the lights out on the world's largest Coca-Cola sign. "This is the year I was going to get rich, pay off all the debts," he told them. "Now I feel like Napoleon in [Russia] waiting for the other guys to surrender. Then this soldier comes in and tells me they haven't given up, they're still out there, and besides it's starting to snow. And now for the bad news: there's no food."

By the end of 1974, Turner had given up his big gas-guzzling Lincoln cruiser and was complaining about the wastefulness of the country. He himself was tooling around in a small white fuel-efficient Toyota. Although he had projected a total sales growth of $2 million for his Atlanta television station, Channel 17 profits had dipped to $732,340 from the previous year's $1 million plus. The one development that did cheer him up was a notification from Brown, the school that had unceremoniously booted him out six months shy of a diploma, that he had been selected for a special honor.

On November 1, Turner attended the Fourth Annual awards dinner in Providence, where, along with thirteen former undergraduate sports stars, he was inducted into Brown's Athletic Hall of Fame. Ted was one of two "special" inductees, having distinguished himself largely after his undergraduate days. The toastmaster for the evening, appropriately enough, was his former classics professor and one of the few teachers at Brown for whom he had ever had any respect, Dr. John Rowe Workman.

Decked out in a blue blazer and gray flannels for the occasion, Turner, who would be thirty-six in less than three weeks, seemed to have matured, and the additional gray hairs he had collected over the summer contributed to that impression. But Workman had no need to check the name tag Turner wore to recognize him. The two chatted away happily together before dinner.

One of the subjects Workman loved was disasters, and he had hundreds of books in his private library that dealt with this theme. Turner remembered how Workman had often told his students that it took the sinking of the *Titanic* to bring about a stricter maritime code requiring sufficient lifeboat space for all passengers. And it was the tragic loss of life in the Coconut Grove fire in Boston that had brought about important improvements in fire regulations. Given the "disaster" Turner himself had just been through with *Mariner,* Workman's sympathetic attitude had to be especially appreciated. Perhaps something good might come of that, too.

When it came time for the toastmaster to make his award, Workman, who was as fond of ice hockey as he was of disasters, stood at the flower-bedecked head table and described Turner as "the Bobby Orr of his profession," and "probably the top ocean racer in the world." Then he went on to cite his two Yachtsman of the Year Awards, and some of his other sailing honors. Ted was enjoying himself. It was too bad his father wasn't there to hear this.

But a sailing friend, big Bob McCullough—Brown, class of '43—was there and heard everything. In honor of the occasion, he had driven up from New York that afternoon. McCullough, who would be elected commodore of the New York Yacht Club the following year, headed the *Courageous* Syndicate that had just successfully defended the America's Cup. The older man and Turner were fond of each other, though they differed about what it took to win in Cup racing. In a lively postprandial exchange between the two men, McCullough said, "One of the main reasons we walloped the Australians was that we had better sails. It wasn't the material. Their sails just weren't cut right."

Turner, still smarting from his Cup wounds, defensively insisted that neither skipper nor sails was the crucial element in success. "You can always change skippers if they are hacking things up. You can even change the cut of your sails, as *Courageous* did in July while fighting *Intrepid* for the right to defend the Cup. But you can't change the design of the boat."

Although it was true that *Mariner* had been a lemon, Turner had hacked things up badly and knew it. He also knew that very few who do so ever get a second chance at the helm in the high-stakes gamble that is modern Cup competition. Thinking ahead to 1977 and the next America's Cup, Turner had already approached McCullough with an offer to buy *Courageous,* but he had been turned down. He was determined to redeem himself somehow. But as he shook hands with McCullough and parted company that night, he could have had no notion that there was a *Courageous* in his future.

CHAPTER 11

ONE DAY IN LATE 1974, Ted was once again at work at WTCG, pacing back and forth in his office and shouting orders as if he were still on the deck of *Mariner*. Will Sanders, his chief financial officer, was accustomed to having his boss away from Atlanta 30 percent of the time during any given year, and sometimes more. He didn't mind, but there were some employees who didn't care for the Turner style of long absences followed by bursts of hyperactivity when he returned and began stirring up the pot. Those staffers were soon gone. "They just couldn't take it," Gerry Hogan recalls.

Tom Ashley, who was the station's national sales manager, admitted that it was tough working for Ted "but we all loved it. If you were producing, pulling your weight, he pretty much left you alone. You had to be in the office by 8:30 every morning." Jim Trahey would arrive by 7:30. "I couldn't wait to get in there. You were just so pumped up. I couldn't wait. It was something about Ted Turner. A lot of the other people felt the same way. Gerry Hogan was my boss. He'd be in there before I was."

While Turner was away sailing in Newport, Sanders says, "we would communicate on a regular basis by phone. Then when he was in town and in the store we were meeting every day. Essentially, he ran the company in a way that nobody really made any major decisions without his approval."

On this day in 1974, Ted was on the verge of a major decision—one of the biggest of his career—but he didn't know it yet.

He was expecting Sid Topol, the president of Scientific-Atlanta, who had called and was on his way over with a man by the name of Reese Schonfeld. Schonfeld was in the television-news business and had just arrived in Atlanta from New York. Topol thought he was a very bright guy. He also thought that Schonfeld, with his interest in the new technology that Scientific-Atlanta was selling, might be useful to both Turner and himself. Eventually, it would prove to be an important meeting for all three of them.

In late 1974, the newest thing in the air was satellites, and Scientific-Atlanta made satellite dishes. A few months earlier, on April 13, Western Union had placed its *Westar I* in orbit, and RCA was scheduled to send up its own communications satellite the following year. Schonfeld, a graduate of Dartmouth and Columbia Law School, had first become interested in the use of satellites for the television industry in 1972, when the *New York Times* did a major story on satellite technology. By 1974, working for TVN, a scrappy independent television-news company, Schonfeld was faced with a problem: how to get his news packages to his customers, the small television stations not affiliated with any network. The AT&T telephone lines were too expensive. Microwave didn't cover the country. Schonfeld thought satellites might be the answer. It was for that reason he had come to Atlanta.

Topol met him at the airport. Schonfeld turned out to be a big man, about six-foot-three, with glasses and an authoritative manner. He seemed to know what he was talking about. On the way to Scientific-Atlanta, Topol suddenly announced that he had to make a stop. He wanted to introduce Schonfeld to Ted Turner.

The small WTCG-TV station on West Peachtree didn't make much of an impression on Schonfeld. The physical plant was shabby, the equipment poor. It was like a dozen other small independent TV stations he had visited while promoting his news service. The year before, in fact, he had met the general manager of WTCG, Sid Pike, who had laughed at Schonfeld's offer of news. "We'd never do news," Pike had told him. "I've got more movies on my shelves than I could run in a hundred years. This crazy Ted. Every salesman that comes in here [with a movie package] Ted buys it. We're never going to do news."

As far as Turner was concerned, "No News is good news." News programs were much too negative for his taste. Better *The Three Stooges* and *The Mickey Mouse Club* than Vietnam and

Watergate. Whatever news the station did carry (in order to satisfy the FCC requirement) was run primarily at three in the morning. It was handled by Bill Tush, who, according to his own description, sat at a "chintzy plywood desk in front of a blue wall" and presented, with Ted's enthusiastic support, a goofy program that was more entertainment than news. Tush did a "rip and read" from the wire-service reports and threw around a lot of pies, usually lemon meringue. Turner called him his "low-budget Walter Cronkite."

The only office in the building where a tall man like Schonfeld might not feel claustrophobic was Turner's. Recalling their first meeting, Schonfeld says, "It was very strange. Ted was sitting in a not-quite-clean yachtsman's cap behind his desk, and Topol wanted me to tell him about satellites."

Ted had one question on his mind, and he didn't beat around the bush. "What's it gonna cost?" he wanted to know.

"It'll cost you a million dollars."

He couldn't believe it. "A million dollars to reach *everybody* in the country?"

"That's right. But you understand, they've got to put in their own dishes [to receive your signal]."

"Oh, I understand that," he said. "I understand that. But do you mean that for one million dollars I'm gonna be able to put this out all over the country?"

"Yeah, absolutely," Schonfeld said. His figure was based on information he had gotten from Western Union and RCA. "Maybe a million-one. Ninety thousand a month, something like that."

Although Turner had seemed merely curious about what Schonfeld told him, Topol was encouraged by the meeting. The owner of Scientific-Atlanta felt that his prospects for a future sale were good, and as the two visitors left, Topol was delighted. Schonfeld, on the other hand, was wondering who in their right mind would want to watch a dinky little family station like Turner's, which seemed to carry the worst kind of independent programming imaginable. Looking back, Schonfeld says, "I thought the idea was crazy. I thought he was just a fool. I had no interest in him at all."

It was in the latter part of 1974 or early 1975 that Ted's friend Andy Goldman received a call. Goldman was now vice president of marketing for Teleprompter, and working in the company's New York headquarters on West 44th Street.

The call was from Gerry Levin of Home Box Office, a regional pay-television service that offered movies at home by subscription. Levin wanted to discuss his plan to transmit HBO programming to Goldman's cable customers by means of the new RCA satellite. It was going to be a cable-TV first, and a major milestone in the history of television.

After talking to Levin, Goldman picked up the phone and called Atlanta.

"You know, Ted," Goldman told Turner, "I think you ought to look into going up on the satellite. Long-term, it'll make a helluva lot more sense distributing by satellite than it will by terrestrial methods."

"What's an HBO?" Turner asked.

Hardly anyone knew what HBO was back then, much less how satellites worked. Goldman knew, because HBO was one of his early customers, and he explained. "Look," Ted said, "I'm a regional television station. What the hell do I need a satellite for?"

"I really think you ought to look into it," Goldman repeated, "because long-term it'll allow you to connect from one point to multiple points a helluva lot more efficiently than you're doing right now." Ted said he'd think about it.

He did, and decided that he'd like to find out more about *Westar*. At Sid Topol's suggestion, he arranged to pay a visit to Upper Saddle River, New Jersey, to see Ed Taylor, vice president of marketing for Western Union. Taylor remembers that when they first met, Turner's knowledge about satellites was small. "*Very* small," he says. "But he's a quick learner. He's the type of guy who seems to talk all the time, but he gets what he needs to know out of the person he's with."

Although Turner had known little about microwave systems when he began in television, he had learned. Microwave depended on 500-foot towers erected every 20 miles or so that carried a television signal from point to point to point. It was an expensive system to install and maintain. But it was allowing WTCG to go well beyond its 40-mile broadcast range to distant cable systems in neighboring states. Still, microwave was limited. Satellites, however, held the promise of national distribution.

Communications satellites had been whizzing around the earth since the 1960s, but it wasn't until the first one was sent up 22,300 miles to a geosynchronous (stationary) orbit above the equator, that satellites became practical for television use. Taylor told his visitor that with the aid of a powerful transmitter—an uplink—WTCG would be able to send the Atlanta Braves skyward to a satellite such as *Westar,* and in one-fifth of a second they would be beamed down to dish-shaped receivers all over the United States. It seemed like a kind of magic. And because the signal traveled perpendicularly through the atmosphere, it was free of static and "snow." Ed Taylor also mentioned that it was relatively cheap.

If the cost-conscious Turner wasn't interested before, he was now. With microwave, Ted had moved from being a local to a regional player. Now, he could be national. This was his chance to send little WTCG's programs to television sets all across America. The technology was new and untried for television, but assuming it worked, he could expand his business, perhaps radically, in a single decisive move.

There was just one problem, Ed Taylor recalls. Inexpensive as it was, "Ted didn't really have the money to pay for a transponder [one of the transmitter relay slots on the satellite]." So Taylor worked out a plan for him to finance the operation, which involved charging cable companies ten cents a subscriber, and then the two of them went to talk it over with Andy Goldman in New York.

"I'm gonna get one of those transponders," Ted excitedly told Goldman, after introducing the salesman from Western Union.

Goldman stood up. "Mr. Taylor, would you mind excusing us for a few moments?"

As soon as the door closed, Goldman said, "Look, Ted, you're about to rent a transponder on the wrong goddamn satellite. HBO is going up on RCA."

The problem was that Goldman's company, Teleprompter, would need two different earth stations in order to receive the signals from two different satellites. Back then, the only downlinks (that is, receivers) the FCC was approving were huge dishes—standing 30 feet high and measuring 10 meters in diameter—that cost between $80,000 and $100,000 each. "And I'll be goddamned," said Goldman, "if I'm going to put in an $80,000 earth station to receive you and another one to receive Home Box Office. Go talk to the people from RCA."

In short order, Ted had signed on with RCA and applied for FCC approval. As usual, once Turner's mind was made up, he acted quickly, decisively. He was soon on the phone to Topol.

"Sid," he said, "I wanna buy an uplink. Send a salesman over, and I'll get you a check." Remembering the call, Topol laughs and says, "It was a funny way of negotiating." But Turner was never one to quibble about price when he really wanted something—even though the electronic transmitting equipment he had just ordered, a Series 8000 Satellite Earth Terminal, was priced at about three-quarters of a million dollars, money he didn't have.

To chief financial officer Sanders, and others in the Turner organization who possessed less of an appetite for risk-taking than he did, Ted seemed to be acting recklessly: "He was always playing 'You Bet Your Company,'" says Sanders. "Every project he took on had the potential to sink him. No sooner did you feel like you were comfortable and able to breathe a little bit than we'd take on some other impossible task." And unlike most businessmen, Ted rarely seemed to worry; he wasn't afraid to be embarrassed by failure.

Sanders had not had much luck in raising cash for WTCG from the Atlanta banking community. In Atlanta, Sanders explains, there was no tradition of lending to the broadcasting industry. The value of a broadcast company such as WTCG, with very little in the way of tangible assets, was primarily in its FCC license. "Fortunately," he says, "there were forty or fifty major regional, superregional, and money-center banks that learned that broadcasting was a cash-flow business. That it was a good industry to lend to." Banks such as First Chicago, Chemical, and Chase specialized in broadcasting and already had a billion-dollar investment in the industry. With the help of these banks and insurance and pension funds like TIAA and Home Life in New York, Sanders was able to find enough money for Turner to start Southern Satellite Systems, the firm that would handle satellite transmission of his programming to cable companies.

But Turner quickly discovered that the FCC would not allow him to be both a programmer (creating shows) and a "common carrier" (distributing them). There had to be a middleman between Ted and the cable-system owners. It was around this time that Turner hired the Washington law firm of Corrizini & Pepper to help guide him through the FCC regulatory maze. In an effort to separate himself from Southern Satellite Systems, Turner offered the company to his friend Andy Goldman, who turned him down. He then offered it to two or three others. Whether they didn't like the risk, the concept, the burden of assuming the company's debts, or the prospect of working with Turner—or perhaps because they disliked all four—they also turned him down.

In February 1975, Ted invited Ed Taylor and his wife to Atlanta to make Ed a proposal. Taylor recalls, "It was a typical Turner whirlwind weekend. We went out for dinner and ran around, but we were always talking business." Will Sanders took part in the discussions. Attempting to insulate himself from FCC regulations, Turner offered to give Taylor his programming—letting him, so to speak, capture it and then sell it to cable viewers for a few cents a home via Southern Satellite Systems. The potential financial return was enormous. And all he'd have to pay for Turner's satellite company was one dollar. Although it was still a gamble for the Western Union executive, in that it meant giving up his $65,000-a-year job, Taylor found it an offer he couldn't refuse.

"So I gave Ted a dollar for the company," he says, and thereby took over the financial responsibility of his contracts with Scientific-Atlanta and RCA. In effect, Taylor was freeing Turner from the cost of an earth station and a transponder that he couldn't legally use. The major advantage for Turner was the happy prospect of thousands of new WTCG viewers and a huge increase in his advertising income because of them. Of course, he'd have to get more women down in the station basement to handle the overflow mail for Ginzu knives, bamboo steamers, and fishing tackle, which by 1977 would generate annual gross revenues of more than a million dollars.

Turner assigned Will Sanders to help raise the initial money for Ed Taylor's new corporation. It wasn't easy. They were turned down by fifteen different groups. In the interim, RCA launched *Satcom I* on December 13, 1975, and two months later, while a frustrated Turner remained grounded, HBO was up on the bird. Under such circumstances, Turner is about as patient and tender-hearted a boss as the pre-Christmas Scrooge. One observer at the time recalls, "Ted was horrible to Will. And not one time, but again and again."

"Amazingly," says Taylor, "it took almost a year to get the money." Finally they found a small group of investors in New York, fewer than a dozen, willing to put up $20,000 each for

1 percent of the company (though some bought more) and raised about $300,000 to help Taylor get started. Reflecting on these early hurdles, Taylor reports, with obvious satisfaction, "They got their money back in eighteen months, and, in effect, that 1 percent is today worth one-and-a-half million." (Approximately five years ago, Taylor merged his one-dollar company with Tele-Communications, Inc. (TCI), the nation's largest cable company, receiving stock worth $50,000,000.)

Ted Turner, when asked if he'd ever had second thoughts while trying to get his station up on the satellite, snaps, "Of course I did. I'm not an idiot. I knew the risks. But I had to move fast, without a lot of people knowing what was up."

The equipment for Southern Satellite Systems, Inc., was located in a remote wooded hollow about 10 miles away from WTCG's transmission tower on West Peachtree. Turner had put up a high chain-link fence there and topped it with three strands of barbed wire. No Trespassing! Inside stood a microwave antenna positioned to pull in Channel 17's signal from downtown Atlanta. Next to it and tethered to SSS's ghostly white electronics trailer, was a 30-foot-high transmitter-receiver, its pale orbicular face tilted skyward ready to beam up *Leave It to Beaver* or *Father Knows Best* to *Satcom I* the minute the FCC gave Turner clearance.

To anyone stumbling upon it by accident, this emplacement had the air of a secret military installation. There was no reason to alert the competition to what was going on. But, typically, Turner couldn't resist passing the secret around like a family photo. While trying for almost a year to get this earth station set up and FCC-approved, he had been talking to people, and there were many in the broadcast business who already knew all about it. He wasn't called "The Mouth of the South" for nothing. The only real secret was whether Turner's bet on satellite transmission would pay off.

Ted's man in charge of special projects such as satellite transmission was the affable, twenty-five-year-old Terry McGuirk. He had worked summers for WTCG while still an undergraduate at Middlebury College and then come aboard full-time in 1973. Of all the people who worked for Turner, according to Gerry Hogan, it was Terry who was closest to him. Some even speak of a father-son relationship between the two. In 1975–76, there was no project

McGuirk was working on that was more special, more dear to the heart, more headache-provoking for his boss, than trying to send Channel 17 by satellite to cable companies all over the South.

"For a while," says McGuirk, "even Ted wondered what he'd gotten into. It was so impossibly complicated. We had committed millions of dollars not knowing what the FCC would do, we'd had to separate from SSS, we'd hired people on at big salaries. A lot of the time nobody understood what was happening except Ted, and people would say to me, 'I hope he knows what he's doing, because nobody else does.' I was still new at the time, but one day about eight-thirty in the morning I got a call from Ted. 'Come right over.' When I got there, he was really down.

"'Am I crazy?' he said. 'Am I absolutely out of my mind? Is this whole thing going to collapse around me?' I spent about three hours sitting in his office listening, and then he just walked out and started abusing people in the halls. I waited for a while and then I got up and went back to my own office. I don't think he even knew I was there that morning."

In the summer of 1976—seemingly out of the blue—Turner was invited by Congressman Lionel Van Deerlin, a Democrat from California, to testify in Washington before his House Subcommittee on Communication, which was looking into the future of cable television. Van Deerlin was especially concerned about the failure of the cable industry to grow more rapidly. Turner rushed to the capital, eager to argue for a lessening of federal restrictions and to go on record as one of the few broadcasters in the country who supported the cable owners. It was a typical outsider's position for the maverick broadcaster. It was also a case of enlightened self-interest. Network broadcasters were eager to restrict cable's growth, because it fragmented their audience, but cable viewers already represented more than half of Turner's customers. Called to address the committee just before their luncheon break on July 20, Turner stepped into the Washington limelight for the first time and really seemed to enjoy himself.

Seated at the microphone in the small subcommittee hearing room, the witness looked up at the handful of congressmen arrayed behind their imposing desks. "I am unique, I think," he told them, "or fairly unique in being friendly to the people in the cable industry, and I am considered a bit of a traitor to be here on behalf of the cable industry this morning, although I do feel that I have a pretty reasonable position."

Turner on this occasion had brought a written statement, which was very unusual for him. Even more unusual was that it had been written by someone else—Don Andersson, one of his employees. Turner acknowledged his indebtedness in a wry, self-deprecating reference to the fact that he was not a college graduate. He then proceeded to read the prepared statement, in which he explained what he, as an independent broadcaster, had done for cable sub-scribers in small communities outside Atlanta.

He began by telling the committee that a UHF independent station like his had to do "a lot of scratching" to attract an audience and advertisers in Atlanta, a market where there were three VHF network-affiliated stations. One thing he had done, he said, was to invest a great deal of money in acquiring what he claimed was "the best programming available," and that included the television rights to three of Atlanta's major-league teams. In addition to Braves baseball, he had bought the rights to televise Atlanta Hawks bas-ketball and Atlanta Flames ice hockey. He had also established a close relationship with cable operators, visiting their association meetings and talking with them. As a consequence, WTCG was currently being seen via microwave by more than four hundred thousand subscribers in five states on nearly one hundred different cable-television systems.

Turner pointed out to the committee members that outside Atlanta there were communities that had few if any independent stations, and the only commercial television available came from the three New York networks. In his opinion, the one opportunity for many people in small towns in the South to have any choice in programming was to let cable operators bring in an increased num-ber of distant independent signals, like WTCG.

Then, like some idealistic Frank Capra hero fighting entrenched power and desperate to protect the little guy, Turner said, "My hat is off to the cable people, and I am pulling for them. I am, I really am. They are providing a real public service for the American people. There should not be any distant-signal-carriage restrictions for cable TV. To limit the number of television stations available to cable systems is to shortchange the American people and

perpetuate a broadcast monopoly. Especially this year of all years,"
he concluded, alluding to the Bicentennial, "we should be promoting and fostering the widest possible freedom of choice."

Congressman Charles J. Carney was impressed. "I would say
that the gentleman from Atlanta, Mr. Turner—he said that he is
without the benefit of a college education. Maybe that is why you
had to learn to think, Mr. Turner."

Though Ted may have liked the compliment, he was quick to
set Carney straight. "I did go to college. I just did not finish."

"And maybe that is why you had to learn to think," Carney
repeated, insisting on his little irony. Even Ted knew enough to
let him have the last word on that subject.

Next, Tim Wirth, a young first-term Democratic congressman
from Colorado, wanted to know whether unlimited importation
of distant signals would inevitably lead to the development of a few
superstations, endangering the effective coverage of local news and
events. "Mr. Turner," he said, addressing the witness, "your station
is known, as I understand it, as Super 17, correct?"

That was more or less a joke, Turner explained. "I got that
name six years ago, when we were losing eighty thousand dollars
a month and were watched by no one. We had a young girl down
in promotion who did not last very long, but she had one great
idea. I said, 'We need to jazz this place up a little bit,' . . . And she
said, 'Why don't we call this place "Super 17"?' I said, 'That's a
great idea!' You know, in other words, it was a real tongue-in-
cheek thing, and it was long before the superstation concept."

Wirth asked, "Do you have any reaction to the notion of a
superstation?"

Much to the delight of his audience, Turner made no attempt
to conceal his feelings or ambition. "Well, I would love to become
a superstation. I would love desperately to create a fourth network
for cable television, producing our own programs, not just running
I Love Lucy and *Gilligan's Island* for the fifty-seventh time. And I
intend to go that way, if we are allowed to."

He turned to Wirth: "You have to remember there are three
supernetworks who only own four or five stations apiece that are
controlling the way this nation thinks and raking off exorbitant
profits, and most of these local stations that everybody is crying
about are just carrying those network programs that are originated
out of New York. They have an absolute, a virtual stranglehold,

on what Americans see and think, and I think a lot of times they do not operate in the public good, showing overemphasis on murders and violence and so forth.

"So if we do become super," he said, leaning into the microphone to make sure they heard this, "it will be another voice. Perhaps it might be a little more representative of what we think the average American would like to see. A little less blood and gore on television and more sports and old movies and that sort of thing, that we think might encourage children not to go out and buy a gun and start blasting people, like in *Taxi Driver*."

Turner's enthusiasm and candor easily won over the members of the committee. By emphasizing the nastiness of the networks, the importance of diversity, and the need to provide wholesome entertainment for the average American family, he had championed removing restrictions on the distant importation of signals and neatly avoided the question of protecting the small local stations. *Taxi Driver* was the enemy—which, of course, was not a television program at all but a movie that had only just opened and had never been televised—and he had slyly smeared its "blood and gore" all over the networks.

Congressman Wirth was delighted with the witness. "Mr. Chairman," he said, "I personally find that a very healthy and productive approach to this. I wish that [approach] were reflected all across the board." Ted left Capitol Hill greatly pleased with himself. Though other broadcasters condemned him as the quisling of the broadcast industry for his testimony, cable owners embraced him as a brother. Turner was sure that he had done his cause some good, and he had.

On December 17, 1976, after an investigation of almost a year, the Federal Communications Commission determined that Taylor's company, SSS, was independent of Turner Communications and approved its petition as a common carrier. That same day WTCG-TV went up on *Satcom I*. Down below in Atlanta on West Peachtree, the telephone operator now answered Channel 17 callers with the cheery greeting: "The superstation that serves the nation, good morning."

The FCC issued another decision that day that would also have important repercussions for Turner's future. In an effort to encourage the growth of the cable industry, it announced that receiving dishes no longer had to be 30 feet high but could be half that size. Overnight their price dropped from more than $100,000 to $50,000. And within two years, the size of dishes as well as their cost had shrunk again by half. As prices fell, the demand increased. When HBO first went up on satellite, there were only two dishes in the country available to receive it for cable distribution. Two years later, there were 181 receivers in operation, with some 800 more expected by year's end, and applications were flooding into Washington at the rate of more than two dozen a month. The demand proved to be so great that by the close of the decade, tiny Scientific-Atlanta, Sid Topol's electronics boutique, had grown to become a billion-dollar bonanza.

As for Turner's superstation, within about a year it was being received from Atlanta to Alaska, as well as in Canada and Hawaii. Via microwave, Channel 17 had been on cable networks in five states. Now by satellite, Turner proudly announced, it was being seen in twenty-seven states. But in the beginning, availability didn't necessarily mean income for WTCG-TV.

For all the crazy antics and hysterical hijinks during those early superstation days, when the staff thought of themselves as "Ted Turner and his merry men," there was a good deal of free-floating anxiety as well. Not only did the boss and his chief financial officer have doubts about the company's future, but so too did its salesmen. How were they going to convince local Atlanta companies such as the Sunshine department stores or Rocket wines ("Put a Rocket in your pocket") to advertise to a national audience, most of which might never visit the state of Georgia? And how were they going to convince major national accounts (Ford, General Foods) to advertise to an audience that was minuscule by broadcast standards? WTCG was now positioned as an almost impossible sell—it was too large and too small. "I mean, this had never been done before in the history of broadcasting," Jim Trahey pointed out. "And, you know, it scared me to death, with a little girl at home."

Turner, fully aware of how tight money was and of the necessity for a major effort on the part of his sales team to keep WTCG afloat, called a special meeting early one Sunday morning. It was held at the Stadium Club before a Braves game that was scheduled for later that afternoon. Trahey recalls that in addition to Turner and himself, three others were there—Gerry Hogan, Bob Sieber, and Don

Lachowski. It had begun raining the night before, and it was still raining when their meeting began. Trahey was sure that the game would be rained out, but Turner seemed to be all optimism. Talking for two hours straight, he told them that they were going to have to work harder than they'd ever worked before. The superstation meant problems, big problems, but problems, as Turner viewed them, were nothing more than opportunities waiting to be seized.

"It's gonna be tough," he told them. "It's really gonna be tough, but we'll do it. Someday you four guys are gonna make a whole lot of money."

Outside it was still pouring when they had finished, and the game was finally rained out. But despite the dreary weather, Trahey and the others were positively upbeat when they left, and determined to do whatever it took for the superstation to survive. There were few who could motivate better than Turner.

Ted was, of course, also a salesman himself, selling the superstation to advertising agencies, cable companies, anyone who might eventually make them some money. Trahey thought he was great at it. "I'd go and do the groundwork. Get beat up and beat up." Advertisers would say, "When you're in a third of the homes in America, come back and see us." Trahey told himself, "It ain't gonna work." Then he'd bring in Ted. Trahey marveled at Ted's charisma, his knowledge of rating books. "But you know," he says, "a lot of [our] sales people, they didn't want [to go out with] him. They were afraid of him." Perhaps it was, as Trahey suggests, because "he cast a giant shadow." Others felt scared of Ted because he was so unpredictable. He rarely prepared for a presentation, preferring to wing it. They never knew what he might do next.

Gerry Hogan recalls Turner repeatedly nagging him to get an ad from Rich's, a blue-chip Atlanta department-store chain. It was a very staid, very proper old-line Southern company, and the woman who bought advertising for it, Mary Jean Meadows, personified the Rich's image. "She was probably at that time about fifty," says Hogan, "very prim and proper." Somehow he managed to get an appointment with her, and Ted, eager to win the account, went with him.

Hogan still shudders at the memory of that meeting. "He just put on a performance that was like Jerry Lewis. You know, like the wacky professor—the wacky station guy. He was all over her office. He couldn't sit down. He was walking around, picking up stuff off her desk, and finally he got down on his knees, and he said, 'Mary Jean, you gotta tell me what I can do. I'll do anythin', anythin' to win your favor.' And he held her hand like he was sort of wooing her, like he was a suitor in a John Barrymore kind of scene from some movie, trying to impress this woman. She couldn't believe him. She was not even *slightly* amused." Hogan adds dryly, "Needless to say, we didn't get the business."

Turner would indeed do anything to sell his station to advertisers. He'd jump up on chairs, on desks, on tables, on anything that didn't move, and shout at the top of his larynx. If he met really serious resistance, he might even drop to the floor as if he'd been shot and cry, "You're killin' me!"

It was during this period that he began to hold an annual fall review of "new" programs for advertisers. Newsman Bill Tush sums up these new schedules as "essentially rearranging reruns." There was one party, according to Tush, that was especially memorable. It was a luncheon held at Atlanta's Hotel Sheridan, and the room that afternoon was crowded with well-dressed Southern ladies, all advertisers, chatting quietly among themselves. Standing in front of them, Ted was eager to make the women feel at home, so he began his opening off-the-cuff remarks by saying, "You know, I grew up with advertisers. And my daddy taught me how to treat advertisers." He smiled brightly, showing them the gap in his front teeth. "My daddy said, 'If advertisers want a blow job, you get down on your knees.'" The women stared at him unbelievingly. It was a luncheon that none of them was likely ever to forget.

Encountering all kinds of problems at the start, the superstation found its advertising revenue slow to increase. The nervous, chain-smoking Don Andersson and Terry McGuirk visited every cable company in America trying to sell them Channel 17. Ed Taylor, realizing that his future was inextricably linked to Turner's regardless of what the FCC might think, went with them. They often ran into a stone wall of resistance from cable managers who doubted the satellite would work. Local managers would ask, "What the hell do people in Denver (or Tupelo or Fairbanks) want with a station from Atlanta?" On occasion, when faced with an especially tough prospect and desperate, the three of them would decide, "We've got to sic Ted on him" and put in a call to the home office.

In time, cable systems in outlying areas began to sign up, and things started to look better. "Look what's happened," Andersson gloated, after a year. "In terms of cable homes, we've done in one year on the satellite what it took seven years to do with microwave." His optimistic prediction was that by the end of 1978 WTCG would be in two million cable homes, more than double the number of viewers they had in the Atlanta market.

A major problem for Turner was that *TV Guide* declined to list cable stations, and it was almost impossible to build an audience if there was no way for viewers to find out what WTCG was offering and when. Ed Taylor says, "Terry and myself would go to meetings and plead with *TV Guide.* . . . It took us a year and a half to get them intrigued enough [to decide] that they had to become a cable guide as well. But at that point cable was so small that it wasn't in their self-interest to list the cable programs. HBO was paying them some for buying ads; so they wanted us to pay them. In those days, it was pay them—with what?"

Taylor remembers well how difficult the early satellite years were and how everyone who had any connection with the super-station seemed to be gambling. Sid Topol was gambling for a long time that he'd ever get paid. RCA was also gambling. Taylor confides, "When the first month came, they sent me a bill, and I couldn't pay them. And the second month, I couldn't pay them. And the third month, I couldn't pay them. We ran ninety days behind on our bills to RCA for four damn years."

But the biggest gambler of all, of course, was Ted Turner. And the very same year that he had invested so heavily in creating his superstation, he decided to buy yet another new toy with the potential of being a major-league bleeder.

CHAPTER 12

THE ATLANTA BRAVES were a losing team when Turner first acquired the rights to televise their games in 1973, and they were still a loser at the end of the 1975 season. But win or lose, cable owners were eager to have baseball for their customers. Sports coverage was important to Turner's business. "We televise all the Braves' away games, and we got the Hawks [the Atlanta Hawks, of the National Basketball Association] away from ABC. Big bucks. It costs us money to televise the Hawks, but it's bringing in viewers. Big bucks. It was a bold move," Ted says, pleased with himself. "But a faint heart never won a fair maiden."

Although Ted himself—despite his limited athletic ability—had always preferred an active rather than a passive relationship with sports, he began to attend the baseball games at Atlanta's Fulton County Stadium, just to watch what was going on. It was nothing good. The people who had been fascinated in 1974 to see whether Hank Aaron would hit 715 home runs and break Babe Ruth's record were less thrilled a year later by a nearly last-place team that had traded Aaron to Milwaukee and seemed to be just going through the motions. Turner had to admit, "The team really stank. Ugh!"

One night in 1975, he looked around the stadium and saw that there were only a couple of hundred people in the stands. If something wasn't done to improve the Braves soon, even cable audiences might not want them. His salesmen had been packaging the Braves as "America's Team," which worked fine for large areas of the

southeastern United States where there was no major-league baseball, but if they continued to play next year the way they were
playing this year, they could well end up as the team without a
country.

While he watched the game and worried about selling ads for
the Braves for the coming season, Turner had a beer. It was good,
and he had a few more. Then he went up to talk to Dan Donahue,
the team's president, who was sitting in the glassed-in owners' box.

"Hey, Dan," he said, "what are we going to do to get these
Braves going?"

"I don't know what *you're* going to do, but we're bailing."

"Oh my God!" Turner almost had a heart attack. There had
been rumors about the Braves leaving Atlanta. He reminded Donahue that the Braves had a five-year contract with Channel 17. Then
he asked, "Who're you going to sell it to?"

"To you," Donahue said.

"To me?" Turner was startled. "For how much?"

"Oh, about ten million."

"Yeah, well, how much is it losing this year?"

"Oh, about a million bucks this year."

"What?" Turner looked at him in disbelief. "I'm going to pay
you that much so I can lose a million a year, too?"

Turner told Donahue that he'd think it over. Although he didn't
think very long about it, he did think very hard. What to many at
the time might have seemed a foolhardy gamble was to him a
logical, if somewhat risky, defensive move. His repeated nightmare
was that one day the Braves might move out of town. Toronto
was eager to buy the franchise. Turner loved the idea of snatching
the Braves away from the competition and saving them for Atlanta.
The role of savior, which had long been emotionally important to
Ted, was playing an increasingly significant part in his business life.

Another consideration was that even if the Braves remained in
town, the cost of TV rights would almost certainly go up in 1977,
when Turner Communications' five-year contract with the owners
ran out. But if he owned the team, he could count on controlling
both the continuity and the cost of his television programming—
and programming was crucial in his decision.

Jim Roddey, formerly the president of Turner's company, had
done the calculations for him. According to Roddey, "He could
lose five million dollars a year on the Braves and still break even."

As Roddey explained to Ted, "There are 162 games. They're three-and-a-half hours long. And by the time you do the pregame and the locker-room show and everything, you're talking about a four-hour show times 162. You'd have to lose a lot of money in sports to offset what it would cost to buy that amount of programming." And later on, the same line of reasoning would explain Turner's interest in the Hawks when they were on the auction block.

Years before synergy became fashionable in business, Turner was exploiting it. And with the help of his financial experts, he would soon discover the advantageous tax loopholes that his sports teams provided to help offset their losses.

Turner was determined to buy the Braves. Once his mind was made up, he was almost impossible to stop. Meeting with his board of directors, he laid out his case, and there was no question what the outcome would be. "This is something we *have* to do," he told them. "We need to do it. If we don't do it, we might lose the rights. Who knows? And then how are we ever gonna be a major player in this business?" Will Sanders says, "Ted presented a very persuasive case," stressing the advantages, and how acquiring the television rights to the Braves games had already given Channel 17 a new respectability.

Turner's CFO was worried about the price Donahue was asking and didn't know where they would get the money to pay it. He had learned that the Braves were having trouble meeting payrolls. And as the team had not been playing very well, their attendance had plummeted. But knowing how impulsive his boss could be, Sanders was especially worried that Ted intended to buy the Braves "no matter what," which would give him no bargaining position. Sanders cautioned him about that, but his caveat fell on deaf ears.

Turner went back to see Donahue and told him to tell the other owners—the "Chicago twelve," as they were known in the local press—that he'd do it. "What the hell!" he said casually, as if he'd just flipped a coin.

Sanders recalls trying to protect Turner from himself. "I was involved in that negotiation. It was essentially little or nothing down—maybe a half a million dollars, and a note for eight million dollars payable over ten or more years at six-percent interest [plus some debt]." Their discussions lasted several weeks, and then after all the details had been agreed to by both parties, everything nailed down, the Braves' negotiator suddenly demanded Turner's *personal* guarantee for the total amount.

Although upset at this new last-minute condition, Sanders was sure that it wasn't a deal breaker. He met with his boss to discuss the matter at the Midnight Sun, a Scandinavian restaurant in Atlanta's Peachtree Center. Over dinner he explained what the other side wanted and told Ted not to do it. There was no need to put up his home or the superstation as a guarantee. The other side would cave in soon enough. Or if he liked, he could guarantee a portion of the asking price—say, a million dollars at most—rather than the entire amount.

But when Turner came face-to-face with the Braves' owners, he had neither the patience nor the cunning to be cautious. He wasn't going to lose the team *now*. Despite all of Sander's advice, he stuck out his hand. "Fine," he said, accepting total personal liability, "I'll do it." Sanders shrugged, realizing that there was nothing more to be said. It was Ted's decision. According to Sanders, "He gets to a point where he doesn't negotiate very carefully or thoroughly and leaves too much on the table." But in this instance, as things turned out, the price was a bargain.

Sanders knew that among the Braves' assets were earnings from season-ticket sales and hot-dog vendors, money that was still trickling in. But after purchasing the stock of the Atlanta Braves National League Baseball Club, Inc., from the Atlanta-LaSalle Corporation (the parent company that owned the team), Sanders made a discovery. The exact amount of money they had acquired was somewhat larger than expected. He rushed to tell Turner the news: "The company has cash in the bank in excess of a million dollars."

His boss was delighted. "So I bought it using its own money," Turner says, "which was quite a trick." Indeed, for in buying the Braves he got back twice as much cash as he paid—a nifty investment for a team that today is probably worth between $150 million and $200 million. "In retrospect," Sanders notes, with quiet satisfaction, "it turned out to be a pretty good deal."

On January 6, 1976, a press conference was held to announce the sale of Atlanta's Braves and introduce their new owner to the public. Seated sandwiched between ex-owners Bill Bartholomay and Dan Donahue—two beefy-faced smoothies in dark, custom-tailored suits—the lanky, dimple-chinned, thirty-seven-year-old Turner, who was decked out in a tacky tattersall vest and a souvenir Hank Aaron 715 tie, looked like an innocent being led to the slaughter.

Ted began his typically rambling, extemporaneous remarks to the reporters crowded into the press lounge at Fulton County Stadium by saying, "We're gonna operate, ahh, freely and, ahh, openly. Atlanta is my home, and I love it here. I've been all over the world racing and stopped in Tahiti on the way back from Australia which I read a lot about as a little boy, and I decided the closest thing to paradise on this earth is Atlanta."

He told them that even though financially it was a dumb move, he was buying the Braves for Atlanta. He said that the only way he could afford to purchase them was on the installment plan, because as of that moment he was "Tap City" and "we're gonna owe the Atlanta-LaSalle Corporation quite a bit of money!" He said nothing about the fact that if he couldn't come up with the money, Atlanta-LaSalle would not only get back their team but Channel 17, too. That kind of thinking was for losers.

He emphasized again, "It's not an economically wise move to buy the team, but, ahh, money itself is not the prime motivation here. I believe I'm doing it primarily for the city and the southern part of the country, believe it or not." Two years later, when asked by *Playboy* why he had paid so much for the world's losingest baseball team, Turner seemed to have changed his emphasis, if not his mind. "Well, I got a deal: nothing down and twelve years to pay. We already had the Braves on our television station, and it just made sense to buy, instead of paying six hundred thousand a year for broadcast rights."

As usual, Turner at the press conference was taking songwriter Johnny Mercer's advice to accentuate the positive, eliminate the negative, and definitely not to mess with Mr. In-Between. He warned the reporters, "I don't wanna see any more headlines in the *Atlanta Journal-Constitution,* bless their souls, that call Atlanta, 'Losersville, U.S.A.' I wanna see 'Winnersville.'" He promised that he'd bring a World Series to Atlanta in five years. He promised that the buck stopped at his desk. And he made one more promise. After admitting that "I really don't know as much about baseball as I should, I'll be quite honest with you," Turner said, "but I intend to learn and learn as fast as I can with everyone's help." And that was the one promise he almost kept.

The fact is that Turner was really an ignoramus about the game. But the National League owners couldn't have cared less about that when they approved the sale of the Braves to him. According to Turner, all they really wanted to know was "what my financial backing was." As a businessman, he understood their priorities. "What do you need to know about baseball?" he asked, shortly before buying the Braves. The answer was simple: "Both sides have ten guys." His education in the game would have to be a cram course.

In February, Turner and Terry McGuirk went down to Florida to watch spring training. "He'd never done anything with a ball when he was growing up," McGuirk says. "Not even a stickball. He had no hand-eye coordination and he really didn't know what was going on." That year, faced with a dispute over the reserve clause, major-league owners were conducting a preseason player lockout, but that didn't stop Turner. He worked out informally with his players, running wind sprints, and on March 14 in Sarasota, he organized an unofficial game of nonroster players with Bill Veeck, the well-known baseball maverick, and his Chicago White Sox. It was televised on WTCG-TV. "With all the negative stuff going on," Turner said in defense of their unsanctioned exhibition, "this is something positive."

Janie came down to see the game, and the new owner and his wife sat in the stands in their dark glasses and Braves hats, clapping and cheering their heads off. When Ted first told her that he was buying the Braves, Janie thought he was kidding, but she knew better. Then she thought that at least now he'd be home more, but again she was wrong. The only way she was going to see more of her husband was by tagging along after him.

The owner-player dispute was soon resolved, and the Braves' spring-training camp officially opened in West Palm Beach. The Braves' manager, Dave Bristol, had been given his orders by Turner at their first meeting. Roughly paraphrased, they were "Win or die!" Bristol says, "When people would ask me about Ted, I'd always tell 'em that he just burns a different fuel than the rest of us, and that's true."

It took Turner a while to learn the difference between a balk and the infield-fly rule. Jim Trahey, no doubt exaggerating for effect, says, "It took me a year to teach him what a hit-and-run was." But Ted hung around the coaches and the players and he did

learn something about the game. One thing he picked up quickly
was chewing tobacco. Turner says, "I was down at spring training
the first year and all the coaches were chewing and somebody offered me a chew. They were teasing me, so I took it and chewed
it. They all stood around, waiting for me to get sick, but I fooled
them: I liked it." At the cost of high cleaning bills, Ted became
one of the boys. Janie hated the tobacco stains on his clothes. Trahey
remembers one time riding in the backseat of a jeep with Turner
at the wheel. He was talking excitedly and chewing Red Man, his
favorite. Casually spitting over the side, he covered Trahey with
tobacco juice from head to toe.

Back in Atlanta, Turner called up Will Sanders and told him
he was on his way over. They were going to play ball. In a little
while, he appeared on Sanders' doorstep with a couple of mitts, a
bat, and a ball, all ready to play. "We went over to Chastain Park,"
Sanders recalls, "which is right down the street from my house,
and pitched the ball back and forth." As they played, Turner talked
business. "Hey Sanders, what do you think about buying the
Flames?" Sanders couldn't believe his ears. Was he going to buy
Atlanta's hockey team, too? Though Ted was trying hard to play
ball, there was something pathetically awkward about his movements. He would never be a ballplayer, but then he really didn't
have to be. He owned the team.

And Turner was going to sell it for all he was worth. He made
a singing commercial with his players, in which he swung his autographed Louisville Slugger and urged the locals to "come on out
and see the Braves at your Atlanta tepee." He promised to produce
a winner and make baseball fun. He called himself "the frightened
new owner of the Atlanta Braves." Advance ticket sales soared to
three times what they had been the previous year. "Awwriight!"
shouted Turner. Roaming the halls at Channel 17 with his bat on
his shoulder and his Braves cap on his head, he psyched himself
up, getting ready for opening night, the first home game of the
season against the World Champion Cincinnati Reds.

There was a crowd of more than thirty-seven thousand in the stands
on that Tuesday evening, April 13. They had come out to see the
1976 version of Atlanta's Braves and the nice-looking local boy
who was their new owner. Turner, in his Braves tie and bell-bottom
pants, was easily as visible as his team. Unlike the previous owners,

Ted (bottom row) in 1950–1951 at The McCallie School in Chattanooga, Tennessee, where he was the only seventh grade boarding student. *(Courtesy The McCallie School)*

Ted's father, Ed Turner, in front of his house in Savannah. *(Courtesy Jane Greene)*

Ed Turner (in light suit) and his second wife, Jane Dillard Turner (with arm raised), entertaining friends on the rear porch of their plantation house in South Carolina. *(Courtesy Jane Greene)*

After about two years at McCallie, Ted began his metamorphosis from class cutup to leader.
(Courtesy The McCallie School)

Ted in his junior year at McCallie was Sergeant of Company C and winner of the "Neatest Cadet" award. In his senior year he distinguished himself by being promoted to Captain of Company E.
(Courtesy The McCallie School)

The Debating Club at McCallie and its director, history instructor Elliott Schmidt, with Ted at his left. In 1956, the four-man varsity debate squad won the Tennessee State Championship.
(Courtesy The McCallie School)

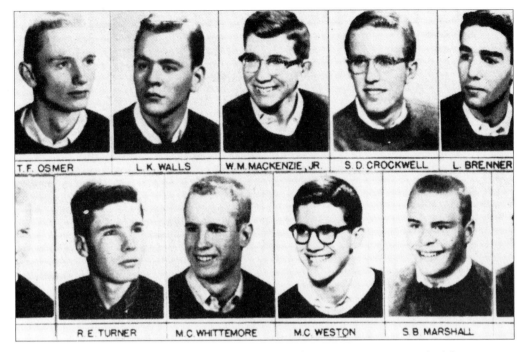

T.F. OSMER   L.K. WALLS   W.M. MACKENZIE, JR.   S.D. CROCKWELL   L. BRENNER

R.E. TURNER   M.C. WHITTEMORE   M.C. WESTON   S.B. MARSHALL

Ted (bottom row) with his 1957 Kappa Sigma pledge class at Brown University. To his left, Yacht Club teammate Mal Whittemore.
(Courtesy Brown University Archives)

Elected Commodore of the Brown Yachting Club in 1960, Ted, when not racing in regattas, could be found on the Seekonk River sailing with a girlfriend.
(Courtesy Brown University Archives)

Father and son at Ted's wedding.
(Courtesy Judy Nye Hallisy)

Julia Gale Nye (Judy) and Robert
Edward Turner III on their
wedding day, June 22, 1960,
in Chicago.
(Courtesy Judy Nye Hallisy)

Ted and Judy (pregnant with
their first child), plus family pet
Blackie on a visit to the Turner
plantation in early 1960s.
(Courtesy Jane Greene)

At the Outdoor Advertising Association of America convention in 1963, the 24-year-old Ted was one of the youngest participants, registering himself as R. E. Turner III.

On November 1, 1974, Turner was inducted into the Brown University Athletic Hall of Fame. The Toastmaster for the occasion was his former Classics professor, Dr. John Rowe Workman.
(Courtesy Brown University Archives)

In the opening home game of the 1976 season, the Atlanta Braves' new owner astonishes players and fans by running out onto the field to congratulate Ken Henderson, who has just hit a home run.
(Bud Skinner/Atlanta Journal and Constitution)

Ted entertaining Braves fans with an ostrich race. The number on his uniform was to remind them to watch the team on his TV superstation, Channel 17.
(AP/Wide World)

Ted and one of the Braves ball girls cavort around the bases at Fulton County Stadium after the team breaks its 13-game losing streak in May 1976.
(UPI/Bettmann)

At the helm of *Courageous,* Turner defeats the Australian challenger in the 1977 America's Cup competition, going on to win four straight races. *(Copyright © Christopher Cunningham 1977/Gamma Liaison)*

Turner, learning that baseball commissioner Bowie Kuhn's 1977 one-year suspension of him has been upheld in federal court, indicates he'll keep his mouth shut.   *(UPI/Bettmann)*

Turner with unidentified Newport admirer in September 1977, celebrating his triumph over *Australia*.
(UPI/Bettmann)

The victorious skipper of *Courageous* being escorted by two of Newport's finest through a throng of well-wishers.
(UPI/Bettmann)

Playboy cover girl Liz Wickersham was Ted's girlfriend in the late seventies and early eighties, working for him as WTBS and CNN hostess on such programs as *The Lighter Side, Showbiz Today,* and *Good News.*
(Courtesy Playboy *magazine, copyright © 1981 by Playboy)*

who had sat behind glass in their private luxury box high up in the stadium, Turner sat right behind the Braves dugout, where everybody could see him, guzzling beer and rooting for his players. "I never could understand why owners like to sit up behind bulletproof glass sipping martinis," he told a sportswriter for the *Boston Globe*. "I sit in the front row."

That night Turner shook hands, slapped backs, signed autographs, threw kisses. Georgia governor George Busbee was in the stands. Atlanta mayor Maynard Jackson was there, too. Turner seemed to be in his element. He ran out onto the field, and one of the biggest crowds ever to see a home opener in Atlanta stood up and cheered. They liked the fact that their team was now owned by a local boy, and they approved of his positive attitude. Waving his hands, the new owner led the crowd in singing "Take Me Out to the Ball Game." This was followed by a five-minute welcoming speech that Turner stretched to twenty-five minutes, unable to let go of his audience. He was having fun.

Back in his seat, he watched the first, scoreless inning go by, spitting tobacco juice into a paper cup and cheering his pitcher, Carl Morton, out of danger. Then came the bottom of the second and Atlanta's Ken Henderson smashed a ball deep into right, the team's first homer of the season. In the center-field stands, team mascot Chief Nok-a-homa bobbed and weaved, doing his ritual home-run dance in front of his tepee.

Turner was delirious. In a flash, he had leaped to his feet, jumped over the fence separating the viewers from the viewed and become part of the action. Braves shortstop Darrel Chaney couldn't believe his eyes. "Here comes Ted out of the stands, and I thought, What's this guy gonna do?" Henderson, rounding third, saw the new owner clutching his Braves cap and racing him to home plate. The next day, a photograph of the disheveled Turner shaking the hand of the startled Henderson as he crossed the plate made papers all over the country.

The game was tied 1-1 until the eighth inning, when Morton was replaced by Pablo Torrealba. "Come on, Pablo, throw strikes! Throw strikes!" Turner shouted, yelling himself hoarse. Torrealba, following instructions, threw one to Reds outfielder Ken Griffey, who promptly smashed a blistering liner between first and second, driving in two runs. Turner buried his head in his hands and moaned. "Oh no! I don't wanna look up." Then Torrealba went back to the mound. *"Buena suerte,* Pablo! *Buena suerte!"* called Turner, still screaming encouragement, but it did no good. The Reds won 6-1.

Turner hated to lose. Later in the clubhouse, he picked up an onion and hurled it furiously at the wall outside Dave Bristol's office. The manager came out to see what had happened and noted the onion stuck to the wall. ("I think it's still there," he says.) Turner told his players, "I'm proud of you all, anyway." Then storming out, he kicked the door so hard that he nearly broke his foot. The Braves' publicist during this period, a man with the high-visibility name of Bob Hope, was witness to more than one such Turner outburst. "He has an incredibly intense and volatile personality" is the way Hope described his new boss, "and really doesn't have total control of it at all times." The next day, Turner showed up at his office limping and leaning on a cane.

"That first year," Turner remembers, "I was really active. I devoted half my time to the Braves, which is a huge amount. I went around to every National League ballpark, met all the general managers, discussed ticket pricing, how to develop players, even how you print up a program. I didn't have any idea how complicated it all was. It was a nightmare." Hank Aaron, whom Turner later brought back to Atlanta as director of player personnel, was impressed with the amount of baseball knowledge Turner had picked up since they first met, a year earlier. Aaron recalls that at the players' draft, when he and Braves executive Bill Lucas and Turner were discussing trades, "Ted surprised us at how much he already knew about the available players—[he'd] put a tremendous amount of time into studying this thing."

Initially, Turner couldn't keep his hands off his team, wanting to meddle in every aspect of the organization. He demanded change, and with the sole exception of Rowland Office in center field, the Braves' starting lineup was completely altered. He wanted upbeat reporting from Skip Caray, whom he put into the broadcast booth to handle the 1976 Braves' sixty-six-game television schedule. Caray says, "He went through a phase early in the year where he wanted me to sugarcoat the pill a little more than I thought necessary . . . making everything positive on the air." Unable to restrain himself, Turner actually donned a uniform and cast himself as a batboy and then as a manager, but even he had to recognize the improbability of convincing anyone that he was a player.

It was when he began to shake up management that he ran into trouble. Donald Davidson, a vice president, had been with the Braves organization for thirty-eight years and was well liked by sports writers. Davidson was only four feet tall. Though Turner didn't give a damn how vertically challenged the guy was, he did care very much about the free and easy way Davidson had with company money (*his* money!), insisting on VIP suites at hotels while traveling with the team. "It's over, Davidson," Turner snapped, after a few months. "Get your bags packed and get out of town!"

As Turner quickly discovered, it's not easy to look like the wronged party when you're six-foot-two and picking on a dwarf. Frank Hyland, a sportswriter for the *Atlanta Journal,* called it "a chickenshit thing to do," and wrote in his column, "Ted Turner has money. Ted Turner has a lot of money. Ted Turner has enthusiasm. Ted Turner has his own baseball team. Ted Turner belongs to some of the world's most prestigious clubs. But Ted Turner has no class." Recalling this episode, Turner readily admits, "The sportswriters almost killed me."

The firing of Davidson was followed by a thirteen-game losing streak. Not surprisingly, there were some who saw this as divine retribution. Similarly, many discerned a relationship between the bad press that the new owner had gotten and his draconian announcement a few weeks later that local sportswriters would no longer receive free food and drink at the stadium. The *Journal's* sports editor, Jesse Outlar, wrote, "I rarely dined in the press lounge. The food there is almost as bad as the chow you purchase at the concession stands. More important, if the Braves continue to perform as they have thus far, Turner can't afford to give away anything."

Not inclined to halfway economy measures when the team was losing money hand over fist, Turner got rid of Eddie Robinson, the team's general manager, too. Like Davidson, Robinson also had a taste for VIP suites and first-class travel, and he drove a Cadillac that he charged to Turner as a company car. "Give me a break," Turner told him. At a time when Turner was leasing Chevrolet Novas for his Channel 17 executives and he himself drove a stick-shift Toyota, Robinson's Cadillac infuriated him. "When you screw around with his money," says a Turner business associate, "you're playing with one of his vital organs. That he does not tolerate."

Turner could be penny-wise and pound-foolish, but contradictions are typical of the man. In his seat behind the Braves dugout, he'd watch with a sinking heart as foul balls sailed up into the stands and were scrambled after by fans. "Hey, throw that ball back," he'd shout, hoping to save a few bucks. But when it came to big money, he could be utterly reckless.

The million-dollar contract he gave to free agent Andy Messersmith was his way of telling players and public alike that the Braves were under new management. Turner justified the money by saying, "The Braves had a reputation for being cheap, and I just wanted to let them know I would pay what it took to get a quality player." As things turned out, Messersmith had a mediocre season for the Braves the next year, and Turner was forced to dump him, selling his contract to the Yankees.

Not particularly good at contracts, he brought in his friend Mike Gearon, a retired realtor, to help. Gearon recalls Turner coming out of a meeting with a player's agent, who was loudly demanding the amount of money he wanted for his client. Turner responded by offering more than was asked for. "Take it or leave it!" Ted said. "One of my jobs was to protect him against this sort of thing," says Gearon. "With [Braves pitcher] Phil Niekro, Ted would say, 'I wanna do this for Phil. I wanna do that for Phil.' I'd say, 'Ted, let me take care of this. You're doing enough.'"

If at the beginning Ted knew little about baseball, he knew a great deal about promotion and advertising. Like his fellow owner Bill Veeck, who reminded him that they were really in the entertainment business, Turner was intent upon making a day at the ball game fun. Though he couldn't produce an instant winner, he could at least sell a good time. "The first thing I did was spend a million dollars on a giant TV screen over the scoreboard," says Turner, once again emulating Veeck, one of the premier promoters in the business. In order to draw crowds, Veeck had also instituted Gourmet Days and Bartender Days and had brought back Ladies' Days. Turner, with the help of Bob Hope, came up with even zanier events, determined to change the boring image of the Atlanta franchise and put people in the stadium's empty blue seats.

Soon there were ostrich races and home-plate weddings, Easter egg hunts and dollar-bill scrambles, mattress stacking and motorized bathtub competitions, free halter-top giveaways and aerialist Karl Wallenda in a Braves hat doing a skywalk 200 feet above the infield. All this plus fireworks, wrestling matches, ball girls in hot-pants, and belly dancers! Turner said he'd do anything to bring people into the stadium, and he clearly meant it.

But most interesting—and characteristic of the near middle-aged Turner—was his eagerness to participate in these events, and his willingness to look like a damn fool in the process. For the ostrich race, the superstation mogul wore a tiny jockey's cap and a custom-made silk uniform with the number 17 sewn on his arm, reluctant to miss any opportunity to advertise. His principal opponent was Frank Hyland, his outspoken critic on the *Atlanta Journal*. The race wasn't even close. Hyland won going away in a strut. Defeated, Turner would later complain bitterly that he hadn't been given enough time to practice. "One lap?" he screamed, his competitive juices boiling. "How the hell can you determine the fastest ostrich in one lap?"

After the Great Mattress Stacking Championship, a squashed Ted turned to Hope and said, "Oh, God, why do I do this type of thing, man? I'm gonna end up killing myself someday." While pushing a baseball with his nose around the infield in a race with Phillies relief pitcher Tug McGraw, he very nearly scraped off half his face. Afterward, as if it were terribly important to him, Turner crowed, "I beat Tug by a mile." Though he did win, he bloodied his nose and forehead so severely that the scabs took weeks to heal. "When you're little, you have to do crazy things" is the way he explains himself. "You just can't copy the big guys. To succeed you have to be innovative."

Turner's firing of Donald Davidson, coupled with the team's long losing streak, had led to angry public rumblings of discontent, signs reading "TRADE TED" sprouting in the bleachers, and falling attendance. But whether it was his omnipresence at the games, his high visibility, his obvious involvement, his crazy antics, or all of these, somehow "Teddy Ballgame," as the irrepressible new owner was labeled by Atlanta sportswriters, gradually managed to win the fans back. In July, one of them came up to him during a game and said, "I just want to thank you, Mr. Turner. I want to thank you for the great job you're doing for the city of Atlanta." Obviously touched, the Braves' owner turned to his wife and said,

"Isn't that great?" It was, and especially given the fact that the team lay buried in last place. More remarkable still, by the following year a sports column appeared in the *Atlanta Constitution* with the headline "TURNER, ATLANTA TOGETHER IN A COMMON LOVE."

When her husband bought the Braves, Janie Turner told a local reporter that up until then the family had been "fairly anonymous" in Atlanta. Ted's yachting accomplishments had gone "almost completely unnoticed" locally. Perhaps, as she claimed, that was because Atlanta wasn't a sailing city. Now that he owned the Braves, Janie was afraid that they were about to lose their splendid isolation, and she was right.

Unlike sailing, which has never been a major sport in this country, baseball is played and seen and loved by millions of Americans. Baseball, after all, is homegrown, as Southern as grits, as American as popcorn. And Ted Turner was wonderful copy. He could make even the most outspoken of the other new major-league owners, such as Ray Kroc of San Diego, Brad Corbett of Texas, and George Steinbrenner of the Yankees seem like Marcel Marceau. Easily accessible, Turner began in 1976 to become a public figure, his name and face constantly in the papers. "Before I got into baseball," Ted told popular TV interviewer Tom Snyder, "hardly anyone knew who I was."

Turner relished the attention. Whether consciously or not, the image he began to project was, in many ways, evocative of Charles Foster Kane, the hero of Orson Welles' classic movie. Turner, by his own Falstaffian accounting, has seen *Citizen Kane* more than a hundred times. Given his obsessive tenacity, he obviously must have felt a kinship with Kane, who was also a man willing to do whatever it took to make his business a success.

Like Kane, too, Turner saw himself as trying to give the little guy a square deal. "I'm the little guy's hero," he told one reporter. "They love me. I run the team the way they think they would if they owned it. I come to all the games. Sit in the stands. Drink a few beers. Even take my shirt off. I'm Mr. Everyman to them— their pal, Ted."

During one Braves losing streak, he impulsively picked up the public-address microphone beside his seat and announced, "Nobody is going to leave here a loser tonight. If the Braves don't win,

I want you people to come back tomorrow night as my guests. We are going to be in big-league baseball a long time, and we appreciate your support. We're going to beat the hell out of all these guys who are beating the hell out of us." And the handful of suffering fans cheered wildly.

If few private lives have been more public than that of the fictional Kane, the factual Turner's is one of them. Just as the millionaire newspaperman was "more newsworthy than the names in his own headlines," so, too, was the millionaire superstation owner. Turner's name and face soon started popping up in newspaper stories and on magazine covers from coast to coast. His outrageous antics began to capture the public's imagination. *Time* printed a picture of a cigar-chomping Turner, a Braves hat on his head and one of his attractive Braves ball girls on each arm, and called him "Terrible Ted." *Sports Illustrated* emphasized his eccentricities and summed him up as "an exciting guy." *Playboy* quoted him as saying, "I don't care what a ballplayer does, if it makes him happy. Just as long as he wears something over his cock, you know." No matter how an article read, Turner seemed to enjoy having people read it aloud to him in the presence of others. Perhaps it was the narcissist's fondness for having laundry aired in public, regardless of whether it was clean or dirty, as long as it was his. Given his behavior, it was only a matter of time before Turner ran afoul of the baseball hierarchy entrusted with protecting the sanctity of America's national pastime and its unimpeded cash flow.

Charles (Chub) Feeney, the popular president of the National League, liked a good cigar and a good joke. What he didn't like was trouble. The square-faced extrovert was rarely seen without a smile on his face, but in 1976–1977 it seemed as if any mention of the name Turner could drain the blood from his lips. Rumors had reached him that the Braves' owner was in the locker room playing poker with his players, taking showers with them, eating meals with them. He seemed to want to be one of the boys.

Feeney summoned him to the league office, which was then in San Francisco. Since the notorious Black Sox scandal of 1919, baseball has been hypersensitive about gambling. "Owners don't play poker with players," Feeney explained. Turner snapped back, "Why not? Is there some kind of goddamned rule against collaborating with the enemy? On my team, I am *part* of the damn team."

Turner couldn't believe that there was "this double standard where all the players have to bow their heads and say, 'Yes, sir,' and 'No, sir,' to the owners. Maybe to the older guys, but why me?"

Feeney flatly ordered him to quit gambling with them. "So I did," said Turner. "But the players still like me." Then he added with almost touching innocence, "Even Pete Rose of the Cincinnati Reds told me one day, 'I wish I had an owner like you.' *Strong!*"

But playing cards was not the only thing Turner was doing that the league president didn't care for. Ted had put nicknames on the backs of his players' uniforms, and on the back of Andy Messersmith, who wore number 17, was the nickname "Channel." Turner had made him a walking billboard for his superstation. The nicknames had to go. Ted had promised his players a $500 bonus for each game the team finished over .500. The illegal incentive bonuses were out! No more using the center-field TV screen for instant replays to call into question an umpire's decision. No more running onto the field to greet home-run hitters. Turner says Feeney told him that "the things I was doing were not endearing me to the establishment." Feeney warned him to clean up his act.

For the moment, the Braves' owner was apparently chastened. Noting how different he had become since the start of the season, one of his players said, "You could see the change. He didn't come into the locker room after every ball game and such, but I still think that after a clutch hit or a close winning ball game, he's going to keep jumping over the fence and coming out on the field to congratulate us when it's over—he's just that way."

While fond of ritual and tradition, Turner hated formality. None of this "Mr. Turner" stuff for him. "Call me Ted," he told his players. He was, after all, one of the boys. According to team manager Bristol, there had never been such harmony among his players. They were all one big happy family. It was during a series against the Mets in New York that Turner had a brainstorm. He would take his entire baseball family for a visit to the New York Yacht Club. Imagine *Guess Who's Coming to Dinner* multiplied by about forty. Feeney might not like it, but what the hell! On West 44th Street, the New York Yacht Club cringed in anticipation.

Shortstop Darrel Chaney recalls that evening vividly. "We were in New York, and Ted told us after a ball game that he was going to take us all out. He sent us a memo and told us about it and said that the attire was 'coat-and-tie.' Nobody on our club *even wore* a coat and tie very often. Ted finally said, 'Oh, heck, forget the ties—we'll make out all right, let's go.'"

With Turner in the lead, the team marched across town and entered the club's elegant landmark building. Somehow they managed to get in without ties. Ted showed them the Trophy Room, where the America's Cup was kept, and the model boats. The bar had been specially set up for them for dinner. Chaney says, "Even the waiters were first class—towel around the arm and all."

Now, Chaney, thinking that he'd have a little fun, had brought along his invention. It's a device made out of a coat hanger and shaped like a shoehorn, with two rubber bands and a washer in the middle. When wound tightly and released, it makes a loud, rather vulgar crepitating noise.

"Anyway," Chaney says, "at this dinner Ted had for us, they gave all of us the royal treatment—prime rib with horseradish sauce. One of the waiters asked how we enjoyed the meal. 'Oh,' I told him, 'it was a great dinner and all, but *that horseradish* . . . eghh, I'm tellin' ya!'" Chaney leaned over and let the thing go. "'Oh,' he said, 'my God, I'm truly sorry.' Turner got all excited. 'Who's doin' that?' he yelled from back in the rear of the room. 'What's that!!!' He didn't know about the thing, you know. I did it a couple more times, and Ted yelled, 'You guys are SICK!!!'" The exclusive yacht club had, needless to say, witnessed more decorous evenings.

The Braves' disastrous 1976 season under their new owner fittingly ended as it had begun, with a home-game loss, but this final defeat was even more humiliating than all that had gone before. Though weakened by a virus that night, John Montefusco, the young Giants' pitcher, was still able to hurl a no-hitter against the hapless Braves. Turner took it philosophically and threw a party for his players and their wives anyway. Hell, at least they had tried. "These guys are the finest bunch of fellas I've ever been associated with, and I love them," he said.

Despite the fact that total attendance was up over the previous season, it was still relatively low, at eight hundred and twelve thousand, and the team was costing Turner his shirt. He lost more than a million dollars that first year and the next, and by 1978 he acknowledged, "We're losing as much money in baseball as the rest of the company is making." But as long as his ownership of the

Braves allowed him to televise more than sixty games a year, the team was worth its weight in programming.

Having never seen a World Series in person before, Turner headed for New York to watch the Cincinnati Reds take on the New York Yankees in the 1976 version of the national classic. The Reds won the first three games in a row. On Wednesday, October 20, they might have made it four straight had that game not been rained out. Turner was disappointed because he had to be in Tuscaloosa the next day for a speaking engagement. "When I agreed to go there, I forgot that the World Series would be going on," he explained to Jesse Outlar, of the *Atlanta Journal*.

They were attending a party in the Yankee hospitality room at the Waldorf-Astoria, where Turner was staying. Another familiar face at the party was that of Bob Lurie, the rookie owner of the San Francisco Giants. Turner was interested in acquiring one of Lurie's players. Giant outfielder Gary Matthews had been a leading long-ball hitter in the National League for three years and would be a free agent on November 4. The Braves owner had already been fined $10,000 by Baseball Commissioner Bowie Kuhn for talking to Matthews during the season.

Turner, who was on his fifth or sixth vodka, couldn't resist the opportunity to needle Lurie. He invited him to Atlanta that Saturday to attend a party he was hosting at the Braves' new Stadium Club—a surprise party for Gary Matthews. In the family of baseball owners, it was as if Turner were making a pass at his brother's wife. Lurie warned him that this time it might cost him more than $10,000.

"Everybody is going to be there to welcome Matthews to Atlanta," Turner continued, jiggling the needle. "The governor, the mayor, everybody." Lurie asked if Commissioner Kuhn had been invited.

Turner became defensive. "There's no law against showing a player our fine city. You know I'm not going to make him an offer. That's illegal before the reentry draft. I just want Matthews to be aware what a wonderful city he can play in." Then unable to resist and keep his mouth shut, Turner unfortunately added, "When the time comes, I'm going to offer Matthews more money than the Giants will take in next year. Of course it may take much more than that." It was obvious to Outlar, who was a witness to this exchange, that whether Matthews stayed with the Giants or left for Atlanta, he was going to be a wealthy young man.

Over two hundred distinguished Atlantans showed up at the Stadium Club for the Gary Matthews surprise party. Though Turner couldn't talk to the player about contract and salary until the draft, he could try to win his heart. Moving restlessly from table to table, the Braves' owner whispered into the ears of his guests, "Go shake hands with Gary. . . . Tell him what a great city this is." Turner wondered if he was overdoing it. "But if it were me," he said, "I'd be moved by this kind of turnout. I just don't believe money is everything to these guys. Anybody wants to feel wanted." As people came up and shook his hand, Matthews said, "This is nice, real nice."

On November 2, two days before the reentry draft was scheduled to be held in New York, an event occurred that would have an even greater impact on Atlantans than whether or not Gary Matthews chose to become a Brave. In one of the closest presidential races of the century, Americans elected Jimmy Carter, Georgia's very own peanut farmer, as the thirty-ninth president of the United States. Though Ted, of course, was a Republican—like his father— and had supported Gerald Ford, he wasn't sorry to see Carter going to the White House, even if he was a Democrat. He liked the idea of having somebody from his own state as president, the first president from the Deep South since Civil War days. In fact, he liked the idea so well that he had contributed money to Carter's campaign as well as to Ford's. It was a no-lose election for Turner. Later, when in need of favors from the Carter White House, he simply called in his markers. Not long after the new president had been sworn in, the Braves asked him to come to Atlanta to throw out the first ball of the 1977 season.

Turner and Carter genuinely liked one another. They had a great deal in common. In addition to being Georgians, they were both ambitious and idealistic, and they shared a naval background, a knowledge of business, an interest in sports, and a preference for informality.

It was on Saturday, January 15, 1977, that the Braves caravan pulled into Plains, Georgia, for an informal softball game, and the president-elect was going to play. Ted, according to Bob Hope, "was all psyched about seeing Jimmy Carter." Grinning from ear

to ear as he pumped Carter's hand, Ted said, "This is strawwng."
Jimmy's brother, Billy, had arranged everything, including a huge
barbecue, and the entire Carter clan was there including Miss Lillian,
the matriarch of the family and a rabid baseball fan. "Then," says
Hope, "we had a softball game of the Braves against the Billy Carter
all-star team. And Jimmy played. Billy played. Everybody played."
But not that much, because Billy had brought in every ringer in
south Georgia, and they were some of the best players that Hope
had ever seen. Even Ted played, pitching three straight balls before
he was taken out. "Ted's not very athletic," Hope explains.

To top off the day's festivities, on the bus ride back to Atlanta,
and with their boss's blessing, the players and their wives were
treated to a pornographic movie. Though it suited Ted's business
interests to bash the networks for their corrupting influence on
"family values," he was by no means a prude about dirty pictures.

Gary Matthews had been impressed by his visit to Atlanta, and at
the reentry draft, he was completely won over when the Braves
offered him a five-year contract estimated at $1.5 million. Much to
the delight of Turner, he signed on the dotted line. But Ted's joy
was short-lived. Bob Lurie filed a protest against him with the
commissioner. When Bowie Kuhn heard what the Braves' owner
had said and done, it seemed clear to him that Turner was tampering
with another owner's property. The commissioner was beginning
to have serious doubts about the wisdom of his ever having let this
loudmouthed troublemaker into organized baseball. Turner would
have to be punished.

Just before the winter meeting of league owners in Los Angeles,
Ted sat down with Hope and discussed what the commissioner
might do. "Well," Ted said, "there are a couple of things he could
do. First of all, he could return Gary Matthews to the Giants. But
that wouldn't be punishing me. That would be punishing him. Or
he could fine me a lot of money. But I've got a lot of money. If
he fines me a lot of money, he knows I pay it and then I go on my
merry way. Or he could *really* punish me, and at the same time get
me out of his hair, by suspending me from baseball. When we go
to the winter meetings, we've got to make sure he suspends me
from baseball."

It may be, as Hope believes, that because Turner planned to be off sailing during the summer of '77 in the America's Cup, he calculated that suspension would give him an easy, unassailable reason for being away from baseball. Or perhaps he thought that if he were a lightning rod, his suspension would save Matthews for the Braves. Hope says, "You tend to try to rationalize why he's doing things, and after a while you just run out of reasons. You think, Well, that's just the way he is." Whatever his reasoning, when Turner headed west to Los Angeles early in December, it was with more than a little anxiety. Would Kuhn actually deprive him of Matthews for tampering?

Arriving at the Los Angeles Hilton, Turner was notified by his Atlanta office that he had just received a telegram from the commissioner. Kuhn's message was "hold up on Gary Matthews' contract pending further investigation." Why did Kuhn have to send him a telegram, unless he had already made up his mind? They were, after all, both staying at the same hotel. If Turner was worried before, he was hyperventilating now. He tried to arrange a meeting with Kuhn and was told that the commissioner was too busy. Too busy! Turner had to find some way to get his attention.

When Bill Lucas, now the Braves' new general manager, entered the hotel lobby, he couldn't believe what he was seeing. Standing on top of the concierge's desk, where no one could possibly miss him, was Ted. Hope was in the lobby, too, and the sight of his boss reminded him of a street preacher ranting and raving to a sidewalk crowd. Half of the people in the lobby were listening and the other half were cringing as they tried to ease past him. "The commissioner of baseball is going to kill me!" Ted screamed. "Bowie Kuhn is out to kill me. My life is over." Close by was a flashy young woman Ted had picked up in town by prior arrangement. She looked on, acting as if it all made sense.

In the days following, Turner talked about Kuhn incessantly and was quoted in the *Los Angeles Times* as saying, "Kuhn's going to gun me down in this hotel like a dog!" He told a radio reporter for KNX that he was going to get a gun and kill the commissioner before the commissioner got him first. Only a little less riveting than an earthquake, Turner was becoming more and more outrageous as he grew increasingly frustrated at his inability to meet with the commissioner. Whatever plan Turner might have originally had, he was now clearly out of control. Hope says, "I'm not sure he wouldn't have behaved that way even if he hadn't planned the other. Because Ted was just getting carried away. I mean . . . he just was bonkers. . . . It was scary."

On January 2, Commissioner Kuhn notified Turner that for behavior "not in the best interests of baseball," he was suspended for one year. In addition, the Braves would not be allowed to participate in the June free-agent draft of high-school and college players. Turner told reporters, "Just say that for once in his life, Ted Turner was speechless—that he was at a loss for words." The good news was that the Matthews' contract was not voided.

Turner's speechlessness lasted only about five seconds, and the next day he was back in the news. A year after buying the Braves, he had bought the Atlanta Hawks of the National Basketball Association, acquiring a 55-percent share of the team for about $4 million. Again, he claimed that it was to prevent the loss of a local professional franchise, to save the Hawks for Atlanta. "I may be spreading myself a little bit thin," he joked at the press conference. "Maybe if they carry me off to Milledgeville [the site of the state mental hospital], I can watch the game on television." Surprisingly, he failed to add, "on Channel 17," where the Hawks games were a regular feature and would now be guaranteed to continue without interruption.

Turner would later liken his buying the Hawks to taking over the Confederate Army on the steps of Appomattox Court House. In quick succession, he had acquired what he called "the world's two worst sports franchises." Since coming to Atlanta from St. Louis, the Hawks had been below .500 six years out of nine. "But the thing that's a disaster for a basketball franchise," he pointed out, "is not to have any season-ticket holders, like we don't. That and having a whole lot of hacks under long-term, no-cut contracts for big bucks."

The first thing he did was reduce his payroll by one-third, getting rid of top management. "We had some guys around here who could fuck up a two-car funeral. And they were wasting my money." Turner admitted to Tom Snyder that "it was strictly a survival thing." Initially, he had appointed Mike Storen, a former commissioner of the defunct American Basketball Association, as the Hawks' president and general manager. But Storen didn't know how to defer to Turner, and he was soon replaced by Mike Gearon, whom Ted had first met at a Hawks' Booster Club meeting.

In the process of reducing his overhead, the new owner eliminated not only top management but middle management as well. There was, for example, a young, relatively inexperienced woman in publicity who was fired in memorable fashion. Because she was voluptuous and getting more money than anybody else in the Hawks' office, Ted may have assumed that there was something going on between her and Storen. "But," Gearon says, "she was an innocent victim."

"Is this girl still working here?" Ted asked him one day, and she was. "We oughta let her go. Bring her in here."

When she was brought in, the young woman was clearly upset, not knowing what to expect but fearing the worst. Suddenly, Turner launched into a tirade. "We don't have any money around here. In fact, we're even thinking of turning these offices into offices for doctors and dentists and closing down the Hawks' offices. . . . The Hawks—we're sort of like Napoleon in the battle of Russia. We're just bleeding and beaten. . . . We don't really need a PR person. This organization is like a boat that's takin' on water. You gotta learn around here to put the biggest people over first. In this case, the ones that are makin' the most money. . . . Now, maybe if you could get some other source of income. You know, maybe get a part-time job out on the street. Maybe even walking the streets." Turner went on in this exorbitant fashion, speculating at length about her new career. When the young woman left, she was in tears.

It was hardly a secret among people who worked for Turner that he could be verbally abusive. On one occasion, Will Sanders put together a long, detailed business report, and Ted, glancing through it, found a mistake. He hurled the study across the room, pages flying in all directions, and began screaming at Sanders as if he had questioned his manhood. Even those like Mike Gearon, who were genuinely fond of Ted, found him difficult to live with. "He could be abusive," Gearon admits. "Horribly abusive."

Ira Miskin, who later worked for Turner Broadcasting, recalls the way Ted would dig in his heels once his mind was made up. "He could be very cruel. He'd just tell you, 'Up yours! Do what I say.' He was loud, tough, mean." Turner would say, "Goddamnit, this is my fuckin' company! I'll do as I goddamn well please! If you don't like it, get the fuck out!"

Farrell Reynolds, who worked in sales for Turner, also acknowledges that "Ted came down hard on people." Reynolds, however, makes the point that all anyone had to do if the boss became abusive was simply stand up to him and tell him to stop. "And he stopped," says Reynolds. But not many who worked for Turner had the nerve to do that.

At the press conference announcing his purchase of the Hawks, Turner mentioned in passing a cordial telephone conversation he had had that morning with Bowie Kuhn. "He was very friendly. I've got a meeting with him [on January 18] to discuss the suspension. I don't know how restrictive it will be." Kuhn was allowing him to continue his activities with the Braves pending that meeting. "I'm just glad that he didn't order me shot," Turner said.

Turner was responding to Kuhn the way he had learned to respond to his father: Authority figures were to be either placated or challenged, whichever led to his getting what he wanted. Earlier, he had promised the commissioner to be "an extra special good boy." That hadn't been enough. Perhaps he should let Kuhn beat him with a coat hanger. When he appeared in the commissioner's office to plead against his suspension, Turner spoke in a strange, extravagant fashion, as if he were the head of a defeated Indian nation rather than a baseball team and wanted to make peace.

"Great White Father," he addressed Kuhn, "I am very contrite. I am very humble. I am sorry. I would get down on the floor and let you jump up and down on me if it would help. I would let you hit me three times in the face without lifting a hand to protect myself. I would bend over and let you paddle my behind, hit me over the head with a Fresca bottle, something like that. Physical pain I can stand."

The commissioner was not impressed. Earlier in the day, Kuhn had been visited by an Atlanta civic group headed by Mayor Maynard Jackson, who presented him with a petition in favor of the Braves' owner signed by ten thousand people. The group pleaded with the commissioner to either drop Turner's suspension or make it less severe. Jackson said afterward, "The thrust of our argument is that we think Turner is important to baseball, to Atlanta, and the entire Southeast." But nothing they did would change Kuhn's mind, and Turner knew it. He had earlier made preparations to turn over temporary control of the team to a triumvirate consisting of Hope, Bill Lucas, and Charles Sanders. Now casting himself as one of history's great martyrs, he told the *New York Times,* "The world has gotten along without Abraham Lincoln, John Kennedy, and Jesus Christ. Baseball can get along without me for a year."

Turner, however, was not ready to leave the stage quite yet. Where others saw only half-empty cups, he saw ones that were almost brim-full. Atlanta newsmen and sportscasters were flocking to his support, and even the local Burger Kings joined the crusade to back him in his struggle against the all-powerful czar of baseball. Loving nothing better than a crisis, Turner managed to turn this one into a publicity bonanza. He even secretly hired two people in the commissioner's own office to copy the names and addresses of those who sent in letters of support for him. The Braves would later mail them season-ticket information and, Hope says, "amazingly that became our little computer database of people who loved Ted Turner."

Unwilling to give up the struggle against Kuhn as well as the limelight, Turner decided to challenge him legally, and on March 8, 1977, he did so in federal court in Atlanta. Despite the fact that all twenty-six major-league owners had signed an agreement that they would be bound by the decisions of the commissioner and had waived all rights of recourse to the courts, Turner contended that Kuhn had so exceeded his legitimate powers in this case that the no-suing provision should be discounted.

While most Atlantans backed Turner in his battle with Kuhn, his fellow owners and the baseball hierarchy definitely did not. They were furious that this dispute was being aired in public. Bob Howsam of the Reds said, "I think any member going to court is wrong. They shouldn't be in it [baseball] if they have to do that. They should get out." George Steinbrenner of the Yankees accused Turner of trying to undermine the authority of the commissioner. Furthermore, the owners didn't care for the fact that they would have to share in the court costs for Kuhn's defense, which could amount to several hundred thousand dollars.

Turner must have known that his chances of winning in court were slim to none, but he seemed to be happy as long as he could keep stirring the pot of public opinion. An interesting insight into his behavior is provided by something he told Hope at that time.

Turner said, "Hope, if you wanna get to the top, you gotta argue with the top. If there's a big guy and a little guy in an argument, if the big guy will argue with him, the big guy doesn't come down to his level. The little guy rises up to *his* level. Now I'm in a fight with Bowie Kuhn. He's big and he's important and he's commissioner. I'm gonna fight him for all he's worth as long as he'll fight back, so I can rise up to his level." Then, as if to clinch his case, he said, "Think of Jesus Christ. Jesus Christ was just an itinerant little preacher until the Roman Empire decided it was gonna attack him. And they went after him"—Turner paused, flashing his gap-toothed smile—"and look where *he* ended up."

Appearing in U.S. district court before Judge Newell Edenfield, Turner was flamboyantly outspoken. He admitted that he was "scared to death" of the commissioner, charging him with being a despot. And as long as he had their attention, he faulted baseball executives for their "mumbo jumbo" and "doom and gloom." He explained to the court that he had been drunk when he made his remarks to Giants owner Bob Lurie on October 20, and that anyway they were made "in a jesting manner." Under cross-examination, he proved to be a generally uncooperative witness, answering questions with questions. At one point, under heavy pressure from Kuhn's attorney, Richard Wertheimer, Turner, believing his honor was in question, suddenly exploded, "After this is over, you keep that up and you'll get a knuckle sandwich."

When Kuhn took the stand the next day, he defended his right to suspend the Braves' owner, as a restless Turner looked on. Kuhn said, "Ted Turner's statement on October 20 violated the collective-bargaining agreement. That's wrong. Everybody in baseball knows it's wrong." He insisted that Ted knew it, too. Kuhn feared that remarks such as Turner's, with their promise of big money, could upset the competitive balance of major-league baseball. He pictured Ted as "baseball's bad boy," who had to be punished harshly. Judge Edenfield gave the attorneys until May 9 to file final written pleadings, and meanwhile Turner remained free to run his team as he pleased.

At the time, Ted wasn't pleased with very much that his team was doing. They were, once again, back in the basement of the National League. After yet another loss, he called Hope to his office and said, "Hope, I just don't understand. Why are we doing so bad? We have such a good team. You always tell me the truth whether I wanna hear it or not. So what's the problem?"

Frustrated with losing, too, and tired of excuses, Hope said, "Ted, one of these days you've got to sit down and realize we just got a horseshit team." Turner went crazy. He began to run around and scream, "You can't say that. That's like telling me my kids are ugly. You can't say that. You're through! Come in on Monday, and we'll work out a deal. You're gone!" On Monday, when Hope arrived to pick up his final paycheck, Turner seemed to have no idea why he was there. Hope told him. "Oh, I was just kiddin'," said his boss. "Couldn't you tell I was just kiddin'?" It wasn't the first time that Turner had peremptorily fired someone and then forgotten all about it.

By the middle of May, the Braves had lost sixteen games in a row, and Ted could stand it no longer. Shipping Dave Bristol out of town on a scouting trip, Turner donned a Braves' uniform and became their new manager. The Braves lost their next game by only one run. Turner thought he noted an improvement. When word about what he had done reached the National League office, Chub Feeney was horrified. He ordered the Braves' owner to stay out of the dugout. Turner watched the next game, a victory, from the stands, but he was soon back in uniform and once more ready to save his team. This time it was Kuhn who stopped him in his tracks, with an angry telegram. When Turner called him to find out what was wrong, the exasperated commissioner seemed to be at his wit's end. He had had enough of this thorn in his side. There was almost a note of pleading in his voice when he asked, "Why can't you be like everybody else?"

Judge Edenfield's ruling came down on May 19. While upholding Kuhn's authority to discipline the Braves' owner, he prevented Kuhn from taking away the team's first-round draft choice. Turner would be banned from baseball for approximately ten more months, his reinstatement timed by a whimsical fate to coincide with April Fool's Day. Edenfield wrote in his decision, "In their encounter with the commissioner, the Braves took nary a scalp, but lived to see their own [scalp] hanging from the lodgepole of the commissioner, apparently only as a grisly warning to others."

Turner was barred from the stadium except as a paying customer. His lawyers told him to keep his mouth shut about the ruling. Sadly, he stored away his cigars and his baseball shoes. For the Braves' players, it was as if they had lost a teammate.

CHAPTER 13

TED TURNER MAY AT TIMES have behaved foolishly, but he was far from stupid. He understood that he wasn't merely up against the commissioner but the other baseball owners as well. His superstation's television coverage of the Braves threatened to cheapen the value of organized baseball's network-TV contract. Turner says, "I had as much chance of winning that case as Czechoslovakia had against Hitler."

And when it came to whether he'd ever again be chosen to skipper an America's Cup defender, he was well aware that the odds against him were similarly daunting. In such back-to-the-wall situations, Turner almost always took the offensive. In 1974, it was right after being dismissed by the New York Yacht Club's Selection Committee that Turner immediately had attempted to purchase *Courageous:* he was already thinking three years ahead to the next Cup competition and, in effect, attempting to buy his way into consideration. But the Kings Point syndicate, which owned the boat, turned him down in favor of Ted Hood, who had just sailed *Courageous* to victory. Once again Turner had lost out, but events soon conspired to bring him back into the picture.

The well-liked Hood asked an old friend from the New York Yacht Club, Alfred Lee Loomis Jr., to manage his 1977 America's Cup campaign. Not only did Hood want to sail a new aluminum boat, but he wanted to design it himself and make its sails. Internationally known as a sail maker and the only skipper to have won yacht racing's "Big Four," (The America's Cup, The Mallory Cup, The Bermuda Cup, and The Southern Ocean Racing Circuit), Hood was the Leonardo of sailing, a Renaissance man who could do it all. *Courageous* would be used as a trial horse for his new boat. The question was who would sail her. Dennis Conner, despite his obvious sailing skills, was not a wealthy man, and Loomis felt that the success of their syndicate depended on money. Though Ted Turner meant trouble, he had what they needed, and Loomis decided to risk inviting him to join their team.

"No, thanks," Turner said. He didn't care for the terms, which were that when trials began in June, Hood would get the boat that proved fastest and he would get the also-ran. Turner had another idea. He proposed that he'd either raise or put in $350,000 of his own money, but come June 1, *Courageous* would be an equal partner and his to sail, regardless of how the other boat did. Loomis agreed, and both men accepted Ted's friend Perry Bass—a wealthy Texas oilman and sailing aficionado—as arbitrator in the event that Turner was to be dismissed.

Loomis had yet one more condition. Turner would have to use Hood sails exclusively. "I told him to count me out," said Turner. He wanted to be able to choose his own sails. Furthermore, he disliked the idea of using the Cup to further anybody's commercial interests. He told Loomis, "It would be nice to have one amateur out there who wasn't part of the factory team, who was doing it like it had always been done." Turner's outburst was a refreshing plea for amateurism, but in fact since 1958, when the 12-meter formula was first adopted for Cup racing, Hood had provided some sails for every defender.

If Loomis had any doubts before that Turner would be a headache, he was sure now. "OK," he grudgingly gave in. He would later say, "I got a lot of advice before I signed Turner on, and all of it concluded I should keep him on a tight leash. He had a strong father, a military-school background, and I was told he reacted well to discipline." Loomis certainly hoped so, because it was generally acknowledged that the high-octane Southerner had gotten his reputation as a free spirit the old-fashioned way—by earning it. But on the plus side, Loomis was also getting a good sailor, a good manager, and a good crew, in addition to a hefty infusion of cash.

It was his hope that Turner would provide the sort of high-level competition that by September would make Loomis's pal Hood an unbeatable America's Cup defender.

The third candidate vying to defend the Cup that year was the West Coast sail maker Lowell North, who was part of the *Enterprise* syndicate. *Enterprise* was the newest of the three boats and represented the latest thinking of 12-meter designer Olin Stephens. From sails to rudder, it was the most computer-analyzed 12-meter ever built. Although Hood used computers as a general guide to sail design, he tended to rely more on intuition and trial and error than North did. As for Turner: "We didn't use tank tests or anything. We didn't even have an onboard computer—except for a thing we got toward the very end just to keep track of where we were in case it got foggy, which it never did." He left his sails to young Robbie Doyle, his tactics to equally young Gary Jobson, and delegated the rest of the detail work and basic preparations to his crew; meanwhile, he focused all of his own sailing attention on winning.

New boats are always given the edge over old boats, and in 1977 there was no question which of the defenders was the underdog. Free of the high expectations of 1974, Turner, the have-not, was like a pig in hog heaven. What he did have was a proven 12-meter in *Courageous,* an experienced crew—seven of the eleven having sailed with him in his previous Cup races—and a fierce desire to win. From the security of hindsight, Turner insists, "I mean, there was no way we could lose, we wanted it so much after 1974. We were the best sailboat crew in the history of the planet, and we knew it."

On April 1, Turner took a break from his losing baseball team and his pending suspension and his superstation and his money problems and his personal headaches to fly up to Boston to meet with his crew and other members of the Kings Point syndicate. At the Boston Yacht Club, he went up on the roof with Gary Jobson, and Jobson recalls staring down at the two boats, *Courageous* and *Independence,* with his mouth open in amazement at these two beautiful white sailing machines, their powerful masts and graceful curves dazzling in the sunlight. Finally, Turner broke the silence. "You know," he said, "when you were fifteen years old, did you ever think you'd be sailing on a twelve-meter?" Jobson said that he never did. "Neither did I," said Ted. "Isn't it the *greatest* thing that ever happened to you?"

The first observation trials began in Newport in June. It was Turner's understanding that he would be able to buy sails from Lowell North, but the West Coast sail maker reported that he was unable to follow through on his promise. His syndicate had said no. Turner was livid. He called North "a no-good liar." He called him "bush." He called him anything that came to mind, and all summer long he kept riding him. The manager of the *Enterprise* syndicate, Ed du Moulin, couldn't believe Turner's harassment of North, his outbursts and public name-calling. "I am tolerant of Turner's shenanigans," he said, "but in this case he went far beyond the realm of good taste." On and off the water, the relationship between the two men grew heated. So much so that North told a reporter that he hoped "the Australians beat Turner and take the Cup home with them." It was not the sort of thing one expected from a New York Yacht Club member.

Though Turner may have loudly proclaimed his confidence on the day of his first race in June, his insides were churning as if they hadn't gotten the message. He threw up three times on the way to Bannister's Wharf, the dock where *Courageous* was tied up. There was no question how much was at stake for him in these early races. In his aviator glasses and his lucky railroad engineer's peaked striped cap, he sat behind the wheel at the stern of the boat beside Jobson, his young tactician, barking orders at everyone in sight. "Trim, damnit! Trim the main," he yelled, his voice leaping up an octave. Then, "Get the main in. Oh, for cryinoutloud!" Then, "Nice goin', Mr. O'Meara."

If there was poetry in long-distance ocean racing for Turner—a world of flying fish, spouting whales, and stars—point-to-point match competition had its own kind of poetic imagery: arms flashing in unison as two men winched in a line, a team of dangling crew members in hiking straps as *Courageous* heeled around a mark, the synchronized ballet of a tacking duel. The crucial element for the skipper of *Courageous* was to make all eleven men dance to the same tune—*his* tune.

"The most fun that you ever have as a man," Turner told an interviewer the following year, "is in doing men's things. Men's things are primarily getting a bunch of guys together and going out and conquering a country, fighting a war, winning a big fight, putting a baseball team together." Or, he might have added to his list, defending the America's Cup. "But first of all, you got to get a good bunch of guys together and do it, whatever it is. And then you have to get them all excited and motivated so they'll bust their ass. People have the most fun when they're busting their ass." Turner claims that what makes him a successful sailboat racer is that he has executive ability and can inspire people. "I can make eleven guys work harder and longer than anybody else."

There was no question in Gary Jobson's mind about the truth of that statement. He and Ted talked sailing constantly. On long walks together, they discussed tactics and grew to respect one another. Turner, it seemed, had the loyalty and respect of his entire crew. Jobson told a reporter, "He can really psych us up. He'll get us all together in the morning, start waving his arms, and then he'll say, 'Hey, isn't this a great day to be alive? Today we've got the opportunity of a lifetime, the chance to go head-to-head against the best sailors in the world. Let's go have some fun.'"

By the end of June, *Courageous* had won seven out of eight races, *Enterprise* had won five out of eight races, and *Independence* had won none out of eight. Turner was jubilant. Although trying to be a good boy, he as usual found it a strain and couldn't resist the opportunity to rub it in on the competition. He told the media, "They came here intending to beat people with boat speed, and when they didn't have superior boat speed, they didn't know how to mix it up in street-fashion sailing." Turner saw his own style, apparently, as rough-and-tumble, that of a street fighter ready to shoot from the hip or the lip. Despite the fact that there were still July and August trial races to be sailed, some knowledgeable onlookers feared that the handwriting was already on the wall. Donald McNamara, a former Cup skipper who had voted against Turner's admission to the club, predicted, "If he's ever selected, he will be the first skipper in Cup history to appear at the starting line wearing a muzzle."

When the July issue of *Sports Illustrated* came out at the beginning of the month, Turner's picture was on the cover. He was thrilled. In time, the photo would be framed and hung on the wall outside his office in Atlanta with a growing number of other magazine covers—including those of *Time, Newsweek, Forbes,* and *Broad-*

casting—as part of his homage to himself, a personal pantheon to remind visitors that they were entering on hallowed ground.

Turner willingly autographed the *Sports Illustrated* cover for anyone who asked. He even signed and sent one to Bowie Kuhn, who most definitely hadn't asked for it. Not satisfied when he failed to get a response from the commissioner, he wrote to make sure that Kuhn had received his gift. "I told him if he hadn't, I would send him another one." Still up to his old tricks, Turner couldn't resist trying to get a rise out of anybody with whom he had crossed swords.

When he arrived at Conley Hall in Newport earlier that summer, Turner had brought with him his own $35,000 earth-receiving station, so that he could keep track of the Braves on Channel 17. Putting it in the backyard, he ran a line up the side of the building to the TV set in his room. Turner says, "I could watch the Braves playing anywhere in the country." The team was not doing well. In the hope that his June winning streak might be contagious, he invited the players up to Newport one day to watch him sail. Unfortunately, his success wasn't catching. By the end of that baseball season, the Braves were still the losingest team in the National League.

Whether it was the notorious jinx of appearing on a *Sports Illustrated* cover, or their problems with sails, or the improvement of the competition, *Courageous* in the July trials experienced an alarming reversal of form. Turner's four-year-old boat lost six straight races to North's *Enterprise*. He insisted that they needed a new jib. Robbie Doyle, who had created seven of *Courageous*'s new sails, says, "We never lost our confidence. We didn't panic." The sail that finally made the difference, Doyle believes, was not the new jib but the new full-cut, lightweight mainsail. But it wasn't until the end of the month that the gloom lifted.

One night, in an effort to cheer themselves up, Turner and his crew went to the movies and saw *Rocky*. On the big screen, the two-bit, ex-fighter-cum-meatpacker hooks and jabs at slabs of beef, because he can't afford to train with decent gym equipment for his once-in-a-lifetime shot at the champ. Though he doesn't seem to have a chance, Rocky does go the distance. After that evening, the theme from the movie became *Courageous*'s theme song, and Turner and his boys were suddenly all street fighters. Once again, they were the underdogs, battling for their lives. As Turner saw it, theirs was not an elite sport for the privileged few but open warfare, a basic struggle for survival.

"I mean, there's not a single really wealthy kid on my crew," he claimed. On the other hand, there were few indigents. Richie Boyd, for example, had gone to Princeton, Robbie Doyle had gone to Harvard, and Bunky Helfrich had studied architecture in Paris at the École des Beaux Arts. These were not ex-pugs that Turner had found in some meatpacking firm's freezer. All of them could afford to take four months out of their careers to indulge their passion for sailboat racing, and in doing so they received free room, free board, free clothes, free watches, free cocktail parties, and all the girls they could find to amuse themselves with in Newport. Not a bad trade-off.

It was unquestionably important for Ted to feel that the cards were stacked against him. Rightly or wrongly, he had convinced himself that Lowell North had betrayed him and that even his own syndicate head, Alfred Loomis, was his antagonist. A heroic effort was called for against such unfair odds. It was the sort of challenge that often brought out the best in Turner. But Loomis had very little sympathy with his conspiracy theories. He told Roger Vaughan, "Don't let Ted sell you the story he was a poor boy being plotted against. Part of his act is to be the underdog."

The final day of July racing dawned with the sun hanging in the heavens like a gold medal on a sky blue blazer. It was *Courageous*'s last chance to reverse her fortunes before the decisive selection trials of August. But even in the prestart maneuvers, North got the jump on Turner, and *Enterprise* appeared to be just out of reach.

"Boy, did we blow that!" cried the frustrated skipper, chewing a mouthful of tobacco furiously as he turned the wheel.

Up the first leg, *Enterprise* had superior boat speed. "Help me, help me!" Turner called desperately.

"That's what I'm trying to do," Jobson shot back.

The noise aboard *Courageous* was astonishing, given that it was being driven by wind rather than internal combustion. There was the jangle of metal on metal, the flapping of sails, the pounding of the hull on the water, the excited shouts of the angry captain. As far as Turner was concerned, "It was to the death." If he'd had guns aboard, he'd be trying to blow North out of the water.

After trailing *Enterprise* for 20 miles, Turner began to grind

North's lead down on the last weather leg. "We got 'em!" he shouted, as *Courageous* came up alongside *Enterprise* and passed her. Overjoyed, Turner cried, "It's all over, baby. We won!" Rushing forward, he gave his trimmer a hug. "Thatta way to go, Robbie. The new jib sure helped." Then smacking his hands together, he began to hug everyone in sight. "Awwwright! Awwwright!"

Turner, the family man, had brought Janie and the kids with him to Newport. They were going to spend the summer together. The children were placed aboard *Tenacious,* a sleek 60-foot sloop he had recently acquired, which could sleep ten. Jimmy Brown was, as usual, there to keep order. Although Turner had little time for them, he'd occasionally come by to see what was going on. Dropping in one morning unannounced, he pointed to the floor and snapped at his teenage daughter, Laura Lee, "Don't just stand there. Let's keep this place shipshape. Get a broom and take care of that dust." He seemed more comfortable barking orders at her than sitting down for a chat. No child of his was going to have it easy just because her father happened to be a millionaire, not by a long shot.

Though Janie was staying at Conley Hall with her husband, she, too, saw little of him. Ted was constantly in motion. His life revolved around his boat and his crew, and he was out on the water practically every day.

New York Times reporter Steve Cady called his behavior that summer almost "monklike." But Newport is no monastery and Turner no monk. *Time* magazine was closer to the mark when it reported: "During the Cup eliminations, he flirted with every girl in sight, crawled pubs with his crew, got tossed out of chic clubs and restaurants for boozy behavior and turned Newport's blue-bloods positively purple."

When passing a boat with a cargo of pretty girls, Turner yanked up his white *Courageous* shirt and yelled, "Show me your tits!" It was one of his crew's patented nautical greetings. Even his own crew members knew enough to keep their wives and girlfriends away from him and at a safe distance.

Drinking got him into trouble repeatedly. In this respect as in so many others, he was his father's son. While Turner claims to be a quick and easy drunk ("I have a very, very low tolerance for alcohol"), it has never stopped him from drinking copiously. After a loud disagreement with the owner of the Black Pearl, a popular *boîte* on Bannister's Wharf, he was barred from the place. Turner explains the incident this way: "The guy was acting like a king of the mountain, crowing about his money and being a pompous ass. Look, I had to put him down. I apologized later in the summer, but I can't stand phony airs."

Then there was the night he was escorted out of Newport's ultraexclusive Reading Room for "boozy bellowing." Next came the incident at Castle Hill, a fancy local restaurant, where a diner who sat down at the table next to Turner's was wearing one of the large buttons that had been handed out by *Enterprise* supporters. It read "Beat the Mouth." Turner, who had been drinking, was so irate that he lunged at the man, ripping the button from his lapel, and challenged him to come out to the parking lot, where he intended to beat him to a pulp.

And then came the most unpleasant incident of all. As recounted by Jonathan Black in his article "Blackbeard Among the Bluebloods," Turner appeared one evening at an exclusive private club on Bailey's Beach and, after having a number of "summer coolers," announced in a loud, belligerent voice to the assembled guests, "The trouble with these stiff bitches is that they really need to be *fucked*—and I'm the guy to do it." The following month, when asked about the story on Tom Snyder's television interview program, *Tomorrow*, Turner replied angrily, "That was a downright lie, a downright lie."

Snyder seemed bemused. "Never said it?" he asked.

"Nope. Or *anything* even faintly resembling it. It was an absolute untruth."

A year later, upon reflection, Turner told *Playboy*, "First of all, I wouldn't *want* to screw old bitches. It doesn't make sense." It was hardly an ironclad defense. Then he went on to explain his side of the story. He and his wife had been invited for cocktails to the house of a couple from Atlanta they knew only casually; this was to be followed by dinner at the Spouting Rock Club, at Bailey's Beach. "When we got to their house, they pulled out ten or fifteen copies of *Sports Illustrated* with me on the cover and wanted me to sign them, which I did."

Turner pegged his hosts as "celebrity hounds," and even after three or four vodka tonics, he didn't care for them any better. At the club, he felt that his hosts were showing him around "like a prize bull," and he was getting mad. So he had some more to drink.

Then he happened to meet this fine-looking young woman in her twenties who was with a man old enough to have been her grade-school principal. "So I asked her, I just said, 'What are you doing with an old guy like that?' And she said she was with him for his money. And I said, you know, 'Have you been laid lately?' . . . And she said, 'I'm horny as hell.' And I said, 'Well, we might be able to get that taken care of,' and that's all I said." Clearly it was enough. Turner may have thought so, too, because he walked out on his hosts without waiting for dinner, leaving his wife behind in his eagerness to get home and in bed before he got into trouble. But it was too late.

Rumors about what had gone on that night spread quickly throughout Newport society, turning stomachs like tainted quiche. Turner's hostess was making it known all over town that her guest was no gentleman. When Alfred Loomis heard what had happened, he was seething. He strongly considered getting rid of the trouble-maker. With the support of his syndicate, he demanded that Turner write a letter of apology to John Winslow, the president of Bailey's Beach, and Turner realized that he had no choice if he wanted to remain on *Courageous*. His letter read:

> It has come to my attention that my conduct at the party, July 2nd, at the Spouting Rock Beach Association may have been bothersome to some of your fine members. If this is the case, I wish to apologize profusely because I certainly did have a couple of drinks too many that Saturday night.
>
> I, as all of the America's Cup competitors, am tremendously appreciative and respectful of the wonderful hospitality of your fine club.
>
> Please accept my sincere apologies.

While willing to make amends to achieve his goal, there were limits to how cooperative Turner was going to be with Loomis and the New York Yacht Club's hierarchy. Once again, Martine Darragon's arrival in Newport stretched those limits almost to the breaking point. By way of identifying herself, Ted's French girl-friend has said: "I'm the one the New York Yacht Club raised all the hell about." Martine and Ted had never made any secret of their fondness for one another. Explaining what happened, she says, "Well, the board of governors of the club up there is very conservative, and they got all upset that Ted was . . . running around in public with a Frenchwoman. But he told them to go to hell. That's how he is."

Despite all his carrying on, Turner somehow managed to get away with it. The wonder is that he could still concentrate on his sailing. "I'm gonna do something that nobody's done before," he announced at the beginning of August. "I'm gonna enjoy the final trials. Win, lose, or draw." But Turner had no interest in being an also-ran, as he plainly revealed when he added, "Because that'll improve our chances of winning." And perhaps it did, for *Courageous* by the end of the month had won all but one of its races.

On August 29, *Enterprise* defeated *Independence,* and Hood was excused from further Cup competition. As Turner watched Commodore Hinman and the Selection Committee go through its ritual of elimination, tears of empathy welled up in his eyes. He liked Hood and knew firsthand how painful it was to be axed after so many months of hope and effort.

On the following day, there was a crowd of people and photographers around Bannister's Wharf as *Courageous,* about to do battle with *Enterprise,* pulled away from the dock. Only the bearded commercial fishermen at the Aquidneck Lobster Company, on the next dock, seemed to be indifferent to what was at stake. From the starting cannon, *Courageous* was ahead, and she led around every mark, winning the race by one-and-a-half minutes. The members of the Selection Committee saw no reason to delay any longer. Bob Bavier and the others had no doubts about *Courageous:* "She showed no weakness. She eloquently proved herself." Thanking the red-shirted *Enterprise* crew for their good effort, Hinman dismissed them. Then he approached Turner and his crew. "Gentlemen," he said, "congratulations. You have been selected to defend the America's Cup against *Australia* in the twenty-third Challenge Match."

The cheer that went up from *Courageous*'s excited supporters split the air as crew and captain celebrated. The underdogs of spring had emerged as the defenders of August. Turner praised the genius of Robbie Doyle for his sails, the skill of Gary Jobson, and the teamwork of his entire crew. But with no more modesty than an Odysseus or an Achilles, he knew, of course, that none of this would have been possible without him. With a cigar in his mouth and a can of beer clutched in his hand, he told UPI reporter Terry

Anzur, "There will never be a time in my life as good as this time. I can't believe all this is really happening to me. I'm so hot I just tell my guys to stand by me with their umbrellas turned upside down to catch the stuff that falls off me and onto them."

Though Turner may have been an embarrassment to the New York Yacht Club, he had sailed brilliantly in the trials and caught the attention of the public as well. There were buttons, banners, even T-shirts emblazoned with his photograph and the logo "Captain Courageous"—T-shirts the making of which he himself had funded.

If the amateur had temporarily preserved the Cup from the dreaded commercialism of the two sail makers, Hood and North, there were unquestionably business advantages for Turner in his selection. Will Sanders, for example, was able to invite loan officers from the First Chicago Bank and Home Life Insurance to spend a few days aboard *Tenacious* observing the races as Ted's personal guests. It was the sort of stroking that businessmen appreciate. There was also the obvious usefulness to his salespeople of being able to present themselves as working for the defender of the America's Cup. And finally Ted even decided to market himself, selling his "Captain Outrageous" image to Tiparillo cigars and Cutty Sark, the scotch for those "who still follow their gut feelings."

The day after being selected, Turner, who had had his fill of that "stubborn bastard" Alfred Lee Loomis, made a secret trip to Kings Point. He went there to discuss acquiring *Courageous* for himself in 1980, before Loomis could get it for Hood.

It was an act that thoroughly outraged the syndicate manager. Loomis condemned it as "complete disloyalty," but with the Cup defense about to begin in two weeks, there was nothing to be done. Confronting Turner, he asked, "I've been fair to you, haven't I? Hasn't it been all right?" Never one to hide his true feelings, Turner replied, "Lee, I hate to tell you, but it hasn't been perfect."

After learning of Turner's selection, President Jimmy Carter sat down and wrote him a personal note:

> Congratulations to you and to the crew of *Courageous*.
> I am proud of you all, and all Americans—Yankees and Southerners—wish you well in the coming races.

A few days later, an invitation from the White House arrived for Mr. and Mrs. Ted Turner to attend a special dinner on September 7, 1977, for Western Hemisphere leaders to celebrate the signing of the Panama Canal Treaties. Turner was thrilled. He was going to the White House to break bread with the president of the United States and other world leaders. He seemed to be on some incredible lucky roll. Although he didn't like leaving his boat and crew at a time like this, he was truly excited about being asked. At the local airport, Ted paced nervously around the small, frail-looking, single-engine plane, kicking its tires to see if he could trust them, as he waited for the pilot who would take him the short distance from Newport to Providence. Once there, he boarded a jet for Washington. Janie, who had gone home to get a dress for the occasion, was flying up from Atlanta.

The dinner was held in the State Dining Room of the White House. It was a glittering affair. In addition to the visiting foreign dignitaries and the Panamanian leader, General Torrijos, the honored guests included former president Gerald Ford, Lady Bird Johnson, Mrs. Martin Luther King Jr., and Muhammad Ali. Following dinner and before the entertainment, President Carter had a few informal words of welcome for his visitors on this historic occasion, and he introduced his special guests.

"There's another man that I would like to introduce," Carter said, after presenting Ali. "I've been a very close reader of the sports page for the last several weeks, because we have a very distinguished Georgian who has, I think, come forward with a great deal of enthusiasm and skill, a great deal of understanding of the elements, the oceans in particular. He's exemplified, I think, the name of his boat. He's a very courageous man—Ted Turner. We are very proud to have you here tonight. And as you all know, he will represent us in the America's Cup races very shortly, having overwhelmed his opponents much better than has been the case with his baseball team, the Atlanta Braves." Turner happily joined in the general laughter. Even his lowly Braves couldn't dampen his spirits on a night like this.

Of the would-be foreign challengers for the prized America's Cup—the Swedes, the French, and the Australians—it was the new aluminum 12-meter *Australia,* the early favorite, skippered by Noel Robins, that ran away from its competition in their best-of-seven series. *Australia* was a formidable challenger, regarded by some as "the most dangerous in recent years." When told that if he lost the Cup his head would be cut off and placed under glass at the club on West 44th Street, Turner said, "I sure as hell don't want that to happen. But even if I lose I don't think they'd take my head. They don't allow mustaches yet down at the New York Yacht Club."

Both sides seemed equally confident, and there was some good-natured psychological warfare as each jockeyed for position in the days leading up to the first race in the best-of-seven finals. One day, Turner and Alan Bond, the head of the Australian syndicate, were in a Chinese restaurant and Ted casually announced, "I'll show you how hot I am." Picking up a fortune cookie, he cracked it open. "A precious possession will soon be yours," he read aloud, with a straight face. Not surprisingly, shortly thereafter Ted discovered that his lucky railroad-engineer's cap had been stolen. Like most sailors, Turner is not untouched by superstition. Together with Gary Jobson, he immediately ran out and bought a replacement, plus a couple of dozen extras just in case.

On September 13, when *Courageous* left Bannister's Wharf for its first head-to-head confrontation with *Australia,* the crowds were enormous. Reporters and cameramen were everywhere. Spectator boats churned the harbor. Overhead flew the Goodyear blimp. Turner threw kisses to the crowd, and they waved back. The unpredictable skipper had helped to convert an elite international yachting competition into a popular media event.

Noel Robins' strategy was to get ahead at the start and stay ahead. It looked like a good plan when *Australia* immediately jumped into a twelve-second lead. But no sooner did the two boats approach the first mark than *Australia* fell behind. Although the challenger could hold her own on the reaching leg in the light to moderate air, she couldn't overcome the advantage that the defender built up on the wind. *Courageous* crossed the finish line one minute and forty-eight seconds ahead of the Australians. At the press conference that followed the race, Turner, when asked for his analysis of what happened, looked into the cameras and said, "I'm happy to be alive and be from the United States and be able and healthy and fortunate enough to be participating in this great competition with my good friends on the *Australia,* and it's a little overwhelming to see all the nice people that have been so kind and everything—"

Interrupting him, someone said, "Thank you, Ted, for that insight into the race today." There were howls of laughter, and Turner beamed. He was not about to say anything critical of his competition, which could appear on their bulletin board and be used to psych them up. There were, after all, three more races still to be won.

The second race took place three days later, under a threatening sky. Though starting slowly, the pace gradually quickened. The race ended with an exciting tacking duel, *Australia* gaining on the front-running defender but still coming up short. After the victory, while Ted puffed satisfiedly on a cigar, his mother, Florence, who was in Newport to watch her son race, came aboard *Courageous*. Tall and serious, wearing white slacks and a multicolored cardigan, she walked in stately fashion to the stern of Ted's boat and, getting down on hands and knees, kissed it for good luck. It was just the sort of theatrically flamboyant gesture typical of her son.

With friends and family there, there was only one thing missing for Ted at that victorious moment: "Sometimes I think my father is somewhere watching all this," he said. "Watching me make the big time. I wish he could come back and see, you know? Like the father in *Carousel?* The dead father, when he comes back to see his daughter at graduation? Damn! We were good friends. I wish my father could come back just for a day."

The next day, Saturday, the third race began at precisely 12:10 P.M. Responding well to the starter's cannon, *Courageous* leaped ahead, gaining a twelve-second advantage. Throughout the race, she was never behind, winning by a lopsided two minutes and thirty-two seconds. Turner cautioned his crew against counting their chickens, but the only question that seemed to remain was how much they were going to win by on Sunday.

Later at the press conference, Turner refused to touch it. But then someone asked him if he'd win even if he switched boats with the Australians. Turner made clear that he considered it a dumb question. "It's like saying is Noel Robins' wife better than my wife. I like my wife." Perhaps thinking he was out of microphone range, he then added softly, "She ain't much, but she's all I got." If Janie heard, she gave no indication. But the audience heard and laughed loudly, much to Turner's delight.

While national interest in the outcome of the America's Cup had grown exponentially over the summer, local interest was at an all-time fever pitch for the fourth, and what was expected to be the final, race in the series. Thousands mobbed the wharves. Cameras were everywhere. Zigzagging behind the two 12-meter racers at a respectful distance, pleasure boats followed them as they moved elegantly out of the harbor toward the starting line. The spectator fleet was enormous. From overhead came the clatter of TV and Coast Guard helicopters.

That afternoon there was a wind to *Australia*'s liking and both boats crossed the starting line side by side. However, Turner and Jobson always seemed to make the right decisions. By the first mark, *Courageous* was forty-four seconds ahead. Then, little by little, she pulled away. According to Bob Bavier, once having built a commanding lead Turner would sail conservatively, which was exactly what he did now. Rounding the last mark, *Courageous* was two minutes and thirty-five seconds ahead. One hundred yards from the finish, Jobson says, "We started hugging and telling each other how great we were. But Ted made us wait until we crossed the finish line before we broke out the six-packs." With the final cannon blast signaling victory, not only six-packs but all hell broke out. After a summer of fierce dedication, the captain and his crew had earned their celebration.

Turner recalls that it took about an hour and twenty minutes for *Courageous* to be towed into Newport. In that time, the first thing that happened was a boat came alongside and someone threw a couple of cases of beer aboard. Then as they approached the harbor another good Samaritan delivered five bottles of champagne. Amid the drinking came cheering and horns and whistles and cannon fire and general bedlam, as an armada of yachts and ferries and boats of every size and shape followed them toward Bannister's Wharf. Coast Guard fireboats shot towering cascades of water higher than the screeching gulls. Turner, who was not especially looking forward to getting thrown into the filthy polluted water near the wharf in the traditional baptism of the victors, took a few more guzzles of champagne to immunize himself.

Much to the delight of the crowd, Ted was among the first to be tossed in. There were more than three thousand well-wishers jamming the docks and cheering from rooftops, car hoods, and masts. Drenched and disheveled when he finally climbed out, Turner began running around on the dock and spraying water on everyone. The air was getting cold, and as he stood around accepting congratulations, he began to shiver. "Somebody stuck a bottle of aquavit in my hand," Turner later told Tom Snyder, "and that's the last thing I remember."

One reporter present recalls seeing him go up to the quarters above the wharf offices to dry off and stand there in front of a mirror in his wet clothes, clutching the bottle of aquavit. "You," he said, admiring himself in the glass as he lifted his bottle in tribute, "are the fucking greatest!"

The postrace press conference was held in the National Guard Armory on Thames Street. The small stone ivy-covered building resembles a toy fort, but it has a large basilica-like wooden shed in the back, where hundreds of reporters and photographers and television cameras were packed in. By the time Turner finally arrived, he was carrying two bottles and was plastered. The crowd cheered thunderously and Turner, beaming, sat down on the dais next to Gary Jobson.

It may have been Sid Pike, Channel 17's general manager, or perhaps someone else trying to protect Turner, but just before the television lights went on, his two bottles disappeared under the table. Turner exploded and dove under the table after them. As the cameras rolled, he popped up on national television grinning inanely, a bottle in each hand and took a long slug. It was the sort of image to bring the New York Yacht Club to its well-pressed knees.

As Turner lit up a cigar, Noel Robins accepted his defeat graciously. "I think we were up against the best defender that's ever been given the job of defending the Cup." Then his crew sang a rousing chorus of "Dixie," in honor of the hero of the day. Bill Ficker, who was moderating the press conference, then turned to Ted. It was his turn now.

Rubbing Ficker's bald head affectionately, Ted blew smoke in his face and, with a crooked grin, grabbed the mike. "I never loved sailing against good friends any more than the Aussies. I love 'em," Ted gushed. "They are the best of the best. The best . . . ," he wavered, as if he had momentarily lost the thread, "of the best." Jobson nudged him and pointed to the note he had written. "And I want to thank George Hinman and the New York Yacht Club for the opportunity. We worked hard, we busted our behinds." Then he said he loved everybody in the room. Everybody! And it was then he remembered his crew. As if on cue, the men of *Cou-*

rageous climbed onto the dais en masse and, hoisting their skipper to their shoulders, carted him off.

Turner had no idea how he got home to Conley Hall that night. "I was in bed, they say, by nine o'clock." Janie, of course, had been there to pick up the pieces. Also there to help were Mr. and Mrs. Carl Helfrich—Bunky's parents—old friends of Ed and Florence from Savannah. "My husband, Carl," Mrs. Helfrich recalls, "went up the stairs to help Jane in getting him on the bed. And Ted looked up at Carl, and the last thing he said was, 'Wouldn't the old man be proud of me tonight?'"

Following a farewell luncheon the next day given for the two crews and skippers by Alan Bond, the Australian syndicate head, Turner and his wife left for Atlanta and home. It must have seemed to him that he had been gone from Georgia for a very long time. When he went away in the late spring, he was someone talked about on the sports pages or the entertainment pages of the local newspapers, and now he was returning a front-page hero. Ted was amazed at all he had accomplished in little more than a year and a half. "I had just gone on the air with the superstation when I left for Newport. I had just bought the Braves, got suspended, and bought the Hawks. It was a marvel, really."

The Turners arrived at Atlanta's Hartsfield Airport at 7:15 P.M., the large commercial jet they were aboard landing in a heavy thunderstorm. Despite the weather, there were about two hundred people, plus the mayor of Atlanta and a brass band, waiting to greet the America's Cup champion. Three of the Turner children were among the crowd, and when the youngest, Jennie, noticed a wheelchair being taken aboard, she became anxious. "Oh no," she cried, "don't tell me he broke his leg."

Turner at last deplaned and, much to his daughter's relief, appeared to be in one piece. He seemed dazed by the crowd and the band. "I'm overwhelmed by all this," he told them, smiling broadly. Mayor Maynard Jackson proclaimed the occasion "Ted Turner Day" in Atlanta.

Doffing the blue skipper's cap he was wearing, Ted smiled for the TV cameras. "I wish I'd written a speech, 'cause y'all deserve some good words." About his alleged run-ins with the yachting establishment in Newport, Ted said that a lot had been written that wasn't true. He conceded, however, that there was "a little" animosity between him and some other yachtsmen. "I don't mean to be a loudmouth, but I guess where there's smoke, there's fire. I try to be good," he concluded, in all sincerity, "but sometimes it's hard."

Amid the applause and popping flashbulbs, Turner—the self-styled paladin of the common man—waved at the crowd as if he were running for office, and the band launched into a snappy rendition of the theme from *Rocky*.

The next morning on the front page of the *Atlanta Constitution* a headline proclaimed, "CAPTAIN OUTRAGEOUS IS HOME." The story reported that Turner had received a hero's welcome. The large photograph accompanying it was of the happy skipper with a cute, well-dressed young woman smiling her complete approval.

Janie, who at the airport reception had been standing next to her husband on the rostrum, was angry. "Of all the pictures that were taken at the Atlanta airport when we arrived home after Ted won the Cup, the one they used was of Ted with some strange girl." It seemed to Janie as if she were always destined to be left out of the picture. "In Newport the same thing happened. The press didn't want me in the scene. I suppose it's not too flattering—too colorful—to be pictured with your wife."

Turner, who had been a matinee idol in Newport—women eager to hug him and men flocking to shake his hand—returned to find a desk full of comparatively boring business details to be taken care of and decisions to be made. Not surprisingly, he seemed to be suffering a letdown. He began to yell and scream at people with little or no provocation. Discovering that the dates for the upcoming SORC races had not been entered in his calendar, he informed his secretary in scorchingly emphatic terms that she was a blithering idiot. "It was one of those times," according to Terry McGuirk, "you shut your office door and hoped like hell he wouldn't call." Turner seemed to feel that he had to do something to make up for the time he had been away, and he didn't really enjoy what he had to do. Mike Gearon recalls, "He was the most obnoxious I've ever seen him after he won the America's Cup. He was impossible to live with."

Turner took only one weekend off before he felt that he had to get back to sailing again. A series of victories netted him an unprecedented third "Yachtsman of the Year" award. Then in January of 1978, he brought *Tenacious* down to Florida for the beginning of the SORC campaign.

About one o'clock on the morning of January 28, Turner was in bed sound asleep at a St. Petersburg hotel. He would need all the rest he could get; later that day, he was to take part in a long-distance dash to Boca Grande, Venezuela, and back, one of the major races of the SORC season.

Downstairs in front of the hotel, Martine Darragon had just arrived, after driving her rented Thunderbird from the Tampa airport at 80 miles per hour. The stunning blonde, a lynx coat draped casually over her arm, swept up to the front desk and demanded the key to Mr. Turner's room.

"Madame, it's after one o'clock," the night clerk pointed out, "and I don't know that I should call up there at this hour. I don't have any notice about you."

In her throaty French accent, Mademoiselle Darragon convinced him that it would be all right and, relenting, he handed over the key. It wasn't easy to deny Martine Darragon anything she really wanted. But unknown to Darragon, the bell captain taking her up in the elevator was an FBI agent, who packed a gun under his uniform. The FBI had been following her ever since she arrived in New York.

Approximately two years earlier, Orlando Letellier, who had been the Chilean ambassador under the deposed Chilean president Salvador Allende, had been killed when a bomb exploded in his car in Washington, D.C. It was only the second major political assassination in the capital since Lincoln's. The FBI had discovered that the order to murder Letellier had been sent to this country through a mailbox at Kennedy Airport rented by a Frenchwoman named Martine Darragon. More than thirty agents had been tracking her movements across the country, from New York to Detroit to Windsor, Ontario, to Tampa, to the hotel here in St. Petersburg. Certain that in Florida she would lead them to her American connection, they were tightening the net and about to nail him.

Darragon unlocked the hotel-room door and walked in as if it were her own. The bell captain followed closely behind with her bag, ready if necessary to reach for his gun. It could be Santos Trafficante, for all he knew. The former boss of the Mafia in Cuba, who had once been recruited to assassinate Fidel Castro, lived nearby in Tampa. The nervous bell captain's eyes were riveted on the bed.

Throwing off the covers, a smiling Turner held up a bottle of champagne by way of welcome and waved it in the air. "Hi, honey!" he chirped cheerily. "What took you so long?"

By 7:30 A.M., while Ted, in his skipper's outfit, and Martine were having breakfast, the FBI had checked him out and ascertained that it *was* Ted Turner, the guy who had just won the America's Cup. Rumor had it that Turner had many important friends and a temper "like a hand grenade." The two agents who had Darragon under surveillance feared that he might call their boss, J. Edgar Hoover, if they tried to stop her from boarding his yacht. But they couldn't risk letting her sail with him out of the country.

As she and Turner left the restaurant arm in arm, the agents approached and introduced themselves. Flashing his FBI badge, one of them said to Turner, "I'd like to talk to your friend here, Ms. Darragon, if you don't mind."

Ted didn't bat an eye. "Well, shit," he said. "Is she in some kind of trouble with the FBI?"

"No, sir," the agent replied. "All we want to do is talk to her. We think she can help us in a major case."

"Well, if you gotta talk to her, you gotta talk to her," Ted said amiably. "I gotta go race. You need to talk to me?"

The agent said no.

Ted smiled at Martine. "Well, honey, good luck." Though he was sorry to leave her behind, Martine could take care of herself. Before they were finished, she'd have them cleaning her car. Turning to the agents, Ted winked and said, "When you get through with her, send her back, will you? She's a pretty good gal." Without another word, he hurried away. There was, after all, still one more race to be won.

Taking Martine aside, the agents began questioning her. Did she have any post-office boxes in the United States? She acknowledged that she did—"Three of them, as a matter of fact"—one in New York, one in Miami, and one in Los Angeles. She used them when she traveled in the United States. Before long, it was clear that Darragon knew nobody in Chile, had no knowledge about the use of her JFK mailbox in New York, and was in no way involved with Letellier's murder. Why her mailbox had been used was a mystery. The only reason she had come to Florida was to be with her friend Ted. The embarrassed agents were soon falling all over themselves apologizing. Martine smiled. She confessed that she thought they had come to arrest her for bringing her new lynx coat in from Canada without paying any duty.

When Ted returned, Martine was there waiting. He had placed second in the race to Boca Grande and was eager to tell her about it. He seemed to have forgotten all about the FBI.

CHAPTER 14

BACK IN ATLANTA on West Peachtree, Ted Turner was stalking the halls, and Gerry Hogan was worried. In 1978, according to Hogan, "we were still trying to make a business out of the superstation." WTCG-TV had been up on the bird for little more than a year, and all of Hogan's energy had been focused on getting advertising for it. Now there was something new to worry about.

Turner for some time had been talking about a new idea, kicking it around, sounding people out. While sailing with Terry McGuirk and others that spring, he asked, "What would you guys think if you could turn on your TV any time of the day or night and find out what's happening in the world?" One day he called Don Lachowski into his office and told him, "I'm gonna do a twenty-four-hour news network. Write this down." The format he outlined was four discrete half-hour segments—news, sports, features, and business news.

To most at the station who knew Ted's track record with news, his latest brainstorm had to seem like a bad joke. To Will Sanders and Gerry Hogan, who had some idea of the costs involved in such an undertaking, it simply scared the hell out of them. With any luck, perhaps their boss's sudden infatuation with news might vanish as quickly as some of his other enthusiasms over the years.

But by October, Turner was more excited than ever about the prospect of creating a cable news network. Anxious about whether they could sustain such a project, Hogan warned Ted about the danger. "I urged him not to sink the entire company over the news idea." Rather than being annoyed, Turner asked him to write down what he had just said, and Hogan returned with a large 2-foot-high sign that read:

Turner Be Sure
That the News
Operation
Doesn't Bury Us
Don't Get
Carried Away
Be Absolutely
Positive
It Will Work
At Least Don't
Put Us Against
The Wall

As Hogan recalls, Ted liked the sign well enough to keep it around his office for years afterward. What most people would have instantly recognized as a skull-and-bones to warn away the innocent did not frighten Turner. He enjoyed having the high stakes of his wager published. Rather than flee danger, he seemed perversely eager to court it, to test himself against all challenges in the crucible of risk. His father, after all, had failed just such a test. Though Hogan had intended to give his boss a reality check, Turner saw his contemplated news operation as a great adventure.

The idea for twenty-four-hour television news had been around for a number of years. All-news radio began in the early 1960s, with the advent of Liberty Broadcasting in Chicago and WINS (Group W Westinghouse) in New York, proving that there was a market for news around the clock. Although the television audience for the nightly network newscasts was enormous, the three programs were a mere half hour long, and only twenty-two minutes of that was news. Since 1963 (when these broadcasts first went from fifteen to thirty minutes), the networks had talked about expanding them to an hour format, but nothing had ever come of it. Then in the mid-1970s the development of cable television, along with communications satellites and videotape, began to change the economics of television news.

Robert Pauley, former president of ABC Radio and part-owner of Television News (TVN) with brewer Joe Coors, claims to have been one of the first to try to implement a televised world news service. If so, he was one of many. Reese Schonfeld, who worked for TVN before starting his own not-for-profit news-syndication company in November 1975—the Independent Television News Association—says, "The first person who ever talked to me about a twenty-four-hour cable news network was Gerry Levin [of HBO]. They wanted to do this back in 1977, early 1978. It was all-news. I had long meetings with him and his emissary, Bob Weisberg. I was then head of ITNA. I tried to get my board to go forward and do it, but they refused."

That others should have had the idea for an all-news television operation before Turner is not especially significant. Many people have ideas. What is significant is that in 1978, the thirty-nine-year-old entrepreneur was able to take this idea and turn it into a functioning reality in less than two years. Although Turner likes to be known as a visionary, he is probably best understood as a borrower and a builder rather than an originator, more Roman than Greek. His talent lies less in creating brand-new concepts than in seeing the potential in existing ideas long before others and exploiting it. His success in adopting the concept of round-the-clock television news and making it work despite the formidable obstacles he encountered was a truly brilliant accomplishment, rivaling in ambition and importance anything he had done before.

To many this development seemed to come out of nowhere. "Ted Turner," an Atlanta businessman who had followed his progress said dismissively, "is the last person I'd expect to lead a revolution in broadcast journalism." His entire career, according to the television critic for the *Charlotte Observer,* "had been marked by an unflagging contempt for news." Curiously, even as a youngster he felt hostile to news. Turner remembers, "I grew up hating newspapers, because my father used to come home at night with a newspaper tucked under his arm, [with] editorial after editorial about banning billboards."

News at Channel 17 was really radio news on TV. One day, in order to liven things up, newsman Bill Tush dressed a German shepherd in a shirt and collar, fed him some peanut butter to get the lip action he wanted, and Rex the Wonder Dog became his coanchor. Tush says, "Ted thought it was pretty funny." He called it "tongue in cheek."

From then on, Tush's news program became offbeat entertainment. One week they shot a garage door. The next they'd do updates on the studio's new linoleum floor. Tush on one occasion donned a gorilla suit to announce a guerrilla attack; on another, he cut to downtown Atlanta for a report from the Unknown Announcer, a coworker with a mike and a paper bag over his head. When an outside consultant, brought in to straighten out the station, advised them to operate their newscasts in a more businesslike fashion, Turner fired him.

In his August 1978 *Playboy* interview, Turner describes the Channel 17 newscast flippantly, as pie-in-the-kisser journalism. He makes no mention of his all-news project, which probably at this point reflects the stage of his thinking about it rather than his caution. The Mouth of the South, as noted earlier, was not known for his ability to hold his tongue. "Anything you don't want the whole world to know," he once candidly advised a friend, "don't tell me."

Whenever Reese Schonfeld, who had about twenty-five independent TV stations as his ITNA news customers, would meet Ted at cable conventions and ask if he wanted to join his organization, the answer was always the same. Turner wasn't interested in news. "I ain't gonna do it," he'd say, implying that the reason was that he couldn't afford to do it right. Schonfeld says, "No one ever really believed that." Turner, after all, was a multimillionaire. That year he was making about $200,000 in salary alone, and the next he'd receive $285,000. In fact, though his present wealth was a mere $100 million, he was hoping before long with the future success of his superstation to have a personal fortune of $400 million.

Turner had had enough money to spend $2 million earlier that year to buy the Hope Plantation, 5,000 acres in coastal South Carolina near Jacksonboro, and to install his own earth station there. Everything about the plantation appealed to him, even its upbeat name. The main house, a stately white colonial mansion, was about a mile from the main gate. By reputation, it was one of the best duck-hunting properties in the area. There were also alligators, deer, snipe, wild turkey, and one of the two bald eagles in the state. Giant oaks, dogwood, and azaleas flourished. There were peat bogs 12 feet deep.

When he bought the plantation, Turner said, "Land is a good investment. And when you are a sports owner, things are very, very iffy." Janie thought so, too. She repeatedly told the children that their father could be broke tomorrow. "That's for sure" was Ted's reaction. Just as land had been a symbol of attainment for his father, it also functioned that way for him. Like Gerald O'Hara and his daughter Scarlett, they believed that "land is the only thing that lasts." The plantation was in Colleton County, about a half-hour from Beaufort, where his father had once owned property. Among the Turners' neighbors were the du Ponts, Spauldings, and Donnellys, as well as Barbara Hutton, once known as the richest woman in the world.

Will Sanders wasn't so sanguine about this purchase: "Ted has no sense of how to buy. He knows how to sell, but can't buy. He paid way over for his plantation. Way more than he should. I kept saying to him, nobody in real estate pays the asking price. You negotiate, you bargain. He said, 'No, no.'" The new purchase simply made Sanders all the more anxious about Ted's ability to afford his ambitious plan for a round-the-clock news network. But Turner understood very well that even if he could get someone like Reese Schonfeld to run it, a guy who had spent his whole life working with news outside the networks on a piggy-bank budget, he'd still need very deep pockets to launch such a project. Perhaps Schonfeld was the one who could tell him exactly how deep.

"Dee," Turner shouted to his secretary one morning in November, "get me Reese Schonfeld." When Schonfeld came on the line, Turner explained what he had in mind. "I'm thinking about doing twenty-four hours of news for cable. Can it be done?"

Schonfeld said that it could. He had heard rumors that Turner had had a change of heart about news but had dismissed them as laughable. Now here he was talking about creating a news network, something that Schonfeld heretofore had only been able to dream of.

Turner asked, "Do you want to do it?"

Schonfeld said he was interested. Whatever reservations he might have had about working for Turner—his ability to finance the project, his potential for meddling, his politics, his loudmouth style—he kept to himself. He did mention that he had a contract with ITNA.

"Your contract is your problem," Ted snapped. "Come on down and see me."

A week later, Schonfeld—a high-energy risk taker, too—was on a plane to Atlanta. He recalls that "Turner was going to take me down to his new plantation to woo me, but I didn't want to go. I didn't have time." Besides, Schonfeld had made up his mind. On the back of an envelope, he had already worked out a format and budget for Turner's all-news cable network. "At that time it was easy for me," Schonfeld says. "I had done a similar exercise for Arthur Taylor, the former president of CBS. Taylor, a guy by the name of Ray Beindorf, and I had for a while plotted doing a fourth network that would have been kind of what Fox is today, except we weren't as smart as to aim for that kind of programming. And I did a whole budget for the news division for that group. A thirteen-or-fourteen-million-dollar budget. And all I did was take that budget and rework it again for the Turner budget. Part of what I'd been doing for eight or nine years was budgeting news."

Among the potential advantages of the Turner news operation was that by employing satellites it could—unlike the network newscasts, with their bloated overhead in New York—originate from anywhere; Atlanta would be a much cheaper home base than New York. Georgia was a right-to-work state, and they would be able to operate with inexpensive nonunion labor—a major economic advantage over the three networks. On the downside, however, was the fact that the fledgling news network would have no huge entertainment division to bankroll it. After a few years, it was going to have to be self-supporting or it would die. This was the sort of challenge that appealed to Schonfeld almost as much as it did to Turner.

At the Atlanta airport, Turner himself picked up Schonfeld and drove him to WTCG. His office seemed more scruffy than the last time Schonfeld had been there, with paint peeling from the walls. Schonfeld remembers vividly the first thing Ted said to him. Admitting that news had not been his number-one choice, he announced, "There are only four things that television does, Reese. It does movies, and HBO has beaten me to that. It does sports, and now ESPN's got that. There's the regular series kinda stuff, and the three networks have beaten me to that. All that's left is news! And I've got to get there before anybody else does, or I'm gonna be shut out." Schonfeld thought it was one of the most insightful things that Turner ever said.

Ted had already come up with a name for his new network. "I'm gonna call it Cable News Network! Not the Turner News Network—the *Cable* News Network." It was a gesture to show how committed he was to the project's independence. Even more important, it was a way to encourage the support of the cable owners. Turner said, "This is the one thing where I've got to bring them all into the tent, so I'm gonna call it CNN." Schonfeld approved. Then, as if unable to stand so much humility, Turner summoned in one of his research people to read aloud to his visitor a recently published profile of Ed Turner's boy wonder.

After that, they talked about format. Turner said they should have four half-hour segments and rotate them. One would be news, one sports, one features and women, and one financial. Digging in his heels, Schonfeld immediately replied, "No way! That doesn't work. That's not what we're going to do." Though he didn't mind the elements, he didn't want them in fixed half-hour units. He wanted flexibility; he wanted to be able to go live as often as necessary.

For a fleeting second, it looked as if CNN might not be able to hold these two large egos any better than the small office contained the two large men. The way Schonfeld saw it, "Ted understood that if he wanted me, he was going to have to go along with a different format."

"OK, OK," Turner said, deciding on this occasion to fold his cards. But how much was his twenty-four hours of cable news going to cost him? Schonfeld explained that to make it work they would need bureaus in New York, Washington, Chicago, Dallas, and Los Angeles. They would need a staff of about three hundred, with a liberal use of freelancers and a nonunion shop, and exchange agreements with local stations for news clips, as well as contracts for international feeds. His best estimate was that with $15 million to $20 million in startup costs and $2 million a month in operating costs, they could make it happen. In short, for CNN to succeed Turner would need at least $100 million and would have to hold on for two or three years, until subscription and advertising revenues began to cover expenses.

Turner took a deep breath. It was a huge investment for him

to handle all by himself. Picking up the phone, he called Russell Karp, the head of Teleprompter, which in 1978 was still the largest cable company in the country. He offered to make him a one-third partner in his new news network in exchange for financial support. Karp said no. According to one cable-industry executive who knew both men well, "Russ Karp was scared shitless of Ted Turner. He thought he was a crazy man. He thought he was nuts. Most people thought he was nuts." Then he called Gerry Levin, the CEO of HBO, and made him the same offer. Levin said he didn't think so.

To Schonfeld, "the hardest part of the business was getting distribution," and Turner, had he been able to bring either CEO on board, would have gotten both distribution and financing. Schonfeld thought it was a good idea. "We would have had editorial control and we would have had business control. They would have been passive partners," he claims. Though the attempt failed, Schonfeld, accustomed to dealing with the glacial movement of corporate boards, was dazzled by Turner's ability to take instant action. "It spoils your life for later on. When you deal with Ted, you do things right away."

Continuing their discussion, Turner spoke of stars. What about getting a star as anchor? "I believe in stars," he said.

Schonfeld did, too. He thought of Dan Rather at CBS. Rather was one of several at the network in line for Walter Cronkite's throne, but he was rumored to be miffed at not being the sole heir apparent.

"Well," Schonfeld said, "I think the biggest guy would be Dan Rather."

"Who's Dan Rather?"

Who's Dan Rather! Was it possible that Ted Turner had never heard of the former White House correspondent who had once gone toe-to-toe with Richard Nixon? Could it be that he was totally oblivious of the megastar investigative reporter of television's most popular news broadcast, *60 Minutes*? Turner had to be kidding, but Schonfeld could see that he wasn't. Keeping his amazement to himself, he explained. Turner asked how much Rather would cost.

"Oh, a million a year."

Turner's eyebrows went up. "Just to read the news?" But he knew the value of star power and had even been willing to pay big salaries to a few of his ballplayers, such as Gary Matthews, Andy

Messersmith, and Phil Niekro. Still, a million dollars for somebody he had never heard of just to read the news seemed like a hell of a lot of money.

Right from the beginning, Schonfeld knew that it wasn't going to be easy working for Turner. Before he left, he said, "I've got to have enough money, Ted, so that when you fire me, I don't have to work again." Turner gave him a blank stare. He didn't know what the hell Schonfeld was talking about. Fire him? Why should he fire him?

"Because," said Schonfeld, "we're not going to get along."

Turner paused, frowned, then suddenly asked, "What's your astrological sign?"

"Scorpio."

He brightened. "I'm Scorpio, too. We're not gonna have any trouble."

The estimated cost of getting CNN started was higher than Turner had anticipated and, of course, he had second thoughts. Once again, he turned to Terry McGuirk and said, "To make a go of it I'll have to commit a hundred million! Have I totally lost my mind?" But whatever questions he had were largely rhetorical, for he seemed more than ever determined to realize his idea. He called in Will Sanders to discuss the best way to raise the money. WRET in Charlotte had recently become an NBC affiliate, which, though cutting down on its profitability, ironically made it more saleable than it would have been as an independent station.

Sanders recalls, "We concluded that we needed capital to start CNN—a significant amount of capital—and maybe the best way to raise it would be to sell Channel 36 in Charlotte." Turner gave him the go-ahead to explore the market. "We contacted the thirty or forty group broadcasters who would be the logical buyers," says Sanders, "and eventually conducted a bidding contest. We received offers ranging from about ten million to fifteen million dollars. We felt that the hottest prospect was Group W Westinghouse. I called them up and said, 'Look, we've got some higher offers than you guys, but we think you're probably the logical buyers. You've got the ability to close, and you're a reputable company, so if you just want to close this thing out, raise your offer to twenty million bucks and we'll do it.' And they said, 'OK.'" As events turned out, it would be a long and difficult time before they were finally able to close the deal.

In early December of 1978, Turner flew to California for the Western Cable Television Association Convention. It was held in Anaheim, the home of Tomorrowland, and Turner's revolutionary plan for CNN fit right into that future world. At convention head-quarters in the Disneyland Hotel, some two thousand cable-system operators were registered. No doubt many of them had read of Turner's recent selection by *People* magazine as one of 1978's most interesting people. Now he was about to do something that would make him even more so.

Before a closed meeting of the board of the National Cable Television Association, Turner presented his idea for CNN. There was a great deal of interest among the association's members in pay-cable programming, and all he was asking for was fifteen cents per subscriber per month. Optimistic as ever, Turner was sure they'd be with him. If he could sign up eight million "subs," he'd have $1.2 million a month, the $800,000 balance to be made up by advertising. Turner said he wasn't asking them for any money in advance. He said he was selling his Charlotte station to finance the project. What he needed was their support, and given that, he promised he'd launch CNN on January 1, 1980.

"We're just asking you to agree, now," Turner pleaded, "and if it doesn't work, we'll go out of business and you can quit without paying us. If it does work, your liability is still limited to fifteen cents a month for each subscriber." His voice rose with excitement as he told them that his all-news network was exactly what the cable industry needed. Then he turned to McGuirk. "The contracts, please, Terry." As McGuirk quickly passed them around, Turner explained that these were letters of commitment for them to sign. "If you want to read it, go ahead," he said. "*Take* a minute or two, but I need an answer before you walk out of here."

The board members laughed at his impatience, and asked if they could have a couple of days to think it over. Turner said he couldn't wait. He had to know before the end of the convention. By the time he left Anaheim, he had only a few signed contracts in his briefcase, nowhere near enough to get started. It was a definite setback. The board members may have liked his idea but, as McGuirk said, "They didn't believe he could do it."

The Atlanta Hawks in 1979 were having a great season. In two years, the team had risen from the bottom of the NBA to become the central-division champs. Turner's appointment of Mike Gearon as president of the franchise clearly had a lot to do with it. Gearon says that in general he kept his boss well informed about what was going on with the club, but for the most part Gearon was left alone. On occasion, Turner might call to find out how season-ticket sales were going and to complain that they weren't any higher. "And Ted *always* used to want to know what you're paying guys. But basically," Gearon says, "he would delegate."

Part of Gearon's success was his ability to negotiate contracts and his knowledge of deferred payment and present value. At the time, Ted, according to Gearon, knew nothing about how you could pay a guy $500,000 a year—$250,000 in present value and the rest deferred. Once he understood the practice, he liked the idea. "So," Gearon says, "we structured a lot of contracts that way and captured a lot of free agents. We did a good job. And after ten years of nothing, we took the franchise in two years and turned it into a division champion."

In mid-April, the Hawks met the Washington Bullets in the playoffs. Turner invited his friend Jimmy Carter to one of the games in Maryland, and the president accepted, offering to fly the Turners up from Atlanta as his guests. Janie and Ted boarded Air Force One at Dobbins Air Force Base outside Atlanta and landed in Largo, Maryland. It was a first for both of them. Carter didn't appear until the second half of the game, but even then Turner still remained the center of attention, leaping from his seat at the slightest provocation and triumphantly flailing his fists in the air with every Atlanta slam dunk. He looked as if he didn't have a worry in the world.

While the Hawks, like the Braves and the Chiefs—the Atlanta soccer team that he owned via a surrogate—were hemorrhaging red ink, Turner's superstation was on the rise. Picked up by more and more cable systems across the country, WTCG, which on August 21, 1979 would change its call letters to WTBS (for Turner Broadcasting System), was now in a position to charge advertisers national rates.

Turner would need all the money he could get if there was to be any chance of making his dream of a twenty-four-hour all-news network a reality.

Turner drew up his battle plans. The McCallie graduate had always seemed to thrive on organizing campaigns, attacks, charges against the enemy. Like a military leader, he would send his racing crews into battle with their spirits flaming. Even cleaning the Turner billboards had been handled with teamwork and radio messages, as if it were a commando operation.

Now, in order to sell nationwide advertising, Turner hired a group of national salespeople from New York, most of whom didn't know one another, and invited them down to his plantation in South Carolina for a pep talk to launch the campaign. Gerry Hogan and the rest of the Atlanta sales force were there, too. Hogan recalls that the New Yorkers were met at the airport by chauffeured cars and driven to the plantation. Ted had rigged up a stereo system on the grounds, and as they pulled into the driveway, they were greeted by the theme from *Gone with the Wind*. Hogan says, "All these New York ad salesmen who thought they were in the middle of nowhere were kind of completely overwhelmed by this character." It was all part of their boss's attempt to turn them into a team. And absolutely nobody could motivate like Turner. Some who were there that weekend went away swearing that he could make cripples dance.

Although inspired, the sales force nevertheless found it hard at first to sell national ads. They were asking too much for "spot" ads, and they didn't have the audience to attract network advertisers. Even some of their good customers balked when their rates were bumped up. Delta Airlines, faced with paying over half a million for what previously had cost $100,000, dropped its entire Channel 17 campaign. "We have a retail approach and a localized message," a Delta spokesman insisted. But they were quickly replaced as a sponsor for the Braves by Eastern Airlines, and before the year was out the superstation's list of national clients—including such foreign sponsors as Toyota and Panasonic—grew to over one hundred and fifty.

For years, Turner had been paying Atlanta rates for the programs he bought for Channel 17 (many of them provided by Hollywood-based suppliers). Although he had previously microwaved them as far away as Alabama and Florida, he had not gotten too many complaints from local broadcasters in these states. But now that his superstation was going all over the United States via satellite, he was noticeably cutting into the movie studios' ability to syndicate their programs around the country. The payment of special (that is, higher) rates for the privilege of showing reruns outside Georgia was a bullet Turner had dodged for a long time. Reese Schonfeld says, "He was underpaying by enormous amounts." Soon a number of studios and syndicators refused to sell anything to Turner's superstation.

Then in the spring of 1979 he was asked once again to appear in Washington before the Van Deerlin subcommittee. In rewriting the 1934 Communications Act, the subcommittee had introduced a new provision on "retransmission consent." The proposed regulation stipulated that before a superstation could send a program to a new market by satellite, it had to obtain permission from the owner of that program—and presumably pay additional money for that market. The provision was backed by the Motion Picture Association of America, the National Association of Broadcasters, and the Independent Television Station Association, as well as the three major professional sports leagues. If passed, it might well put Turner's superstation, along with its imitators, out of business. Paramount was one of the studios challenging WTBS. Paramount claimed, "When Turner's signal comes into a market with one of our programs on it, the station there often feels there will be a lessening of the value of the programs."

Seated at the witness table, Turner fulminated against retransmission consent, attacking it as anti–small town, anticonsumer, anti–free enterprise, and "anti-American." This time the subcommittee was not impressed with his bluster. Shortly thereafter, however, he would testify on the same issue before a Senate committee and be more persuasive. He had something special to offer the senators. In the course of his testimony, Turner mentioned in passing, "I've already announced starting a twenty-four-hour-a-day satellite-fed cable news network . . . and we're going to give you gentlemen an opportunity to really air your views up here and spend hours discussing with the American people what's going on,.so the people will be informed." Rarely if ever before had Turner demonstrated a greater understanding of or sensitivity to his audience.

It wasn't too long after this that the retransmission-consent provision was dropped.

It was in May of 1979 that Schonfeld got the call he had been waiting for. "Reese," said Turner, "I'm going to do it *myself.* And you're going to run it." That day Ted's friend Bill Lucas, general manager of the Braves, had suddenly suffered a massive stroke. "Hell," said Turner, "none of us is gonna live forever. Let's do this fuckin' thing."

Turner had already promised Schonfeld that he wouldn't meddle. ("I'm not gonna do anything, Reese. You're gonna run it. I don't know anything about it.") Turner wanted him to come down to Atlanta right away. Since they had last talked, Schonfeld had been asked by people at Scripps-Howard Broadcasting and Post Newsweek Cable to consult about creating an all-news television station, but here was Turner actually ready to go.

They met for lunch at the Stadium Club, and Turner almost instantly produced a contract. Schonfeld explained that he was unavailable until his ITNA contract ran out in six months. Brushing it aside as of no consequence, Turner cried, "Hey, ain't this wonderful!" and slipped him a pen. Following the signing, a jubilant Turner hurried his latest acquisition back to West Peachtree to introduce him to the gang at Channel 17.

"I want y'all to meet Reese Schonfeld," he told them elatedly. "He's gonna be the president of CNN! And I'm gonna be the most powerful man in America!" Turner noticed Schonfeld looking at him and added, "The *two* of us."

If Schonfeld hadn't realized it before, it would soon become obvious to him that his new boss was a whirlwind. Turner seemed to have things all worked out. In two weeks he was planning to attend the National Cable Television Association convention in Las Vegas, where he would call a press conference to announce that CNN, the first twenty-four-hour all-news cable network in the history of the world, was to begin operation on June 1, 1980. This time the cable operators would have to jump on the bandwagon— *his* bandwagon—while it was already moving. Schonfeld, of course, would be there, too. But Schonfeld didn't know whether his board would allow him to go.

"The hell with 'em, Reese! Now, listen, what big name can we get right away, to go out there with us?"

The name that Schonfeld hit upon was Daniel Schorr. At the time, the former CBS correspondent was doing freelance work for ITNA and was available. Schonfeld understood that Turner had to have people with credentials in the news field at his side in Las Vegas in order to give his new network credibility. "I wanted Dan Schorr," he says, "because Ted's reputation at the time was right-wing, and Dan's was left." Turner was apparently ignorant of Schorr's political sympathies, for it was only much later that he came to Schonfeld and innocently inquired whether Schorr was a liberal.

When asked if he would work for CNN, Schorr had reservations. The little he knew about WTBS's lampooning of the news and what he had heard about Turner's drinking and conservative views gave him pause. But he was nevertheless interested. Turner himself decided to call him and put it to him straight. "Look," he said, "I don't have the time to spend going back and forth on this because [on May 21] I'm gonna have a press conference, and I'm gonna announce that a year thereafter we're starting the Cable News Network. And I would like you to be at the press conference with me. If you have doubts . . . whatever . . . I suggest you come out to Las Vegas and spend as much time as you need. If I can satisfy you, fine. If not, you go away and no harm done." Schorr said he'd be there.

Schonfeld, however, was having problems getting permission from his board to go, and the closer it came to the convention the angrier Turner grew. He needed Schonfeld. He wanted him to come to Las Vegas regardless of whether or not he had a release from ITNA. As their lawyers listened in on the conversation, the two men had a furious discussion on the phone, their voices rising in anger.

"Go anyway. Fuck 'em!" Ted shouted.

"No!" Reese yelled, and insisted on playing it his way.

"I'll indemnify you," Ted swore. "Don't worry."

"It's not my money they're going to come after. It's *yours*. They'll charge you with inducement to breach of contract. They'll sue you for millions."

"I'll take care of my own problems," Ted said. "Don't you worry about it."

"Look, Ted, let me work this out *my* way, and I'll get out there."

Ted shouted, "No, you tell me you're coming!"

Reese shouted, "No, I won't tell you I'm coming."

Turner's lawyer, Tench Coxe, got on the phone and talked to Schonfeld's lawyer, and the two lawyers calmly explored the risks involved. "In the end," Schonfeld says, "we did it my way. And doing it your way is never the best thing to do with Ted."

The cable convention opened in Las Vegas on Sunday, May 20, and Reese Schonfeld was there. For miles around, the desert was in bloom. If it could produce flowers, why not CNN? The night before the press conference, Schonfeld met with his future boss in the hotel cocktail bar for drinks and to discuss what they'd say the next day. Schonfeld remembers that when he arrived, he was thinking to himself, "Hey, I'm here!" and feeling buoyant. But it was obvious that Ted wasn't happy. It seemed to Schonfeld that "he was still thinking, You didn't do it my way."

There was a woman sitting beside Turner in the booth. Although the light in the room was dim, Schonfeld could see that she was blond and gorgeous. She looked like a Miss Universe contestant, which, he would later discover, she was. A small-town Texas girl and would-be actress, she had met Ted on a plane trip the year before, and they had become good friends. Ted introduced her as Liz Wickersham.

Then he got down to business and began talking about CNN. He wanted to have a kid's show in the afternoon—a news program for children. He was excited about the idea, but Schonfeld wanted no part of it. "Look, Ted," he said, "I know my business. We're not going to have any kid's show in the middle of the afternoon. That doesn't work."

Turner glowered at him and said that he didn't understand why. He thought it was a great idea. What was the matter with news for kids?

"That's not what this is about. We're not going to do that," Schonfeld said. He claimed that unless he had complete control of three things, he might as well go home. "The things I have to do here are with personnel, format, and content. If you don't think that I can do that, then you shouldn't have hired me."

Liz Wickersham tried to smooth things over, get them to lighten up. "Come on," she said, "you're going to do all these wonderful things together. It's going to be great." But she was having a hard time with them. Their voices began to escalate. Hearing the commotion, people wearing convention badges and milling about the bar turned to see what was going on.

When it finally seemed as if things had settled down, Turner suddenly glared at Schonfeld and snapped, "What makes you think you're so fuckin' smart?"

"Now, Ted," Liz said soothingly.

"In the first place," Schonfeld shot back, "you hired me."

Turner howled with laughter. Schonfeld called it "the best answer I've ever given in that sort of situation." After that, they settled down to discuss what CNN would be like. Schonfeld outlined his plan. They would have world news, local news, Washington news, financial news, interviews, and as much live coverage as they could afford.

Puzzled, Turner said, "But what you're saying, that sounds a lot like . . . it's gonna be like *regular* news."

"Yeah, it is," Schonfeld agreed, "but when the Wright brothers said they were going to fly an airplane, they didn't say that we're not going to have wings because birds have wings. They took what worked. That's how we're going to do this. We're going to take the things that work and turn them into our news." By the time they left the hotel bar, the two men were largely reconciled.

At nine o'clock the next morning, Schonfeld, having warned Dan Schorr what to expect, brought him to Turner's penthouse suite. The sitting room was big enough for a cocktail party; in fact, there was a bar at the far end. Turner and Schorr shook hands. Though the forty-year-old multimillionaire casually introduced Liz Wickersham as "my secretary," Schorr says, "I take it she was something other than, or more than, his secretary."

Throughout their conversation, which didn't last very long, there were people coming and going, and the telephone rang incessantly. It was clear that Turner had a lot of balls in the air and was distracted. He struck Schorr as "a bundle of energy." He never stopped moving. "What are your problems?" he asked bluntly.

Schorr quickly outlined his doubts. He wanted to know more about how Turner planned to conduct a news operation on cable twenty-four hours a day. He was troubled about the "funny" newscasts on WTBS. "You understand," Schorr said, "news is very important to me."

"Put that out of your mind. It's gonna be a serious news operation. That's why I want *you*." Turner made it clear to him that it would be a professionally run, all-news network, and he didn't really care what Schorr's actual role was. Schorr could work that out with Reese.

"Well, let me ask you a couple of other questions," said Schorr. "Would you expect me to read commercials?"

"No, no. Christ no!" Turner dismissed the idea as ridiculous.

"Would you expect me to mention the names of companies in lead-ins to commercials?"

"No, no, no. Nothing like that." Turner glanced at his watch and said, "Look, I have a news conference at four o'clock this afternoon, at which I'm going to announce that starting June 1st of next year I'm going to start the operation of Cable News Network. If you will appear with me, if you want to work with me, let's sign something, *anything,* and I want you to go to the press conference with me. If you can't decide between now and four o'clock, there's no point in the whole thing."

Schorr still had misgivings, but it was clear that the interview was over. He wondered what they were supposed to sign.

"Listen, I tell you what," Turner suggested. "You go out and write down a little statement, which I'll sign. Just say that you're not gonna have to do anything you don't wanna do. OK?" Turner told him to also put in what he wanted to be called. "Like vice president, how'd you like that?"

"You know, Mr. Turner, I don't know if you will understand this, coming from where you come from," Schorr said. "I am not a good administrator. I'll be in Washington, so why don't you call me your senior Washington correspondent?"

Turner shrugged his shoulders and smiled. If that would make him happy, fine.

Borrowing a typewriter, Schorr drafted a letter of understanding that read in part, "Mr. Schorr will not be required to perform any assignment that he considers to be incompatible with his journalistic ethics and standards," and they signed it. Schorr cost Turner about $100,000 a year, which was one of the top salaries paid to anyone at the fledgling network. But Ted accepted the fact that stars cost money, even though he'd never heard of Daniel Schorr before.

The press conference was well attended and rather formal. Turner stood up front, bookended by his two human trophies, who symbolized his serious intention of bringing America an all-news cable network. Reese Schonfeld was president and CEO and would start no later than November. Daniel Schorr had a two-year contract and would be assigned to Washington. It was announced that among the well-known personalities already signed to do brief commentaries were Bella Abzug, Phyllis Schlafly, Rowland Evans and Robert Novak, medical columnist Neil Solomon, and psychologist Joyce Brothers. By the launch date of June 1, 1980, there would be ten national bureaus plus worldwide "overseas sources."

Schonfeld described the centerpiece of their programming as a two-hour newscast from 8:00 to 10:00 P.M., but their special emphasis would always be on breaking news and live, on-the-scene coverage. Schorr said that he was delighted to be with Ted Turner and regarded CNN as a great opportunity and challenge.

Turner announced the price for CNN as fifteen cents per subscriber per month for cable systems already carrying his superstation and that twelve minutes per hour would be available for commercials, though he admitted that as yet he had no sponsors. But for many in the audience, the nuts and bolts of the operation were the least important part of what he had to say. He spoke inspirationally about his CNN, this great new and exciting undertaking. The country was ready to have news around the clock and cable was the medium of the future to deliver it. With satellite technology, the cable news network was about to change the face of American journalism.

After the press conference, enthusiastic cable owners gathered around him, eager to sign up for CNN. If he could make it happen, they wanted to be a part of it. Unlike his departure from California five months earlier following the Anaheim convention, Turner left Las Vegas in triumph, as if the worst were now behind him. But his problems in launching CNN had only just begun.

Turner didn't hide the fact that he knew as little about TV news as he had known about radio and television, baseball and basketball at the beginning. "Can I tell you how many hours of TV *news* I watched in my whole life, before I started my own network when I was forty?" he once asked an interviewer, and typically, without waiting for an answer, Turner revealed that it had been less than a hundred. Based on what this said about his interest in news, the prognosis for CNN looked none too rosy. But if, as *New York Times* editor Howell Raines says, "competition is what journalism is about as it's practiced in America," then Ted Turner was better equipped to start CNN than perhaps anyone else in the country. Competition, after all, had been central to Turner's whole life. Priding himself on being a quick learner, he had always felt he could buy all the specialists he needed. "Experts," he once said, "are the easiest thing there are to hire."

Schonfeld and Schorr were two of the experts he needed to make CNN run. To find a third, he didn't have to look beyond Georgia. When, following the convention, he and Schonfeld discussed where CNN would be permanently housed, Schonfeld suggested an excellent architect he knew in Atlanta. But Ted had already made up his mind to use his old sailing pal, the architect Bunky Helfrich, who had been working in the Savannah area restoring old buildings. "Shit," thought Schonfeld, "another friend of Ted's."

June 2 marked the fifteenth wedding anniversary of Jane and Ted, and there was a party. "The Bunker" was there, much to Ted's delight, and Ted told Helfrich about his plan to find a single location to house both Superstation WTBS and the new Cable News Network. Inasmuch as they had to move in by the following June, there wasn't time to build something new. The next day they found what Turner was looking for on Techwood Drive, not far from the Georgia Tech campus. It was the old Progressive Club, a former Jewish country club on 21 acres of trees and lawns, and it was up for sale. Set back from the road behind a horseshoe driveway encircling a fountain, the main building was a long two-story red-brick structure, its door framed by white columns and a portico. If there was a bit of *Gone with the Wind* about the building that appealed to Turner, there was even more Best Western Motel, of which he seemed oblivious.

The way Turner envisaged the layout, their administrative offices would be on the second floor, the superstation would be on the first, and CNN would be in the basement, where the gymnasium now was. The six satellite receiving dishes could be out back. Helfrich's assignment was to tear out the guts of the building and, working with Schonfeld and others, completely remodel the interior. And it had to be done in a matter of months, a ridiculously short time. Rather than asking if Helfrich could do the job, or, for that matter, if he even wanted to do it, Ted simply gave Bunky his orders.

As things evolved, according to Schonfeld, "Bunky turned out to be terrific. He may be the best architect-builder I've ever worked with. I'm not talking about imagination, but in terms of getting things done and delivering what you wanted him to do at a good price at the right time."

In mid-June, Will Sanders, Turner's chief financial officer, announced that he was leaving the company. Some thought he had serious reservations about his boss's latest risky undertaking, and others felt he had simply been burned out after years of working for Turner. Mike Gearon, who sat on the company board, regarded Sanders' departure at that crucial moment as a significant loss. "Will was as solid as anybody who ever worked for the company, and it was enormously reassuring to have him there. I was very disturbed when he left." Perhaps Reese Schonfeld was even more disturbed when he heard the news. "I would probably not have come down there if Will hadn't been there," Schonfeld says. "Will was a very solid financial guy." His sudden departure was the first of many shocks Schonfeld would have when he moved to Atlanta.

That summer, while Schonfeld was trying hard to put together a CNN staff and solve myriad other problems, Turner was spending more time at the plantation and sailing. For several days in mid-July he took part in the Hobcaw regatta at Mt. Pleasant, South Carolina. He drove into town with Janie and two of his children, his station wagon covered with bumper stickers promoting his teams. Together with his teenage son Rhett, Turner was racing a Hobie 18—an 18-foot double-hulled catamaran—for the first time. "It's just Rhett and me, going sailing, like I did with my father," he told his wife, pleased to be repeating the pattern. "Right, Rhett?" But the competition proved to be stiffer than he anticipated. After being in the lead for much of the three-boat race, they were overtaken and placed second. The way Turner reacted to the loss, it might have been the America's Cup.

A reporter for the Charleston paper wanted to know what new challenges he was facing in his life and Turner replied glumly, "Just surviving." Then, asked the secret of his success, he said, "Don't be a reckless gambler, take risks but only after preparing, and don't

put all your eggs in one basket." These answers would prove to
be prophetic the following month, when he went to England to
sail in the Fastnet Race.

Just before leaving for Europe, Turner purchased the 5,000-acre
St. Phillips Island, near Beaufort, for more than $2 million. St.
Phillips is a barrier island, two-thirds of which is undisturbed salt-
water marshland. The nonprofit Nature Conservancy had tried to
buy the environmentally precious island to protect it from development but had not been able to meet the owner's price. Given
what Ted got for his money, Will Sanders would probably have
said he paid much too much for it.

According to his aides, Turner is not a guy who says, "If I'm not
here, it ain't gonna get done." Turner says, "I ain't gonna be here,
and it *better* get done." Leaving Schonfeld and Helfrich to take care
of business in Atlanta, he flew to England at the beginning of August
to race his yacht *Tenacious*. Accompanying him was his eldest son,
sixteen-year-old Teddy, who would sail as a member of *Tenacious*'s
eighteen-man crew.

The Fastnet Race is 605 miles long. Beginning at the Isle of
Wight in the English Channel, the boats travel along the southern
coast of England to Lands End, then sprint across the Irish Sea to
the turnaround point at Fastnet Rock, and back again the way they
came, with the finish line off Plymouth. There were 303 boats in
the race—many more than usual—and most of them were smaller
than *Tenacious* and had less experienced crews. In 1971, Turner had
won the event aboard *American Eagle,* setting a course record in the
process.

On the second day of the 1979 race, a sudden furious storm hit
the boats with Force 10 hurricane winds, and gusts over 70 miles
per hour scattered them like seabirds. Thirty-foot waves beat on
their decks, blocking out the sky. Five yachts sank, twenty-five
were abandoned; a hundred and sixty sailors had to be rescued, and
fifteen died. Tabloid writers dubbed it "THE FASTNET OF
DEATH."

Reese Schonfeld wanted Bunky Helfrich to design a studio for CNN
that felt like an actual newsroom. He wanted the viewing public
not only to get the news but to get a sense of how it came to be
put on the air. Helfrich says, "It was a completely new concept.

And most people felt that it couldn't be done." Helfrich had never designed for television before and knew little about TV news. But in fact, as early as 1973 in Chicago, WBBM's station manager Van Gordon Sauter, in an attempt to capture the excitement of a newsroom atmosphere, had set-designer Hugh Raisky build a working newsroom, with desks, maps, wire machines, and TV monitors, that could also be used as a studio set.

There was another prototype for what Schonfeld had in mind at CHAN-TV in Vancouver, British Columbia, and in an effort to show Helfrich what he wanted, the two of them made a quick trip to Canada in August. Arriving at the television station, they were met by the station representative.

"Just turn around and go home," he said.

Schonfeld looked at him as if he were crazy. "What do you mean, just turn around and go home?"

"You don't know?"

"Don't know what?"

"You'd better get back on the plane. It's all over. Your boss is dead."

The report from England was that they had lost contact with Turner's boat in the storm, didn't know where he was, and assumed the worst, though as yet no body had been discovered. Helfrich turned to Schonfeld and said, "I hope they strapped him below."

"What do you mean?"

Helfrich laughed. "In times like this, we used to call him Captain Panic, and we would run the ship and strap him down below." He clearly didn't believe that anything had really happened to his friend, but Schonfeld wasn't so sure. He wondered whether CNN would survive if Ted was dead. It would go on no matter what, he decided, believing it was too good an idea to die. "I felt there was enough support within the company for it," he recalls. "We were so committed. I never considered that it wouldn't go forward—with or without Ted."

Mike Gearon was vacationing on St. Martin in the Caribbean. Switching on the shortwave radio he brought with him, he learned that a storm had struck the Fastnet Race. Turner's boat was among the missing.

Although Turner had already committed himself to CNN, Gearon felt there was still enormous skepticism in the company during this early, prelaunch period about the feasibility of that commitment. "Like all his undertakings at the time," Gearon says, "it was heavily leveraged and represented a considerable risk." Hearing the announcement that Ted was feared dead, he began to wonder what to do in the event it was true. "What I was thinking," he recalls, "as a director, my responsibility was to help stabilize the company and get it in good shape for his family, and CNN would probably be one of the first things that would not be done."

At the television station in Vancouver, Schonfeld and Helfrich watched the news feeds, and by the time the six-o'clock news came on, there was Ted, alive and talking a blue streak. "Sailing in rough weather is what the sport is all about," he boasted to reporters, praising his strong boat, his good crew, and their experience. "The people who didn't have those," he added, in an offhand fashion, "went to the big regatta in the sky." Asked about the tragic loss of life, he replied, "It's no use crying. The king is dead. Long live the king. It had to happen sooner or later. You ought to be thankful there are storms like that or you'd all be speaking Spanish."

His wise-guy answers and his flip reference to the fate of the Spanish Armada (which in 1588 also encountered fatal storms), angered the British press. They were expecting sympathy for the victims of the worst disaster in yachting history and were furious when Turner seemed to be faulting the victims for their own misfortune. Once again, he was putting his foot in his mouth. But in Vancouver, Schonfeld and Helfrich were relieved to see that at least he had landed safely and everything was all right. In fact, he had won the race.

Back in the United States that fall, Turner began to sell CNN by courting the media and making personal appearances around the country. In October, he was the featured speaker at the Washington Press Club. Naturally, he drew a crowd. And the fact that he had just been named one of the ten sexiest men of the year by *Playgirl* magazine didn't hurt attendance either. As usual, he was predicting the end of newspapers. In ten or twenty years, he promised that everyone would be getting the news over cable television. He talked about "bad news" and "good news," describing his plans to have CNN report the good news about what's right with America. In the audience, Daniel Schorr winced noticeably, knowing that this would be interpreted as "happy news" or "infotainment"; it could undermine their credibility as a serious news organization. Turner still had a great deal to learn about the business.

But by the time he and Reese Schonfeld attended a luncheon with the editorial board of the *New York Times,* Ted knew what not to say. Whenever someone asked him a specific question concerning how CNN planned to cover the news, Turner would answer, "Well, I think I'll let Reese handle that one." Schonfeld stressed their desire to cover breaking news. "Our philosophy is live, live, and *more* live."

Turner nodded, but then one of the *Times* editors pointed out that the danger of going live was that you could wind up covering a lot of inconsequential news, one-alarm fires that didn't amount to anything. Suddenly, Turner looked as worried as a father whose kid has just been asked a hard one at a spelling bee.

Schonfeld had to admit that this was so. Then he glanced across at the questioner and said, "But until the fire is over, you don't know whether it's a two-alarmer or the fire that burned down Chicago!"

Turner's face lit up. "Awwright!" he cheered. "Strong!"

By October, Schonfeld had hired about twenty people including Burt Reinhardt, a buddy from UPI Newsfilm, one of the independent news services for which Schonfeld had formerly worked. Reinhardt was an experienced news pro, as were such other early staff members as Ted Kavanau, Ed (no relation) Turner, Mary Alice Williams, Bernard Shaw, and ABC's George Watson, who would become the head of CNN's Washington bureau. Schonfeld says, "It was very important that we got George Watson. He lent credibility. He was terrific."

The CNN planning group moved into temporary quarters—a dumpy, turn-of-the-century, two-story wooden building on a bluff overlooking West Peachtree, next door to Channel 17. Out front, a couple of pretentious white columns guarded the glass front door and there was a porch sprinkled with camp chairs. In its earlier life, the building had been a halfway house for drug addicts and criminals as well as a house of prostitution. New hirees would arrive on the doorstep and their hearts would sink. It was impossible to believe that this ramshackle eyesore was the headquarters of a national news network that planned to cover the globe.

Nevertheless, letters of application and résumés poured into 1044 West Peachtree by the score, mostly from young people eager to gain experience in the business. Many were hired at about $12,000 a year. Some reporters were making no more than $18,000. Schonfeld said, "My agenda was to build a great institution," but he had to do it as inexpensively as possible. Bill Zimmerman, who signed on as an anchor, described it as "a pushcart operation" and "built on the backs of slave labor, people just out of school. So at the bottom end it was fairly weak, with people running around who didn't know what the hell they were doing." It wasn't called the "Chaos News Network" for nothing.

At the other end of CNN's nonunion wage scale were people like Mary Alice Williams, who had recently lost her job at NBC and saw being a part of the first all-news television network as a wonderful challenge. "Talk about a chance of a lifetime!" she said excitedly. "And so I took it for half the pay and stayed until 1989. We worked seven days a week, sixteen hours a day for seven months." To experienced journalists, CNN in its formative stage had the feel of a renegade operation. They had joined up primarily because they knew that Schonfeld, a solid newsman, was running the show. According to Zimmerman, "We all said we're working for Reese, not Turner. We felt Reese would insulate us from him."

"But the most employable people didn't want to come," Schonfeld readily acknowledges. He tried to hire David Frost, Geraldo Rivera, Phil Donahue, Linda Ellerbee, John Johnson, and Maury Povich, and they all turned him down. He says, "The great problem with CNN was that I had to hire an awful lot of flawed people. People who had problems elsewhere. Nobody wanted to come to Atlanta." Even many of those who did were uneasy about the prospects. Former ABC anchor Don Farmer insisted that his contract be with TBS rather than CNN, just in case the news network folded during its first year.

In November of 1979, the two Scorpios celebrated their birthdays together. Ted was forty-one years old and Reese, born on November 5, 1931, was about to be forty-eight. Hoisting their glasses to CNN, they drank to its glorious future and tried to forget about the problems. The least of their worries was that CNN's newly hired controller had had to be fired after only a few months. The worst—and it was crushing—was that Turner's financing for the news network had evaporated. His one and only line of credit,

amounting to between $30 million and $35 million from the First
National Bank of Chicago, had been withdrawn. There were rumors that the bank didn't like CNN's prospects or the idea of
Turner's doing news at all. Schonfeld had heard that "Ted had
given another one of his rather unfortunate interviews, and it had
really ticked off the chairman of the bank, who said, 'Do we really
want to be backing this outrageous guy?'"

Whatever the reason, their financing was gone, and they still
needed $50 million to cover CNN's startup costs and first year of
operation. Turner put as good a face on their predicament as any
human being could have, but even he seemed to be pushing an
improbably huge boulder up an impossibly steep hill. "We still
have the twenty million dollars from the station," he reminded
Reese cheerily, alluding to the money they'd be getting from the
sale of Channel 36 in Charlotte, though it had not been approved.
"Don't worry. I'll get it done. We're gonna do it." The hurdles
that Turner jumped in the coming months in order to make the
June 1, 1980, deadline for CNN became acts in a comedy of errors,
and there were several times when the news network seemed
doomed.

By the end of 1979, the only thing Turner had to show cable-
system owners to persuade them to sign on to CNN was a promotional video he had made with Schorr and the various celebrity
commentators. In it, Ted, backed by the bulldozers that were turning the old country club into the Cable News Network, says, "It's
going to be the greatest achievement in the history of journalism."
Though the tape may have thrilled Turner—even after he'd seen it
a dozen times—it was not necessarily enough to convince skeptics
that CNN was a reality, and no matter what he may have told his
top television executives, Turner knew it was still going to be a
very hard sell.

Ed Taylor, whom Turner had made a millionaire with Southern
Satellite Systems, flew to Atlanta just to do his friend a favor. He
owed him one. "I tried to talk him out of doing CNN, because I
was afraid he'd go under. I was also not convinced that he could
do it. It was the best piece of advice he ever said no to."

In his office, Ted would confide to associates, "Nobody thinks
we can do it. Everybody who walks through this door tells me
we'll never get on the air by June 1st, in less than a year. I tell them,

'Tough! We've got to.' And they don't even know the problems."
One of his latest was a drop in the company's earnings amounting
to a net loss of nearly $1.5 million for the year. A sharp decline in
income from his sports teams plus the costs associated with the
startup of CNN were dragging the company into the red. And one
financial problem led to another. The stress Turner was under was
enormous.

His gloomy litany continued: "I based our entire startup funds
on the sale of my Charlotte TV station to Westinghouse—that's
my twenty-million-dollar ante right there. But the sale hasn't even
gone through yet, because there's a group down there that's contesting the license. I've got Bunky Helfrich working full-time
redoing the new headquarters here, and that depends on a ten-
million-dollar bond issue that the city of Atlanta hasn't even approved yet. We haven't figured out a way to bring satellite pictures
to Atlanta, which is where all our news has to come before it can
go back out. Terry McGuirk is trying to sell this new service to
cable systems while all this is going on—without anything to show
them except the promo tape. He's just blowing smoke, a huge
amount of smoke. And we're hiring people right and left. Schonfeld
already signed on about fifty, and we'll have a staff of three hundred
before we actually click on. If any of this stuff goes bad on me, I'll
be up the creek. It's going to be 'Good-bye, Ted Turner, it was
nice to know you.'"

And then, on top of all of the other problems he had to worry
about, he was suddenly blindsided by the mysterious disappearance
of *Satcom III*.

On December 6, 1979, at 8:35 P.M., NASA launched RCA's *Satcom
III* from the Kennedy Space Flight Center at Cape Canaveral, Florida. For NASA it was a milestone, in that it was the hundred-and-
fiftieth launch of a Delta rocket. It was also a milestone for RCA,
in that its satellite was designed specifically—according to Andrew
Inglis, president of RCA Americom—"to increase our service to
the cable-TV industry." RCA's intention was to move the twenty
channels already dedicated to cable TV on *Satcom I* to *Satcom III*
and to add four new channels, so that all twenty-four transponders
of the satellite would be occupied with cable-TV business.

Though the Delta rocket itself was 116 feet high, the satellite
it carried in its nose was only a little more than 5 feet by 4 feet by

4 feet. It resembled a silver refrigerator with wings, and it weighed about a ton. As the rocket slowly rose off its pad and then shot up into the night sky over Florida, Inglis and other top RCA executives were watching. The light from its gaseous trail was "reflected by the water droplets in a layer of clouds, and for a few seconds the whole hemisphere of the sky glowed with a pinkish color which illuminated the earth like an enormous lightning flash." Inglis recalls a great *"Oooohh!"* went up from the crowd of onlookers. There was no thought of failure then. In fact, of the 149 preceding Delta launches, 137 had been successful. Turner, who had reserved one of the four remaining transponders on the all-cable satellite for CNN, never dreamed that it wouldn't work. "I didn't know that satellites failed," he said in all innocence. "I thought this thing was routine."

The launch itself *was* routine, but four days later, when the RCA team fired the small booster engine to send *Satcom III* into its final geostationary circular orbit 22,300 feet above the equator, something went terribly wrong. Whether the satellite exploded, imploded, or was simply struck dumb, all radio contact with it was lost. NBC News, flaunting its independence from its parent organization RCA, concluded its account of the lost satellite by reporting that there were "a lot of red faces at RCA." And at CBS News, the portentous Walter Cronkite was even moved to whimsy:

> *'Twas three weeks before Christmas, and down at the Cape,*
> *The* Satcom 3 *satellite seemed in great shape.*
> *It was RCA's baby that NASA would fling.*
> *As it happens, the one-hundred-fiftieth thing*
> *To be launched by a Delta, a rocket so flyable*
> *That's considered to be—ABSOLUTELY RELIABLE.*

Cronkite's parody concluded with the wry lines "And from somewhere in space comes the seasonal call, / 'MERRY CHRISTMAS, GOOD NIGHT, And you can't win 'em all.'"

Turner got the news of the lost satellite in Anaheim, while attending the Western cable-TV convention. What he had been expecting to get from RCA at the convention was CNN's transponder assignment on *Satcom III*. A worried Turner sought out Harold Rice, RCA's vice president of sales, who over breakfast reassured him that one way or the other CNN would be "taken care of [*on Satcom I*]," even if *Satcom III* was permanently lost.

Pacified for the moment, Turner went back to Atlanta to reassure the CNN staff, who, the minute they learned of the satellite's failure to achieve orbit, felt that CNN might not make it either. "We're dead!" one of them declared. Gloom settled over their temporary quarters on West Peachtree, as if it really were the haunted house that some claimed it was. Turner promised them that everything was under control, but it was far from that. He looked into the possibility of obtaining one of the transponders on *Satcom I* from HBO or Showtime, but they both turned him down. He became worried about the good faith of RCA. Was it possible that the parent company of NBC might try to protect its network from competition by blocking his access to *Satcom I?*

On January 16, 1980, Inglis and Rice met with Turner. The owner of CNN seemed determined to charm a transponder out of them, and Inglis almost believed he could do it. When he wanted to, according to Inglis, Ted could "charm the snakes." Greeting them warmly, Turner described his plans for the use of the transponder they were going to give him on *Satcom I*. The success of CNN depended on it. "He was already facing extraordinary high startup costs," as Inglis recalls, "and he could not afford the additional delay which would result from building up a network of earth stations on [some other substitute satellite]. He was terribly worried about the impact of the loss of *Satcom III* on his plans." Turner told them that his mother had worried so much that she had to have a hysterectomy. He was obviously desperate.

Inglis and Rice hoped to give him his transponder on *Satcom I*. Inglis genuinely liked the gray-haired, youthful looking, dynamic businessman. He considered him "a quintessential entrepreneur" who had been able repeatedly to make a success of businesses "that more conservative and experienced businessmen believed were hopeless." He thought of Turner as "a hell of a man!" And with his superstation, Turner was, after all, a good customer. Unfortunately, there were other customers who had applied earlier than Turner.

"We told him that we were looking favorably on a plan which would give him a transponder," said Inglis, "but that we had to consider the legal risks." Ted said that he "understood" and that he was "encouraged." On this note, their meeting ended amicably.

Unknown to Turner, five days later, RCA Americom's general counsel, Carl Cangelosi, informed his boss that he was seriously worried about the legal problems if they were to assign the remaining four transponders on any basis other than first come, first served. He suggested that they offer all six of their customers transponders on another satellite and then conduct a lottery among them to select four who could be moved to *Satcom I.*

Inglis explained to Cangelosi that if Turner lost the lottery, he'd probably lose his opportunity to build CNN, and Inglis didn't want that to happen. CNN, he believed, would prove to be a service in the best interests of the cable industry. But Cangelosi was adamant, and he had the strong support of RCA's legal department and outside counsel as well. Worn down by the unanimous opposition of the lawyers, Inglis folded his cards.

On February 20, he met with his cable customers and then the press to announce the company's plans for a lottery. Knowing that none of his customers was going to be pleased with the decision, Inglis was sure it would be a hard sell, and he was not disappointed. Although Turner wasn't there, he was represented by his friend Ed Taylor. "After the meeting," Inglis recalls, "Taylor spoke to me privately and predicted ominously that I was in for real trouble with Turner."

When Terry McGuirk heard the news, he regarded it as "a disaster." Cable companies wouldn't be able to receive CNN's signal from another satellite unless they installed an additional earth receiving dish, which would be prohibitively expensive. McGuirk reached Turner by phone in Nassau, where he had been racing *Tenacious* in the SORC, and told him what had happened.

"I'd just got off the boat," Turner recalls. "We'd raced over there from Miami, and we'd only been in for an hour. As soon as I heard, I grabbed Janie and we got the next plane home."

Assembling his support team in Atlanta, including lawyers Tench Coxe and Bill Henry, he immediately flew up to New York to confront the president of RCA Americom. "For Ted, the lost satellite was the worst," Schonfeld said. "He had no bankers. He had no financing. Who would lend him money unless he had assured means of distribution?" Turner had no illusions about just how much was at stake here. If he didn't come out of this a winner, CNN was finished. His mood became darker and darker, and as they shot up more than forty flights in the Rockefeller Center elevator, he appeared to be a storm cloud gathering force and ready to shake out lightning and thunder on anyone unlucky enough to be in the vicinity. The meeting was held in the conference room in the RCA legal department.

Inglis says, "We expected a rough session, and it was even worse than we anticipated." Turner was in a vicious mood, and let loose a flood of vituperation. Terry McGuirk was there, and though in the past he had seen Ted carry on in all sorts of unusual ways, even he was astonished. Ted seemed to "go nuts on them."

Turner shouted, "All you guys get out of here! I want the chairman of the overall parent corporation down here, right now, because I'm gonna break this company into so many small pieces that all of you will be looking for jobs!"

As Inglis recalls, "Turner ranted, he screamed, and he threatened. He accused us of taking this action to protect NBC from competition. He was going to go to the FCC, to the Justice Department, to Congress, and to the public. When he got done with us, he said, NBC would lose all of its licenses."

Turner was absolutely livid. At one point, according to Inglis, he shot up from his chair and, storming around the table, grabbed the RCA Americom president by the shirt. Hoisting him up, he stuck his nose in his face and said, "Andy, do you own any RCA stock?"

"Yeah," Inglis said, "I own some."

"Well, you better sell it," snarled Turner. "When I get done with you, it ain't gonna be worth a dollar a share. We're gonna go after NBC."

But it was Carl Cangelosi, the only RCA lawyer there, who bore the brunt of Turner's attack. "I remember Ted yelling at me for about twenty minutes straight about how he was going to sue NBC for antitrust. How he was going to humble RCA Corporation. He was almost uncontrollable." At one point, the lawyer said, "Listen, Ted, we've looked at your contract with us. You've got no backup. You've got no legal ground. You'll just have to live with it."

Turner nearly exploded. "You're gonna kill me. The networks are gonna kill me. This is my death if you do this to me. This is my blood you're getting." And if CNN went down the drain, he threatened to take giant RCA down with him. He acknowledged, "I'm a small company, and you guys may put me out of business."

He paused for a second, and then at the top of his lungs, he roared, "BUT FOR EVERY DROP OF BLOOD I SHED, YOU WILL SHED A BARREL!"

At this point, Reese Schonfeld tried to say something about satellites, which he knew much more about than his boss, but as soon as he started to speak Ted shut him up. "You don't know about this," he snapped. "This is *mine*. This is satellites."

Turner carried on like this for about an hour. Then Tench Coxe spoke up. "Mr. Inglis," he said, "I think I should tell you something." He announced that he had prepared a lawsuit, which the Turner organization was planning to file in the federal district court in Atlanta the following week. The substance of their claim was based on an old contract signed in 1976, when the superstation first went up on the bird, which they now alleged gave them a right to another channel on *Satcom I*. It was a very thin reed, but they were clutching it hard. And they were claiming damages of $34.5 million. Deciding to let the smoke clear before continuing the discussion, Inglis called time out for lunch.

The participants ate at the exclusive Netherland Club, at 10 Rockefeller Plaza—a rather sedate, private businessmen's club —occupying two tables. As Inglis recalls the scene, Turner was still almost beside himself. He couldn't eat. He alternately picked at his food and paced around the dining room talking to himself. Other diners glanced at him. Inglis says, "I was fearful that he was going to make a scene."

After lunch, the RCA team announced that they wanted to caucus. Behind closed doors, Cangelosi told Inglis that he was confident they'd win the suit Turner was bringing "hands down." Then, like heartburn after a four-star dinner, he added, "But you never know." And if Turner won, he continued, RCA would be liable for enormous damages. For that reason, Cangelosi suggested that they make him an offer. "We would assign him a transponder on [*Satcom I*] temporarily, while the court was reaching its decision as to our obligations under the contract. If, as we expected, the court ruled for us, he would be obliged to move his traffic to [another satellite] on December 1, 1980. If the court ruled in his favor, he would keep the transponder. In return for this, Turner would agree to waive all claims for damages." Inglis approved of the idea, and they returned to the meeting.

The Turner group seemed much more subdued now. Realizing that they didn't have a very strong case, Coxe recalls, "I was sweating blood." No sooner did Cangelosi outline his plan than Turner leaped at it. For him, it was at least a reprieve, a temporary way out. It would be what Cangelosi calls "a friendly lawsuit." They'd let the judge decide the merits. "I'm not saying we threw the case," Cangelosi says, but RCA cooperated fully, agreeing to let it be heard without delay.

Turner was elated. He grabbed Andy Inglis's hand and began pumping it enthusiastically while saying that they really didn't have a conflict of interest after all, and that RCA should hope that he won the case. Eager to protect his client from his own mouth, Coxe immediately stepped in and said, "Ted, shut up!" But his client's silence was only temporary. In the elevator going down, the delighted Turner announced to anyone who happened to get on, "We made a secret deal!" It was enough to cause any lawyer about to file a court case to cringe.

A week later, on February 28, 1980, Coxe filed a suit on behalf of Turner Broadcasting and Cable News Network in the U.S. District Court for the Northern District of Georgia. Charging RCA Americom with breach of contract, he asked for an immediate injunction requiring RCA to give CNN a transponder on *Satcom I*.

With dizzying speed, a few days later U.S. district-court judge Ernest Tidwell granted Turner space on *Satcom I* pending the outcome of his suit against RCA—but for no longer than six months. The judge stipulated that "In no event shall [Turner's] right to retain use of the transponder extend beyond December 1." At a press conference held on March 4 at the Techwood Drive construction site, an unusually subdued Turner, wearing a jaunty yellow hard hat with a CNN logo, admitted, "This is not the solution we would like to have had. But it's the best we could have done. It will allow us to get off on time."

While all this was going on, Turner was out trying to raise financing for CNN. He claimed in his suit to have already invested $34.5 million in the news network, and now they were hiring staff at an accelerated rate of ten a week. Ted flew to Washington to speak to his lawyers there, because the FCC would have the final say on both his *Satcom I* transponder and the sale of his Charlotte station. At lunch with the lawyers on April 8, he was looking more disheveled than usual as he laid his plight on the table.

"This is about it," he said. "The whole deal is crumbling, and when it goes it'll take everything I've got with it. I'm just about flat broke. We haven't got a [permanent] transponder and we haven't got the Charlotte money. The banks are calling in their notes on me, and the insurance company already has. I've got three hundred people on the Cable News payroll, and no money coming in to pay them. I just had to borrow twenty million to tide me over. The interest rate is twenty-five percent. Twenty-five percent of twenty million is five million a year, and there's twelve months in a year, and that's four hundred thousand dollars a month in interest alone. I can't pay it." The lawyers heard him out in silence, advised him against any precipitous acts, and spent the rest of the afternoon lobbying.

Though Turner might be able to hold off the foreclosures on his bank loans temporarily, he needed a large infusion of cash right away. If he could only push through the sale of his Charlotte station, he'd at least have the CNN startup money practically in his pocket. The problem was that a coalition of local black groups was contesting the station's license, because they didn't feel that WRET had done enough to hire minorities.

The longer the delay, the more uncertain Westinghouse was becoming about the $20-million price tag. Turner decided he had to have a heavy hitter with him when he faced the Charlotte black coalition, and the heaviest hitter around was the home-run king himself, Hank Aaron, who also just happened to work for him directing the Braves' farm teams. Turner took Aaron, Tench Coxe, and Mike Gearon with him to Charlotte, where they met with representatives of the black coalition in the coalition lawyer's office.

As it happened, the three people at the meeting who were doing all the talking for the coalition's side were white. When it was Ted's turn, he got up and said, "You know, I don't blame you guys for being mad at me. I'd be mad at me, too." Frankly admitting what a poor job WRET management had done, he was direct and disarming, and the black members seemed to like him. Then he said, "But it looks like you got the same problem I've got in my company. You don't have any blacks in high places either. You got three guys here who are doin' all the talking—and they're all white!"

Gearon says, "He had all the black guys laughing, and the three white guys became uncomfortable." Ted had quickly won them over. Then Aaron told the coalition members that whatever deal they finally arranged, they could count on his boss to do the right thing. After discussing the matter, a list of concessions was drawn up, one of the more unusual (in that it had nothing to do with WRET) being that Reese Schonfeld would meet with a leading black media spokesperson in Washington to discuss minority hiring practices at CNN. But the crux of the meeting, according to Turner, was that "they agreed to stop opposing the sale of the station, and we made some concessions. We made a lot of concessions. In fact, that little trip cost me close to half a million dollars."

Taking stock, Turner wondered what else could go wrong. If there hadn't been so much at stake for him, the repeated crises they had been having would have seemed almost comic. Now with less than a month to go before CNN's launch date and a flood of things still remaining to be completed and problems to be solved, it looked as if they'd never be ready in time. The floor of the Techwood Drive studio was still mud. One wall had all the windows knocked out. The computer system wasn't installed yet. Stored away in a corner, the television cameras were wrapped in thick sheets of plastic. There weren't even any bathrooms, which were designed by Helfrich to be a later addition to the main building. Porta-Johns were out back in the mud, not far from the giant receiving dishes.

Schonfeld and his staff were still trying to figure out how they were going to fill twenty-four hours of television time, seven days a week, with credible news. Turner was still trying to figure out how he was going to pay for it. Most knowledgeable media people didn't think CNN would ever get on the air, and the handful that did predicted Turner would be bankrupt within a year.

On one occasion, he took a brief time-out to welcome a group of young news recruits who had only recently joined his team. "I just want to welcome y'all to CNN and wish you the best of luck," he said. "See, we're gonna take the news and put it on the satellite, and then we're gonna beam it down into Russia, and we're gonna bring world peace, and we're all gonna get rich in the process! Thank you very much! Good luck!"

Any of those listening who happened to have read the recent comments about their boss by J. Christopher Burns, the vice president for planning of The Washington Post Company, might well have thought that Burns was right. "The reason Ted Turner decided to go ahead with it [CNN] in the form that he's doing," Burns said, "may be that he doesn't understand the problem. He's not paying attention. The cable industry doubts that Ted Turner knows his ass from a hole in the ground about news."

When Turner read Burns' statement, he was furious. He swore to friends that that sucker would eat crow when the news network got on the air in June. But even a few weeks before its scheduled inauguration, CNN's startup was not exactly a foregone conclusion.

CHAPTER 15

ON SUNDAY, June 1, 1980, it rained all morning, a dreary gray Atlanta drizzle that fell warm and sticky on the city. In front of the converted country club, workers were slogging around, trying to put the finishing touches on a reviewing stand for the upcoming festivities. This was Ted Turner's big day, the inauguration of CNN, and he intended to introduce his new creation with all the pomp and ceremony that befit a historic occasion. Three hundred distinguished guests had been invited; a military band and a color guard were on hand. "I'm going to do news like the world has never seen news before," Turner boasted. "This will be the most significant achievement in the annals of journalism."

By late afternoon, the rain had stopped, and with the 6:00 P.M. launch of the all-news channel approaching, the skies turned hot and humid and clear. In front of the new headquarters of Turner Broadcasting, three flags rustled in the sultry breeze—the colors of Georgia, the United States, and the United Nations, the latter symbolizing Ted's dream of international brotherhood, not to mention the worldwide scope of his ambition.

Turner's star-studded gala, however, turned out to be a modest affair. The announced three hundred celebrities were a motley group—a handful of journalists, some local Atlanta officials and businessmen, and a few of the "personalities," like Phyllis Schlafly and Dr. Joyce Brothers, who had been hired as CNN commentators. As guests walked across the unlandscaped front lawn, which was still soaked from the morning's rain, their shoes sank into the sod and the mud clung to their heels. To some there, the event seemed more than a little frayed around the edges—like a small-town Fourth of July, with a bandstand and a cut-rate impresario exclaiming, "C'mon kids, let's put on a show!"

Up on the reviewing stand in the late afternoon light, Ted Turner, wearing a rumpled blue blazer, fumbled with the microphone and called the ceremonies to order. Superstation cameras carrying the event live instantly captured Turner at the podium, and bounced his image back inside headquarters, then out to the uplink, up to the satellite circling the earth, and down to TV sets across the country and the monitors scattered about the grounds, where the guests could simultaneously watch Ted in person and on TV. In both places, he was proudly reciting an ode to himself and his new creation, as devoid of modesty as any epic hero is:

> *To act upon one's convictions while others wait,*
> *To create a positive force in a world where cynics abound,*
> *To provide information to people where it wasn't available before,*
> *To offer those who want it a choice;*
> *For the American people, whose thirst for understanding and a*
> *   better life has made this venture possible;*
> *For the cable industry, whose pioneering spirit caused this*
> *   great step forward in communications;*
> *And for those employees of Turner Broadcasting, whose total*
> *   commitment to their company has brought us together today,*
> *I dedicate the News Channel for America—*
> *The Cable News Network.*

Then calling for the presentation of the colors, Turner placed his hand over his heart and, while the band played the national anthem, stood at attention as he had been taught in military school—stomach in, chest out, shoulders back, straight as a statue. The fanfare of the trumpets soared heavenward along with the invisible TV pixels. As the last notes lingered in the Georgia sky, he turned to his audience and let out an earsplitting yell, the yell of baseball fans everywhere eager for the game to begin: "Awwriight!" After shaking a few hands, he strode back inside the CNN building, down the corridor to the office of one of his vice presidents, sat down and flipped on Channel 17 to watch the Braves-Dodgers game from the West Coast.

Meanwhile, on the newsroom floor, a major new journalistic operation was lurching into being. Three of the senior producers were screaming at one another about how to introduce the anchors and the new network:

"They gotta make an opening statement!"

"No, no, no!"

"They should at least say that this is CNN, and we're here to—"

"No, they shouldn't do anything like that! Just let 'em open with the news!"

"We gotta have a *statement!*"

"How about 'Welcome to your window to the world'?"

"No!!"

While the argument raged on, Dave Walker and Lois Hart, seated only a few feet away at the anchor desk, turned to face the cameras. As the final ten seconds were counted down to 6:00 P.M., the voice of director Guy Pepper came booming over the intercom from the control room with one final thought for them as they set off on their historic undertaking: "Just remember," he said, "shit flows downhill."

From the beginning, the Cable News Network resembled its boss, Ted Turner. Both were unpolished upstarts with global ambitions, a bit ragged on the fundamentals but scrambling hard to be taken seriously in the big leagues. Both were easy to laugh at (and frequently would be laughed at, especially in the major media centers of New York and Washington), and yet both had a certain raw energy, a certain drive that held great promise for the future, if only they could survive that long.

In those early days, critics tended to scoff at Turner's new creation, calling CNN "the Chicken Noodle Network." Others—even inside CNN—called it "the children's crusade," for the staff was young, inexperienced, underpaid, nonunion—in short, strictly bargain-basement. Putting on a twenty-four-hour news show live is a daunting prospect to anyone in television, and with such a junior-varsity crew glitches were bound to happen. And they did—frequently. Within its first hour on the air, CNN lost its feed from New York and went black, came back on the air only to cut out in the middle of a presidential speech, started to go live to the Middle East only to find that the Jerusalem feed wasn't ready, and then cut to reporter Mike Boettcher in Key West picking his nose on camera.

Over the following months, there were enough blunders, flubs, and goof-ups for a dozen blooper reels: the time a lightbulb exploded and set Dan Schorr's clothes on fire while he was reporting live from Washington; the time the network ran shots from the Atlanta zoo of a monkey masturbating; the time a cleaning woman walked past Bernard Shaw's desk as he delivered the news and proceeded to empty his wastebasket while America watched; the time weatherman Stu Siroka almost got swallowed up in the revolving panels of his weather map. "Whew," he said, extricating himself, "I suppose I'm lucky I'm still alive." Then the panels spun back around and whacked him again.

More significantly, the network was perpetually short of material, and when they did have material there weren't enough skilled people to put it together. Anchors found themselves on the air fumbling with new copy, mangled copy, or no copy at all. Several anchors took to carrying around stacks of the latest wire-service reports, just in case they were left live on camera with nothing to say. "The writers were terrible, and the editors by and large were terrible," anchor Don Farmer recalls. "Every anchor had his or her favorite anecdote. Mine was—we were always running these pieces on medical advances, 'new hope' for this, 'new hope' for that. One day the writers handed me the copy, 'New hope for crib-death victims.'

"Then there was the time Kathleen Sullivan was reporting this nice little Christmas story about a boy with congenital heart problems who was finally coming home after multiple rounds of surgery. And the story read, 'So this year, the Joneses will be able to open up their Christmas gifts—instead of little Johnny.'"

CNN's content was the least of Ted Turner's concerns. By his own estimate, he was losing a million dollars a month on his new network right from the start, and *Broadcasting* magazine put that figure closer to $2.2 million a month—not just negative cash flow but a river of money all going the wrong way.

CNN, like all of Turner's businesses, was a cut-rate, penny-pinching operation, squeezing out twenty-four hours a day of national and international news on a tiny, $25-million annual budget. But although CNN had the potential to make money two ways—by selling advertising time and by charging cable operators fifteen cents for each viewer—it wasn't doing nearly as well at the outset as Turner had hoped. Bristol-Myers and Sears had bought advertising spots, but they were among the few national companies to do so. And many cable systems seemed to be taking a wait-and-see attitude. The truth was that no matter how much he had bragged, Turner's viewership numbers weren't very good.

"In a sense," recalls Reese Schonfeld, "Ted failed. We were supposed to have five million homes when we went on the air. We kept cutting back and cutting back, from five million to three-and-a-half million. And I remember, right before we went on the air, [the guys who were] running advertising came into my office and said, 'We're going to lose all credibility. We're telling advertisers we're going to have three-and-a-half-million homes, and we ain't going to have it.'" Opening week, the official figure that appeared in *Newsweek* was 2.2 million homes, and even that was an exaggeration. "I refused to believe it," says Schonfeld. "I thought they [CNN's advertising staff] were really being too pessimistic, too negative. But as it turned out, we went on the air with 1.7 million." (At the time, seventy-six million homes watched the network news broadcasts each evening.)

Small wonder Schonfeld had been worried right up until airtime that CNN might not get on the air; small wonder the subsequent weeks and months were filled with anxiety over whether the network would have the funds to stay on the air. Turner had pledged his entire personal fortune to backing CNN, but it amounted to only $100 million, a sum the network news divisions went through in a year. Out at Fulton County Stadium one day early that summer, Turner was observed counting foul balls. As the batter shanked one into the stands, the Braves' owner muttered, "Four dollars." Three more foul balls followed. "*Sixteen* dollars," he groaned.

"The boss may be a rich man in his circles," observed *Panorama* magazine, "but $100 million is less than one-sixth of what The Washington Post Company took in last year, and there are several other huge vultures circling in the sky. No wonder Ted Turner is scared." Commenting on the launch of CNN, former CBS News president Fred Friendly noted solemnly: "Turner will show promise, but he may not be able to deliver. He's the first man on the beach. Unfortunately, the first man on the beach rarely stays around to develop the colony."

It was easy for the experts to write Turner off. With his strange rasping voice that yammered on and on, his clownish antics, his bimbo blondes, his falling-down drunken sprees, he seemed a comical figure—a good old boy with a few too many bucks and an overinflated sense of self-worth parading around comparing himself to Hannibal and Julius Caesar, while running the Kmart of journalism. How could they possibly imagine this joker succeeding at news, that most sacred of broadcasting traditions handed down from Edward R. Murrow and Eric Sevareid? But what CBS and others failed to see in both the man and his rinky-dink network was the potential for growth.

Looking at CNN's newscasts, the media establishment saw only the glitches and the inexperienced talent. Even CNN's exclusive interview with Jimmy Carter on opening day was written off as simply one Georgian helping another. (They didn't know about the memo that three months earlier had circulated in the White House saying, "Ted Turner has been a major supporter of the President and has been very helpful to us politically and financially.") What the media establishment was slow to grasp was the sheer scope of Turner's new enterprise and its two key innovations: first, CNN's ability via satellite to weave together into a seamless flow live remotes from around the world, from Florida to New York to Israel and back again, in order to tell a story; and second, the way CNN redefined news.

By appearing live, twenty-four hours a day, news on CNN had a compelling immediacy. It was no longer a collection of stale facts at the end of the day; now news was an evolving story that happened right in front of the viewer's eyes. And there were plenty of big stories happening that June of 1980: the MGM fire in Las Vegas, the eruption of Mt. St. Helens, the Republican and Democratic Conventions. Whatever the story and no matter how long it took to tell, CNN stayed with it. Although its reporting was usually not as strong as that of the networks, and its anchors were often left killing time, waiting for the next development, CNN transformed news into narrative by doggedly following events from beginning to end, and the appeal could be addictive.

But the media powers were right in one way: all this potential meant nothing if CNN didn't have viewers, and building an audience was Ted's job. He was the carnival barker who had to bring them into the tent. He had to grab the attention of advertisers who would buy his airtime, cable systems that would carry his news channel, and viewers across the country. It was a job for which he was uniquely qualified.

Right from the outset, Turner was shrewd enough to realize that he needed a dual strategy—that he had to sell not only CNN but cable TV as well. At that time, less than a quarter of America's homes with TVs were receiving cable, and if they weren't wired for cable, they certainly weren't getting CNN. So Turner started crisscrossing the nation, preaching the gospel of cable. Pounding the podiums, flailing his arms, he became cable's most visible champion.

"I've got everything I need personally," he told one group of advertisers in Atlanta. "I've got a baseball team, a basketball team, a soccer team, two sixty-foot yachts, a plantation, a private island, a farm, a wife, five kids, and two networks. Ain't never been anybody in the history of the world ever had more than me. So let me tell you what I'm going for. I'm going now for the history books. I'm swinging for the fences. And I don't want to be remembered as a bad guy. And since I'm in the advertising business, I want to be the hero of the advertising business. . . . I'm making these promises to you. I'm gonna deliver like I've delivered in the past. And I'm saying we're gonna be in fifty percent of the homes in this country on January 1, 1985, with both networks. . . . We're wiring the whole damn country, and we're gonna do it together. We're gonna make a ton of money. We're gonna do a lot of good. And we're gonna have a lot of fun."

But in 1980 cable wasn't necessarily the easiest sell. In fact, to many the cable networks seemed amateurish and dull. Turner understood that he needed to stir up some excitement: his plan was to attack the competition, the big three networks—ABC, NBC, and CBS. In typical Turner style, he wasn't satisfied with merely noting how their programming could stand a little improvement. At any given opportunity, in any forum or interview all through the early 1980s, he denounced the major networks as not just boring or bad, but evil.

In front of seven thousand members of the Veterans of Foreign

Wars, he announced that "the worst enemies that the United States ever faced are not the Nazis and the Japanese in World War II, but are living among us today and running the three networks." In *Broadcasting* magazine, he called CBS "a cheap whorehouse [that had been] taken over by sleaze artists." In *Newsweek,* he accused the "bunch of pinkos" running the networks of trashing the work ethic and raising the crime rate: "In the race for ratings, their newscasts dig up the most sordid things human beings do, or the biggest disasters, and try to make them seem as exciting as possible. In their entertainment programs, they make heroes of criminals and glamorize violence. They've polluted our minds and our children's minds. I think they're almost guilty of manslaughter."

Turner was, however, a little short on specifics. The only programs he could cite as "polluting minds" were the innocuous *Dallas* and *The Dukes of Hazzard,* certainly not much better or worse than the shows airing on his own WTBS (*The Caitlins* and *World Championship Wrestling,* for example). But though Turner on occasion sounded laughably cracked, his rhetoric was in fact more or less calculated. Every time he said, as he did on *Donahue,* that "intellectually and morally the networks are tearing this country down. . . . Television is the most powerful form of propaganda the world has ever seen, and in our nation it's being used to destroy us," he was not merely denouncing the competition. He was placing himself on the same playing field, defining himself as an equal adversary, just as he had done with Bowie Kuhn. Rhetorically, Turner was positioning himself as the David who would take on this Goliath for all of us, the knight who would slay the dragon that was enslaving America and American television.

Often his hyperbole seemed ludicrous: "What those networks are doing is making Hitler Youth out of the American people— lazy, drug addicts, homosexuals, sex maniacs, materialists, disrespectful. I mean, you know—mockery of their parents, mockery of all the institutions. It's bad. Bad, bad, bad! They oughta be tried for treason; they're the worst enemies America's ever had." But every time he launched one of these broadsides, he got press coverage, and even as observers laughed and shook their heads they were reminded that there was an alternative to CBS, ABC, and NBC—cable and CNN.

Privately, Turner also attempted to use CNN to attack the broadcast networks. He asked all his employees who were once network staffers to do their part. Anchor Bill Zimmerman recalls, "One time, when Reese was out in Texas, Ted sent a memo around to all the former network people. He said, 'If you've got any dirt on the networks, I want to see it.' Somebody called Texas and told Reese this memo is going around. Reese said, 'Get every copy of it and destroy it.' And they did." Meanwhile, Turner was publicly announcing, "We're going to do a bunch of investigative exposés on the networks. Listen, I'm going after the networks. I'm going to scare the hell out of them."

By now, Turner had begun to endear himself to the cable-systems operators, who realized that he was fighting their fight and who, consequently, were starting to pick up CNN and WTBS. He had also endeared himself to advertisers, who were sick of the broadcasters' high-handed ways when they were the only game in town. And he had endeared himself to a segment of the public who liked his raucous style and his patriotic fervor. Many, however, still wrote him off as a crackpot.

CNN had been up and running for only a few days when Ted Turner packed his bags and flew from Atlanta to New England. Once again, he was heading to Rhode Island, preparing to compete in defense of the America's Cup. Putting on his lucky striped engineer's cap, he stepped aboard *Courageous* and steered the old 12-meter boat out into the steely gray-blue waters off Newport. "I'm up for it," Turner told reporters. "The whole crew is up for it. I've got an awful lot more on my mind than I had in 1977, but we want to prove something."

There was someone else who had come to Newport that summer of 1980 with something to prove—skipper Dennis Conner. Ted's former tactician and frequent competitor had been stoking the fires of a grudge since 1977, when no syndicate would offer him a boat to race for the Cup. Conner had won an Olympic gold medal, a world championship (Star class), and a SORC—in short, he had proved himself one of the finest sailors in the world—and still he had been passed over, while rich amateurs like Turner got to put together their own syndicates and helm their own boats. This time out, Conner vowed, things would be different.

A plumpish man with a soft round face and an easy smile, Conner didn't look all that tough, but he was driven by a powerful no-holds-barred ambition that made him enemies. "Don't turn your back on Dennis Conner," certain members of the New York Yacht Club would murmur knowingly. For the previous two years, while Ted had been building WTBS and CNN, Conner had been sailing—sailing almost every day, sailing under Cup conditions from early in the morning until dark.

Conner was one of that new breed, the professional sailor, that would revolutionize the sport. He had devoted money to every aspect of his boat *Freedom* (raising some $3 million to Turner's $600,000). He had devoted weeks and months honing his crew, and more months still in selecting the perfect sails, with each catalogued according to the kind of wind conditions that might be faced. He had even hired a professional trainer to exercise the crew each morning before they came down to the boat. Conner's philosophy was "No excuse to lose."

For over a year, veteran racing pundits had been predicting that the 1980 America's Cup trials would be one of the great competitions in sailing history—an amazing duel between the tough, no-nonsense pro and the last flamboyant amateur. Ted Turner, who had put in only fifty days on the water in the past year, didn't admit to being all that impressed with Conner's preparation. "He brags that he sailed three hundred and forty out of the last three hundred and sixty-five days," Turner said nonchalantly. "That doesn't necessarily mean too much." Riding into Rhode Island on a wave of media attention, Turner exuded confidence as he smiled for the cameras and signed autographs for the bevy of admirers that followed him everywhere. "We own Newport," he declared.

The first day out, as Turner took his place at the helm of *Courageous,* the salt wind whipping his hair, the boat slicing through the water, the 1977 champion looked as strong as ever, narrowly edging Conner and *Freedom.* But as the Cup trials continued, the much-anticipated duel turned out to be a flop. A series of miscues plagued Turner and his crew—a cracked boom, a broken mast, a jammed halyard 60 feet up. *Courageous* lost not just to Dennis Conner's *Freedom* but also to *Clipper,* a boat that Turner had originally bought himself and then leased to Russell Long as a trial horse. Although only twenty-four years old and just graduated from Harvard, Long had taken Turner's boat, redesigned it, and was now regularly beating him.

Knowledgeable observers noticed that the trial races were being won and lost not so much on Newport's waters but back on land, at Bannister's Wharf. In the prerace hours, as Conner painstakingly examined his choice of sails for the day and his crew stretched and strained and grunted under the eyes of their trainer, Turner could be found at a dockside pay phone yelling down the telephone line to Atlanta. He had so many departments in his business empire to keep running, so many lieutenants to brief: Gerry Hogan, Terry McGuirk, Robert Wussler (his new number-two man)—they all needed to be poked and prodded from time to time. And when he finally hung up the phone, there were two other distractions on Bannister's Wharf—a feisty young brunette named Barbara Pyle, who had come to photograph the Newport races for *Time* magazine, and, just across the dock, a spectacularly athletic twenty-three-year-old blonde named Jeanette (J.J.) Ebaugh, who was accompanying an old sailing buddy of Turner's, Tom Blackaller, a wild-haired San Franciscan serving as assistant helmsman on the *Clipper*. Both women would come to play important roles in Ted's life.

For Captain Outrageous, defending champion of the America's Cup, June and July unrolled slowly, somberly, like some grim morality play in which good old-fashioned hard work beats flash and panache every time. Loss followed loss with deadening regularity. At the end of the first two months of trials, the won-lost record was: *Courageous* 6-20, *Clipper* 9-21, *Freedom* 29-3. Adding insult to injury was the disqualification of *Courageous* for a few days at the end of July: Turner had managed to sneak Ben Lexcen, codesigner of the America's Cup challenger *Australia,* on board his boat as they sailed against *Freedom,* pretending that he was a mechanic—"something akin to having a Russian admiral aboard the USS *Nimitz,*" the press observed. The New York Yacht Club was not amused. "Turner has lost his concentration," said one member. "He's not keeping his eye on the ball. *Freedom* may be a little faster, but that's no excuse for his record this year. There are those who think that Turner is like a schoolboy who wants to get thrown out of school rather than face up to failure."

Turner more than once contemplated pulling out or relinquishing the helm. He even talked it over with his tactician Gary Jobson, but both agreed that he wasn't sailing all that badly. *Freedom* just seemed to have a more powerful hull, faster sails. Perhaps they could turn things around in the final weeks. "Turner's last hope is to do well in the August trials, which get under way Tuesday," announced the *New York Times*.

August 5, 1980, dawned bright and sunny, as Turner took *Courageous* out to battle Conner and *Freedom*. Maybe now, in a splendid, last-ditch effort, he could recapture his former glory. But it was not to be.

Once again, off the shores of Rhode Island, it became clear that Turner's boat was being outclassed by Conner's at every turn. In the final results of their head-to-head competition that summer, *Courageous* ended up with but one victory, that of the first day, to *Freedom*'s eighteen. "It's time Ted realized that he's a middle-aged man chasing a dream," said skipper Russell Long, sounding as cold-blooded as a young Turner. "The level of the game has become higher than anything he's ever seen."

So it came as no surprise when Commodore Robert Mc-Cullough and his seven-man Selection Committee, wearing funereal black blazers and carrying an equally solemn message, came on board *Courageous:* Ted Turner and the crew that had swept the America's Cup series in 1977 weren't even going to make it to the main event in 1980. In fact, they were the first United States boat to be dismissed. "We do appreciate what you've done to defend the Cup," McCullough said gently, trying to ease the pain.

That evening Turner and his crew went down to the Candy Store, a popular local bar, one final time. Knocking back beers and vodka tonics, they toasted the *Courageous* and her skipper and felt more than a little sorry for the passing of an era. As Russell Long had noted, the game had changed. This would be Ted Turner's last America's Cup. "Three times is enough," he said, waving his hand to dismiss the emotion of the moment. Crewman Marty O'Meara confided, "It hurts worse than he'll tell you."

It was only a matter of months before Turner declared publicly—not only declared but shouted in a journalist's face—"I'm through with sailing! I don't even want to talk about it. I'm sick of sailing! I'm sick of the professionalism, the downright cheating. Did you read about the SORC last winter? All three winners were disqualified! They broke the rules. I'm through with all that!" That spring, he called his crew chief, Jim Mattingly, to ask what he thought about canceling their racing plans for the coming summer.

The next day he was on the phone once again: "Scrap the plans," he ordered, his mind made up. "Sell the boat."

Courageous had already been unloaded. With the sale of *Tenacious,* Turner would be out of 12-meter racing altogether. "It's the winningest boat in the world," Mattingly said. "You know he's serious when he puts *Tenacious* on the block. He told me he had no plans ever to return to ocean racing." With one quick decision, Turner had given up competitive sailing, the sport that had transfixed him for most of his life, the sport that had made him famous. But he would need all his energy and all his money for another battle, one that was shaping up in Atlanta.

When Ted Turner bet his personal fortune that he could make CNN work, he knew it would be a risky proposition. What he didn't realize was quite how risky. The all-news network was now carrying a $24-million working deficit, and there wasn't a bank around eager to extend that much credit to a reputed wild man like Turner, especially after First Chicago had turned him down. Money started to get very tight for him.

Staffers remember coming to work in the morning and finding Ted already there, wandering the halls in a well-worn, blue terrycloth bathrobe. He had set up a small cot in his office, and would stay through the night, looking at the figures and trying to make them come out right. There had to be more cable systems that would show worthwhile programming like CNN. There had to be more advertisers that would buy commercial spots on a fine newscast like CNN's rather than those "sleazy, crummy shows" on the networks.

Turner had brought in a new top financial officer to replace the departed Will Sanders. His name was William Bevins, and he was a tall, skinny chain-smoker who had a way with numbers; he could make them sit up and roll over like a trained poodle. Even some of Turner's own board of directors thought that on occasion Bevins was a little creative with his figures—"smoke and mirrors" was the phrase used—but no one could deny that he had a skill Turner valued above all others: he could squeeze money out of concrete. "In scrounging money," Reese Schonfeld recalls, "Bill was absolutely terrific." Running out of cash, Bevins and Turner scrounged marvelously, taking prepayments from the hot-dog and peanut vendors at the Braves' ballpark. Schonfeld says, "We'd take ninety percent of the receipts due—if we could get them *now*—to get us through. It was *that* hand-to-mouth."

Bevins informed Schonfeld that if somehow they could manage to hold on until August of 1981, they'd be OK, because by then TBS would be so profitable that they wouldn't have to worry anymore—TBS could support both itself and CNN. But August 1981 seemed a long way off.

Early in 1981, Bevins came up with an answer: Turner Broadcasting had to issue more stock; the sale would bring in a much-needed $200 million. Bevins had a housemate at the time who worked at the local office of Drexel Burnham Lambert, and Drexel plus Robinson Humphrey would underwrite the offering. It was a neat solution, and it had only one problem: Ted hated it. He had no desire to sell off more stock in his company. His father had always told him to keep control of the business. It was one of the precepts Ed Turner had drilled into his son: Don't give up pieces of your company. But Bevins was insistent: "Ted," he said, "there's no other way." Grudgingly, Turner agreed. A rich prospectus was offered to the public, and the so-called silent period (the time when shares can be bought by investors and the executives of the company must refrain from any public comments on the offering or their company) went into effect.

During this period, Turner suddenly appeared on the *Donahue* show. It was precisely in the middle of the silent period—in fact, it was April Fool's Day, 1981. Seated alone on stage, he leaned back and, staring calmly out at the audience, started off in his usual vein blasting ABC, NBC, and CBS, the evil empires. "They're in desperate trouble," he said. "The networks are like dinosaurs. Most of their money is going to the outrageous costs of Hollywood productions, going to pay for cocaine, so your favorite TV people can do cocaine."

Then, as the bespectacled Donahue hurried up and down the aisles with his microphone looking for questions, Turner made some glowing predictions about CNN: "We'll have enough homes on the news network this fall where we'll be breaking even. We're not that far from breaking even. . . . Between now and the end of the year, the news network's coverage will double. . . . By January 1, 1985, we'll be in half the homes in the country."

The next morning, a pale Bill Bevins conferred with members of the Turner board of directors to tell them the bad news. "Ted's killed it," Bevins said, shaking his head. "The underwriting is off." They were all stunned; no one could quite believe it. Was Ted so completely incapable of keeping his mouth closed?

The answer seems to have been yes—and no. "It was absolutely calculated on Ted's part," Reese Schonfeld contends. "He made all sorts of predictions for the company that were not true, and he killed the prospectus and public offering absolutely deliberately. Because he didn't want to finance that way. . . . Bevins and the banks were telling him that he couldn't get out of it, and Ted was absolutely calculating. He just killed the public offering by breaking the silence on it. (Anything you say on a prospectus has to be true, and if you say, 'I'm going to make a profit in the third quarter,' and it turns out not to happen, then everyone who bought can say, 'I want my money back.')" Deep in debt, yet unwilling to give up a piece of his company, Turner had chosen the *Donahue* appearance as the most graceful way to get out of the offering without alienating Bevins—or, more important, the bankers.

The rationale behind Ted's strategy would become apparent later, but on April 2 Turner's staff, colleagues, and competitors were once again left shaking their heads. Their boss seemed, to put it mildly, unpredictable.

Mary Alice Williams says, "His mood swings were so wild. I mean, he was so loud, he was so high—every normal human emotion was magnified with Ted. He was just so much larger than life. And I suppose I thought at first that he was just crazy as a bedbug."

Ted's lieutenants took to calling themselves "the gorilla keepers." Gerry Hogan remembers, "He was looked at with a certain amount of affection, but almost with a sort of combination of awe and a sense that this guy is absolutely nuts. He was larger than life in such a wacky way. He did such crazy things. . . . He'd offend people. I mean, if a woman had an attractive set of breasts, he would immediately tell her how attractive they were, regardless of whether it was the mayor's wife or a competitor's wife or whoever. He did it out of a sense of enthusiasm, but he was just a little out of control. I think the way people felt who worked for Ted was that he was crazy, but it was fun to work for him."

In fact, "nuts" is a word not infrequently used by his employees to describe the Ted Turner of the early 1980s. "It was in the air at CNN," recalls Bill Zimmerman. "You never knew what direction this guy was going to go in, and there was always that fear or caution you wanted to use around him because you didn't want to kick him off. It was almost like living with an alcoholic and abusive father or something. You always trod very carefully around him."

Those close to Turner at this time noted how, under stress, even some of his admirable qualities were pushed to extremes and became part of the darker side of his personality. He could, for example, be wonderfully frank, but there were times when it was a brutal frankness, lacking in all sensitivity. Several associates can remember instances in which Ted, meeting someone with an affliction, had made remarks like, "Hey, that's a real clubfoot, isn't it?" One notorious story involves an employee who had been operated on for cancer and had to have a piece of his brain removed. When this executive told him what had happened, Turner joked, "Well, since you lost half your brain, I ought to be paying you half as much."

Another curious quality his associates noted was his lack of fear. Ted didn't merely have the courage of his convictions. He seemed to be absolutely fearless. Perhaps it was the result of a fatalism born of his father's suicide and his sister's sickness and early death, but whatever its origin, Ted had not hesitated to risk his personal fortune on WTCG or CNN. Although he had no doubt that his new ventures would succeed, he seemed perfectly capable of writing the whole thing off with a shrug if they didn't. It was as if someone had deactivated the fear mechanism in his circuitry.

"Ted has never been inhibited by fear in a business situation," said Mike Gearon, a member of his board. "He's calculated what he can do. In a sense, you might say he's a gambler, because he knows the odds, but it's not a game of chance. It's a game dependent upon his skills, ingenuity, and—to a great degree—pure perseverance and tenacity. . . . You're always scared of a guy who's a little crazy, who's not afraid, who isn't going to back off. . . . The worst businessmen are bankers and lawyers, because they're looking for all the reasons to be scared. To not take a chance. Patton said, 'You get killed when you start digging foxholes.' Ted is a classic example of how you stay on the offensive."

Although almost always in motion and easily distracted, Turner was capable of great concentration, and when a subject really captured his interest he exhibited a kind of tunnel vision that excluded everything else. Friends at the time spoke about how boring it was to have dinner with him: they might want to discuss politics, the Super Bowl, or a movie; all Ted wanted to talk about was his business. Gearon remembers planning a vacation in Tahiti and asking Ted, "What's it like?" Although Ted had spent four days there racing, he didn't have the slightest idea. "It didn't occur to him to get a car and drive around what was then one of the most unique islands in the world," Gearon says. "He was absorbed with winning the races. He was absorbed with that more than anything else. And then in business—absorbed with making the business successful. Anything that wasn't relevant to that just wasn't relevant at all."

Each morning in those early days of CNN, Ted would explode into work; even before many staffers had arrived, according to some associates, he'd already be revved up "like a hyperactive kid." He'd roar up and down the halls, trying to get his people excited. Although a good deal of his time was spent wooing advertisers and cable-systems operators, his in-house sales work was almost more important: "He was selling *us*," recalls top TBS salesman Gerry Hogan. "His selling ability in advertising is certainly more than adequate, but he's really a great promoter and a great motivator. It's his personality. It's the kind of leadership that the military uses to get young troops to walk into a wall of fire. He had that ability to inspire you and commit you to a higher cause." From the executive offices to the journalists and producers and salesmen to the lowliest researchers in their cubicles, CNN and TBS staffers talked of sharing the same feeling—of fighting for Ted, of being part of a crusade.

Like their boss, the young, underpaid staff also fashioned themselves as buckaroos. They, too, would be out knocking back the drinks at Harrison's, the local power watering hole on West Peachtree, and getting into fistfights—even the senior executives like Bill Bevins and Bobby Wussler.

All through the mid-1970s and early 1980s, tales circulated about improprieties and even illegalities throughout the Turner organization. Drug usage was reputed to stretch from the top on down to the journalists and ballplayers, and law-enforcement officials had investigated and even arrested more than one of his buckaroos on drug charges. Stories about the embezzlement of funds followed a handful of executives. At one point, a disgusted Turner asked his staff to submit to lie-detector tests because of financial irregularities.

If Turner's CNN people were flawed, as Reese Schonfeld claims, so were those at TBS. Ira Miskin, executive vice president of the Turner Entertainment Networks, says, "We were all wounded soldiers, those people who came to Turner from the mid-1970s to the mid-1980s. . . . All of us who worked for Ted at all the various levels were no saints. We were nuts. We were the kinds of kids who should be put in the corner for being pains in the asses—cage rattlers. . . . We all were strident, loud, mean, foul, pushy. It was all part of the internal persona, and it kept the pot boiling. It made that company work—because there were so many crazy, committed, psycho people."

And leading this band of irregulars into battle with his Confederate saber was commanding officer Ted Turner, major general of the misfits. By May of 1981, with sources of funding running out for him, Turner and his troops were fighting on several fronts at once, and starting to look a little beleaguered. "My biggest job is in Washington," Turner told a local reporter, outlining his battle plans. "The entire entrenched entertainment and telecommunications business wants to see me destroyed because I pose a terrible threat to their whole rotten setup. So I've got to protect myself first in Washington.

"I need the cable industry to carry my programs. . . . I mean, believe it or not, as good as Cable News is, over half the cable industry is not carrying it.

"The third most important thing is selling the advertisers and the advertising agencies. But if I get the ratings, if I get more homes, the advertisers have gotta come along."

Turner was convinced that the folks in Washington—the Congress, the FCC—didn't understand the importance of CNN. Their ignorance wasn't surprising, considering that at the time the nation's capital wasn't wired for cable. In an extravagant gesture of goodwill and smart PR, Turner donated a $13,000 satellite dish to the House of Representatives, and wired all the congressmen's offices, so they'd at least understand what he was talking about.

In the spring of 1981, Turner and Reese Schonfeld instituted a legal suit against the White House—President Ronald Reagan, Chief of Staff James Baker, Deputy Press Secretary Larry Speakes, and Secretary of State Alexander Haig—and against the three networks for not including the Cable News Network in the camera pool for White House coverage. This antimonopolistic suit had both practical and symbolic resonance. With its round-the-clock airtime, CNN needed all the hours of footage it could get; but even more important, being included in the White House pool would put CNN on an equal footing with the big three networks.

At a boisterous news conference in Washington, Turner insisted that "the president's staff should not be allowed to deny CNN equal access." He then called for an immediate congressional investigation "to determine whether [network programming] has a detrimental effect on the morals, attitudes, and habits of the people of this country." One reporter in the crowd shouted, "It sounds like you're calling for a witch-hunt." Turner denied it. The networks, he said, were *worse* than witches. "They need to be hunted down and prosecuted."

Turner was in his element, attacking on all fronts. He summed up his position by saying, "We're at war with everybody. Now we have lawsuits against the three networks, the White House, and Westinghouse. And we're winning—or at least we haven't lost yet."

The Westinghouse battle was the latest, and it promised to be the biggest one of all. Throughout the spring and summer there had been distant rumblings on the horizon, rumors of trouble, and on August 11, 1981, the trouble arrived. It came in the form of a news item, a bombshell in *Broadcasting* magazine: "Westinghouse and ABC have combined forces in Satellite News Channel, a joint venture designed to produce two twenty-four-hour channels of advertiser-supported cable news that will be beamed free to cable operators."

For over a year, Turner and CNN had been living on borrowed time, running their cable-news operation with no competitors. The networks had all considered cable news and discarded the idea as too costly. But now Group W (the champion of all-news radio, and one of the largest owners of TV stations and cable systems in the country) and the behemoth ABC were going to team up and crush CNN. The tiny network had been improving month by month; in 1981, it was first on the scene at the Hyatt disaster in Kansas City, and first on the scene after the attempted assassination of Pope John Paul II. But in the pairing of Group W and ABC, it faced a real challenge. Combined, the two companies were worth $4 billion; Turner Broadcasting was valued at only $200 million. How would an already beleaguered network survive this powerful competition? What would happen to Turner's multimillion-dollar gamble when the Satellite News Channel went on the air in less than a year?

Faced with this new and enormous threat, Turner decided to take time out to think things over. He needed some peace and quiet. Leaving Atlanta, he headed off to his plantation in Jacksonboro. Turning off U.S. 17, and following the road as it changed from asphalt to gravel to dirt, Turner drove into another world—a forest that felt somber, ancient, almost primordial, with Spanish moss draped across the oak trees and palmettos and ferns rising out of the mud and ooze. Past faded brick pillars and a worn black metal gate lay his low-country estate. Turner holed up here for a long weekend, walking down by the Edisto River, hiking through the scraggly vegetation, and trying to decide what to do next, trying to figure out some path that would lead to victory, when defeat seemed to be closing in from all sides.

CHAPTER 16

By MONDAY MORNING, August 17, Turner was back in Atlanta, standing in the conference room of his Techwood building, in front of an assembly of CNN staffers, who awaited his verdict with trepidation.

"It's a preemptive first strike!" Turner declared, picking up his Confederate saber and once again slicing the air with it. When the enemy was charging, Turner knew just one response: Attack! If Westinghouse and ABC were going to try to undercut CNN with their Satellite News Channel—its format a repeating half-hour digest of headlines, like all-news radio—then he'd do his own headline service. Reese Schonfeld had in fact developed a budget and a format for just such a network a year before, but there had been no need for it then, and Turner had tabled it. Now he would launch CNN II. And if SNC was going on the air in ten months, he and his staffers would have CNN II operational in four—CNN II would be launched on midnight, December 31, 1981.

It's an immediate counterattack, Turner announced—just like Franklin Delano Roosevelt's after Pearl Harbor. When the Japanese carried out their surprise attack on the Hawaiian base, Roosevelt didn't hesitate; he immediately declared war and went on the offensive. And that's what they would do. "We're gonna offer everything they say they're gonna offer, except that ours will be on the air first and it'll be *better!*" he said, chomping on his cigar. He looked around the room at the excited and fearful faces of his staff, and he plunged on: "They have only money to lose, but we have *everything* to lose. We have to stake out this territory now or we'll lose it forever. This is the time in cable history where you either establish your bona fides or you don't!" Turner's battle cry rang through the Techwood offices. As Reese Schonfeld left the conference room, he was grinning and shaking his head. "Ted's happy," he said. "We're at war again."

Turner soon repeated his battle cry on his own network's program *Take Two*. Attacking Westinghouse and ABC, he once again spoke in the mode of David taking on Goliath, saying, "They're fifty times bigger, but with the superstation and the Cable News Network, we've always been the little guys fighting the big guys, and I really relish the fight. . . . I will do whatever is necessary to survive. *The only way they're gonna get rid of me is to put a bullet in me!*"

The very next day, when he flew up to Boston to speak at the annual conference of cable-system operators, Turner was using the same rhetoric. Aside from his own staff, this was the most important audience he would face: the ones who would decide whether or not to carry his networks—his old cable buddies, who were being tempted away by ABC and Group W and their free service. A standing-room-only crowd had gathered at the Copley Plaza Hotel to hear Turner speak, and as he began, aides recall, he was more than a little nervous: "When I cast my lot with the cable industry ten years ago," he said, exaggerating the number in his eagerness to win them over, "I *dreamed* that all these things would happen . . . and it's not unexpected that the three guys, the second-wave companies, the three major networks, would be in a state of panic and try to figure out a way to cut themselves in and get control of the news on your cable systems, at a belated time, when they see what tremendous damage we are doing to them!"

In point of fact, CNN had (and has to this day) less than one-tenth of the viewers of any single network newscast, but facts weren't going to stop Ted now. He was on a roll. After spelling out his CNN II service, he put aside the few notes he had and leaned across the podium toward his audience: "You know, I've always been a fighter. And I remember when this industry was scrapping for survival. We've always been scrapping. We've had to fight for our right to live against these networks that are now standing in line to utilize your channels. We had to fight for our lives. You'll find that I've spent a great deal of time and money in Washington, helping you fight these battles, because your battles have been my battles ever since I joined the industry."

Having established their common bond, he led the charge against the networks: "The one thing is, you guys and girls and ladies in the cable industry, you are learning. A year ago, you knew very little about news, but you're learning a lot about it. I mean, *I* didn't know much about it either"—the cable audience was laughing along with him now—"but I knew enough to know that the broadcast-news approach was the wrong way to do it. . . . They've turned our people against our wonderful government, and they've turned our people against business. . . . They're antireligion, antifamily, anti-American." Finally he attacked the Satellite News Channel directly, noting that ABC News might not even choose to share its footage with its cable cousins. *"Anybody who goes with them,"* Turner bellowed at his audience, *"is going with a second-rate, horseshit operation!"* Sweeping the room with a challenging glare, he demanded, "Any questions?"

Assembling CNN II, Turner knew there would be certain synergistic economies. Much of the same footage from CNN could be recycled on the headline service. There wouldn't be any need to hire extra reporters either. But all the same, CNN II would need a hundred and fifty of its own staffers—directors, producers, anchors, technicians—not to mention a building of its own and an annual budget of at least $18 million. Schonfeld had assembled a strong team for CNN II under another kamikaze newsman, Ted "Mad Dog" Kavanau, but even so, they had only four months to get on the air.

In his quieter moments, away from the crowds and the speeches, Turner tried realistically to assess his chances for success. If he was under pressure before, he had just turned it up several notches. But his people could certainly crank out the news. That was the one thing he wasn't worried about. His headlong attack seemed like the right strategy; the question was, Would it be a victorious blitz or a charge of the Light Brigade?

"It was basically a defense," Turner told freelance writer Hank Whittemore, "but in defending you often attack. We were holding the ground. We had CNN already, and they were gonna attack with two channels. . . . I figured—knowing they were two big companies and that they were both public corporations, and how slow those kinds of operations usually run—that if their losses were bigger than anticipated, the people in charge of this project would come under criticism. . . . And I knew that if they ran into unanticipated difficulties, there would be friction between the two fifty-fifty partners. . . . So even though we were very, very strapped financially, and they knew it, I decided that we would beat them to the market. We would split the market for that service, so they would not be as viable. I didn't know exactly how long we could last. I think the two of them had resources a hundred times greater than mine. I did know that in a war of attrition we'd lose."

As the construction crews labored and the news staff assembled prototype programs and all the other preparations for CNN II rushed ahead, Turner—much to his distress—found himself embarked on just such a war of attrition, the money leaking out of his company like blood from a wound.

It was at this moment, late in 1981, that Turner went before an audience at Georgetown University. Interested students, faculty, and reporters hoping for outrageous Turnerisms were packed into Gaston Hall for the afternoon lecture, and Ted was in fine form as he lambasted the networks and the Washington establishment. He then turned to his own accomplishments—the business, the networks—and proudly brandishing a copy of *Success* magazine, he held it up so that everyone could see his portrait on the cover. Lifting the magazine above his head, he looked up toward the heavens as if at an unseen audience member, the one who was with him wherever he went. "Is this enough?" he cried out theatrically. "Is this enough for you, Dad?"

Turner loved to collect articles about himself such as the one in *Success*. He understood the importance of PR for his fledgling business; each article was also a tiny measure of validation for a man who had early lost his only real yardstick of validation, the father who had prodded and shaped him. Ed Turner would have been especially proud of the *Success* cover story; he himself had been a longtime subscriber to the magazine and had been thrilled to be mentioned in it some two decades before. Perhaps that's why in the early 1980s Ted would repeatedly reenact this same scene over and over again, holding the same magazine cover up to the sky for approval, repeating the same words, as if they were a mantra.

Yet even as he was proclaiming his success, the forty-three-year-old Ted Turner was, ironically, as close to failure as he had ever been. Like Ed Turner shortly before his suicide, Ted was now stepping up to play in another, bigger league, and by the end of 1981 he had extended—perhaps overextended—himself toward the same precipice. According to several friends, he seemed to be deliberately courting disaster to prove that, unlike his father, he could meet it and not flinch.

In true Turner style, even when faced with a seemingly all-consuming challenge, he somehow found the time and energy to dash off to another corner of the globe. It was Friday, February 12, 1982, when he caught a flight out of Miami and, accompanied by a CNN crew, arrived in Havana as the guest of President Fidel Castro. As Ted stepped into the warmth of the Cuban sunshine with Liz Wickersham on his arm, it may have seemed nothing more than a long holiday weekend, but in fact he was taking a significant stride toward his future.

Turner was certainly a celebrity, but up to this point he had operated primarily in a world of locker rooms and chamber-of-commerce meetings. He had a certain global reputation, to be sure, but he was still primarily a local boy made good. If before this he had any links to high-visibility political power, it was in large part because Jimmy Carter was a fellow Georgian. But now, among the palm trees and the balmy tropical breezes, Turner was making his first foray into the heady world of international politics.

The meeting had come about at Castro's request. Impressed by CNN's uninflected global coverage, and by the network's live reporting of Cuba's 1981 May Day Parade, including a Castro speech—the first live broadcast from the communist nation to air in the United States in some twenty years—El Presidente had sent Turner a note: "I just wanted to let you know that I think CNN is the most objective source of news," he wrote, "and if you ever want to come down to Cuba, . . ."

For all his right-wing views and America-first jingoism, Turner was intrigued by the invitation. "I'm a very curious person," he'd later tell an American reporter. "It was the first time I'd ever been in a communist country, and I was just interested in learning a little about how it worked." To a BBC reporter, he represented himself as an ordinary American tourist: "I just went down there as Citizen Turner."

As the four days unrolled, Turner's excursion into personal diplomacy was an unqualified success. Touring the Havana factories, taking in a Cuban League baseball game, enjoying a day at the beach and a night at the renowned Tropicana nightclub, Turner and Castro hit it off like soul mates. They posed for snapshots together. They clowned around together. Two lusty, larger-than-life characters with a taste for cigars, an eye for women, and a propensity to pace about and lecture everyone in sight, Castro and Turner became friends on that long weekend, even though they could barely talk to each other without an interpreter. They were the ultimate Odd Couple, the communist guerrilla leader and the reactionary Southern businessman.

Of course, it didn't hurt that that week CNN had broken the story of U.S. "advisors" carrying M-16s in El Salvador, a black eye for Ronald Reagan's administration. CNN's coverage had exposed the extent of the government's involvement in Central America—an involvement that was in direct opposition to the White House's stated policy.

And it also didn't hurt that Ted and Fidel both shared a passion for duck hunting. Early Monday morning found the two of them dressed in green-brown camouflage fatigues, crouching in the marshes along the south coast of Cuba, scanning the sky for waterfowl. Ted had some trepidations when he noticed that the birds were Cuban tree ducks, an endangered species, but Castro allayed his fears by informing him that they were the only two people on the island allowed to have shotguns. Castro was a good shot and efficient with his gun—one bird to every one-and-a-half shells, according to Turner—but Ted wasn't so bad himself. Castro awarded him a prize as the best marksman of any guest he had hosted—Ted had bagged the most birds with the fewest shells—and sent him three stuffed and mounted ducks as a memento of the visit. Turner would later exclaim, "Twenty-two attempts on his life by the CIA, and I'm sitting next to him with a loaded rifle! Can you believe that? . . . I could've shot him in the back!"

Not only had Turner and Castro enjoyed each other's company but they had found the basis of all good diplomacy—enlightened self-interest, a way for each to exploit the other. Turner came home with a new visibility on the international scene, a "softball" interview to air on CNN, and even a four-minute promotional piece for CNN by Castro ("When there's trouble in the world, I turn to CNN," declared Castro), which Turner's advisors would beg him not to use. And Castro would get something even better—a friend in the United States during the Reagan era and, if not an ally, then at least a media window to the American public at a time when Cuba was increasingly isolated in this hemisphere.

Turner was none too sophisticated politically. In one discussion with Castro that he recounts, he reveals himself a virtual naïf: "After three drinks with rum, I got up my courage and I said, 'Are you interfering in Nicaragua and Angola?' And he said, 'Yeah, you are, too.' I said, 'Yeah, but we're the United States. We've got every right to be there.' And he said, 'How come?' I said, 'Because we're right, we're capitalists. We've got a free country.' He said, 'Yeah, but what about people that don't agree with that?' I went back and scratched my head. I never even thought there was another side to the picture." But Ted knew what he liked in a man and respected in a leader, and he'd soon be doing some promoting for Castro. "It's hard to understand how we can trade with Russia and not with him," he contended in one interview.

What Turner respected was strength. "That's right," said his right-hand man Robert Wussler, "Ted likes strength, and he doesn't like weakness." Over the years, friends and colleagues recall Ted describing his fascination with another dictator, Adolf Hitler. More than one was startled to hear Turner talk of his interest in the Führer, even reading passages aloud from *Mein Kampf*. Not that Turner had jackboots in his closet. Whether a leader was fascist or communist was irrelevant to him. He simply admired people with enormous power, people who had the ability to get things done. When Turner returned from Cuba and came under attack from betrayed right-wingers who couldn't understand how he had gotten along so well with the top communist in the Western Hemisphere, this longtime friend of the John Birchers, this good-old-boy champion of all things capitalistic, sprang to Castro's defense. "Castro's not a communist," Turner explained to puzzled friends. "He's like me—a dictator."

Back home in Atlanta, CNN's supreme leader found an insurrection on his hands. NABET (the National Academy of Broadcasting Engineers and Technicians) had come down to Georgia and was urging CNN staffers to unionize. The dirty little secret about CNN was that they hired young people and paid them slave wages for interminable hours. Overworked, underpaid, and out-and-out exploited, CNN staffers had had no one to complain to, since Georgia was a right-to-work state. After five months of campaigning, the NABET organizers thought they had started a union groundswell. But despite his newfound admiration for Castro, Turner wasn't about to give the proletariat any power on *his* turf. Unionization, collective bargaining—they could keep that commie stuff back in Havana.

Ted, however, thought he might try Fidel's iron fist. He didn't want the union to win. From where he stood, the only reason CNN was viable at all was that it was lean and mean, with no union rules, hours, or wages. He told staffers that they could organize all they wanted to, but if a union took over, they'd most likely find themselves without jobs. "We couldn't operate here if we were unionized," Ted declared emphatically.

The velvet-glove part of the Turner message was that CNN offered the young, inexperienced workers all sorts of opportunities they'd never have elsewhere—chances to write, to produce. But he never forgot the iron fist. Staffers recall Ted racing down the hall one evening before the election yelling, "Where are the union organizers? I'm gonna kick some ass!"

On the last Friday in February, at CNN Atlanta, the National Labor Relations Board put a voting booth in the first-floor conference room, right next to the newsroom. The staffers filed in, over two hundred of them. And when the vote finally came down, it was 156 to 53 against the union. Schonfeld breathed a sigh of relief. CNN could keep paying the same skinny wages and operating on the same shoestring budget, the only way Turner could afford to let it operate. Afterward NABET accused Turner of exerting undue pressure to sway the vote, and an NLRB hearing was convened. But all that was moot in Ted's realpolitik. Sway the vote? Of course he had. He had stomped out the union just as he intended.

Given Ted Turner's penchant for restlessly mucking about in all of his little fiefdoms, the biggest fear of many of the staffers in the newsroom was that he would single-handedly undermine the editorial integrity of CNN. But from Mary Alice Williams and Dan Schorr to Reese Schonfeld himself, the relieved consensus, after nearly two years, was that—unionizing and financial issues aside— Ted had by and large stayed clear of the newsroom. He didn't really understand the mechanics of news any more than he had understood the mechanics of baseball or basketball. He might occasionally pop up on the floor of CNN and issue strange edicts, like his ban on litter ("I do not want to see one speck of trash on our grounds anywhere. If you intend to litter on our premises, please do us a favor and resign now"), but he wasn't going to demand, say, that one story lead the news and another be dropped.

Or at least that was the way he started. But as CNN began to solidify its reputation and reach some ten million homes by late 1981, Turner slowly started to feel free to throw his weight around internally. A series of small confrontations with CNN president Schonfeld erupted into a full-scale clash of egos, as Schonfeld had long ago predicted. Reese thought that CNN was his—he had shaped the network and been told that he could run it without interference—but Ted knew who really owned it. He wrote the checks.

The first incident occurred in September of 1981, when Turner was set to appear before Congress to testify on violence on TV. According to Schonfeld, "He called me very late the night before, sounding as if he had had a lot to drink. He kept saying, 'We gotta go *live*. We gotta go live.'" But Schonfeld had no intention of tying up CNN with his boss's Washington appearance. "Look, Ted," he said, trying to ease out gracefully, "We can't do it in time. There's no way."

That night, however, Turner marched over to the Washington bureau of CNN, all excitement and enthusiasm. "Hey!" he yelled, "Do you guys know I'm testifying to Congress tomorrow? You're going to be covering it, right?" The staffers got the message and, working overtime, set up the facilities for a live broadcast.

When Schonfeld stepped off a plane the next morning in New York City, he heard his name being paged: "I called the bureau, and I discovered Mary Alice [Williams] almost in tears. They were carrying Ted live on CNN, which was bad enough, but that was the day of the Brink's truck robbery up in Westchester—the famous Brink's terrorist robbery. And Mary Alice is trying to break through to carry the story. And we're covering a speech by Ted. I called Ed Turner in Washington. He said, 'Ted told me I had to keep it on.' And I said, 'I don't give a damn what Ted told you! I'm your boss. I'm telling you to take that down and put New York on! And don't go back to the hearings.'"

In many ways, Reese Schonfeld and Ted Turner were doomed to clash. Both were hard-charging, fierce individualists who would throw orders and invective around to make things happen. Neither one had much patience; both wanted to be sure things got done, their way, *now*. "In several senses, he and I are alike," Schonfeld admits. "In our weaknesses, too."

Fed by rumors from disgruntled aides, Turner in mid-1982 began to have his doubts about Schonfeld's ability to handle a big budget and oversee a large staff: "It came back to me from a number of sources," Turner says, "that he was making all the decisions . . . and really trying to do too much . . . that Reese wanted to make every decision himself. It just wasn't the kind of management style that was going to make the organization strong in the long run." It was a strange complaint, coming from Turner. Meanwhile, Schonfeld felt that Turner was stepping all over his turf and ignoring the promised freedom he had been offered. Prickly in his pride, he resented any intrusion.

It wasn't a good sign that Ted kept forgetting Reese's name. He would call him "Schonberg" or "Schonstein." It was still more ominous when Reese and Ted disagreed over the outcome of the White House newspool lawsuit. CNN had *de facto* won the case when an out-of-court settlement allowed the network to slip quietly in and join CBS, NBC, and ABC in the video pool. But Schonfeld felt that CNN should have pushed on with the suit for a big public PR victory; Turner felt that since they had already spent $1 million on legal fees, enough was enough.

The big fight came over personnel. Anchor Marcia Ladendorff and talk-show host Sandi Freeman were coming up for contract renewals, and Schonfeld had problems with both: Ladendorff wanted nearly double her $45,000 salary; and Freeman, he felt, was burned out, her ratings slipping. Schonfeld was feuding with them and their agent. Besides, he wanted to pick up a Washington, D.C., talk show called *Crossfire*, with political commentators Tom Braden and Pat Buchanan, instead. He had also decided to hire West Coast interviewer Mike Douglas to host CNN's show-business program, paying him $1 million for two years.

All these hirings and firings were too much for Turner. He liked Freeman; he liked Ladendorff. And who was Schonfeld to spend a million dollars of his money without even consulting him? "Ted felt he should have been talked to," Ed Turner recalled. "Not so much to overrule it but, if nothing else, as a courtesy. Reese, on the other hand, felt that he had absolute authority to do what he wished."

The stage was set for an explosion. Ted had been giving Schonfeld some not-so-subtle hints over the spring months—hints that he should cool off a little, try to get along with everybody better. "Business is business," Ted said. "Don't let personalities get involved when you're making judgments." In a friendly way, he took Schonfeld aside and said, "You know, Reese, my father always wanted me to read *How to Win Friends and Influence People.* So I read it, and it changed me." And with that he handed Schonfeld his very own copy of the inspirational book by Dale Carnegie.

Ed and his son may have been deeply impressed with the folksy wisdom of Carnegie's chapters on such subjects as "Six Ways to Make People Like You" and "How to Win People to Your Way of Thinking," and precepts such as "Throw Down a Challenge," "Make the Other Person Feel Important," and "Smile" may have been inspirational for them, but to Schonfeld it was all a lot of hooey, and he never even opened the book. Had he read chapter 1, he would have learned: "If you want to gather honey, don't kick over the beehive."

When Schonfeld tried to hire Braden and Buchanan, Turner declared them "a couple of turkeys" and said he wasn't letting them on his air. Schonfeld responded, "It's not your air; it's mine." So it came as no surprise when, in early May of 1982, Turner fired Schonfeld. "You know," Schonfeld says reflectively, "Ted once said I reminded him a lot of his dad."

"Ted was asserting himself," comments board member Mike Gearon. "He was going to do what he wanted to do with his toys. . . . Ted had to stake out his territory. If you tell Ted somebody in his organization is just brilliant, you're not promoting that guy's success in the organization. Ted'll say, 'I'll get that smartass.' Ted's not magnanimous in that sense. He wants to be in control. It's *his* company."

For his part, Schonfeld hadn't been willing to bend either. "If you can't hire and fire people, you can't enforce discipline and you don't have a company," he says emphatically. Turner's on-again, off-again girlfriend Liz Wickersham tried to mediate the dispute between them, as she had earlier. "Liz said something to me that was very sad and very sweet in a way," Schonfeld recalls. "She said, 'Reese, we're all his whores. I'm his whore. Bunky's his whore. Can't you be his whore just a little bit?'"

Within a few months after Schonfeld's departure, Turner was on the telephone with his ex-president, inviting him to the Braves' Stadium Club for an exploratory conversation about returning to CNN's helm. At the meeting, according to Schonfeld, Turner kept asking, "Why are the ratings down? We want you to come back." Bill Bevins, who was also present, cracked a few jokes about Schonfeld being Turner's Billy Martin (whom Yankee owner George Steinbrenner would constantly fire and rehire). But Bevins was clearly no fan of what he perceived as Schonfeld's fast-and-loose financial style, and was soon arguing with Ted about the advisability of the idea. And while Schonfeld would linger for a while in Atlanta as a consultant to CNN, he would never be rehired.

With Schonfeld out, Turner began to flex his editorial muscles almost at once. The trial of John Hinckley Jr. for the attempted assassination of Ronald Reagan had led Turner to fume about one of his least favorite movies, Martin Scorsese's classic of urban alienation, *Taxi Driver*. The trial merely proved to him what he had always believed—namely, that movies and television exerted a powerful influence on young people's behavior.

Staring solemnly into the camera, Ted recorded a jeremiad for CNN, not unlike one he had once done for WTBS attacking the government bailout of Chrysler. It ran on May 29, and it aired eleven more times on CNN over the Memorial Day Weekend and three times on WTBS. "I am very, very concerned that this movie, *Taxi Driver,* was an inspiration to Hinckley and was partly to blame for his attempted assassination of our president," he told viewers. "Many years ago, I stumbled into that movie almost by accident and was absolutely appalled by the blood and gore. . . . Columbia Pictures and the people responsible for this movie should be just as much on trial as John Hinckley himself. . . . These sorts of movies must be stopped."

It was exactly the sort of thing that Charles Foster Kane and his real-life alter ego William Randolph Hearst had done—they, too, had spoken out from the bully pulpit of their own publications.

Turner owned CNN; why shouldn't he speak out on it? And besides, the issue was clearly so important, so *right*—right?

Not everyone at CNN agreed. Daniel Schorr, with his guarantee of journalistic freedom (and job security) felt most comfortable replying to his boss, and he put together his own editorial. "I must respectfully disagree with any suggestion advocating legislation against movie or television companies," he announced. The rebuttal aired only once, and Schorr felt that he was lucky to get even that. He recalls, "The CNN brass were furious. There were very scared people between Ted Turner and me. . . . They were worried that he would go berserk and blame them."

Turner had decided he liked this editorial business. He always enjoyed sounding off with his opinions, and now he had millions of listeners. He decided to do another editorial, an enthusiastic endorsement of *Gandhi*—the man and the movie. "Take the kids to see it!" he enthused. It was senior business anchor Lou Dobbs who plucked up his courage and tried to persuade the boss that he was subverting the editorial credibility of his own creation. "Ted," he said, "you simply can't do editorials. It's not appropriate. It's not what we're about. It's a terrible risk." Dobbs attempted to explain the distinction between objective reporting and commentary. It was all news to Turner. "To my knowledge," said Dobbs, "no one had ever talked to him about editorials in those terms before." To give him credit, Turner listened and he learned. And he's never attempted a CNN editorial since.

On June 21, 1982, the news war began in earnest, when the Satellite News Channel, with its staff of two hundred and fifty and its 2.6 million subscribers, went up on the satellite. "If it turns into a battle and drags on," wrote *Broadcasting* magazine, "Westinghouse and ABC would certainly seem to have the finances to outgun Turner if they have a mind to."

The next twelve months would be the grimmest time for CNN and all of Turner's companies. Already having trouble making ends meet, Turner's staff now had to cut special deals with the cable operators just to compete with SNC. Instead of receiving ten or fifteen cents per viewer, as at the launch of CNN, they now had to offer CNN II (which would become Headline News Service) free, as SNC had done. Then, with the battle growing increasingly cutthroat, the networks actually began to *pay* for viewers—to bid against each other, spending ten, twenty, fifty cents for each cable-system viewer in a game of financial chicken to see who would flinch first. They would make their money on advertising alone.

Desperate to get CNN II up on the satellite, Turner had given Warner Amex the right to sell all CNN commercials in exchange for a transponder. Gerry Hogan's staff had been pulling in solid sales figures, but when Warner Amex took over, sales plummeted. The movie company knew Bugs Bunny, but they didn't know cable. "We were giving up a million dollars a month," says Schonfeld, "because Warner couldn't sell water in the desert." Producer Ira Miskin remembers coming to work for WTBS in September of 1982: "The very first week I was there, our corporate electric bill was so far in arrears that the electric company was threatening to pull the switch at five o'clock on Friday. They would have shut down CNN, TBS, everything. Well, TBS, if they went off the air, they'd lose a lot of money, but it wouldn't be the end of the world. But CNN was locked in this death-battle for survival— broke, bleeding money all over the place. And you shut down CNN for one day, it's gone. . . . That week Bevins opened up every pocketbook, wallet, and piggy bank to pay off part of what was owed to the electric company." Stretched to the limit, Turner's company relied on the direct-response advertising at WTBS. Gerry Hogan recalls, "If not for the Ginzu knife and the bamboo steamer, Elvis Presley and Boxcar Willie, we would have been out of business."

During late 1982 and early 1983, Hogan, who oversaw WTBS advertising, would frequently find himself on the receiving end of phone calls from Ted Turner or Bill Bevins. "We've got a real cash-flow problem," they'd say. "Do you think you can get some people to pay us in advance for advertising?" And so, while prepaying for advertising spots is an unusual practice, Hogan would gamely troop around the country trying to convince advertisers to ante up in advance. "We'd give them discounts in the twenty-percent range," he remembers. "You had to give them some incentive. Nobody would do it just to be a nice guy. But you had to be careful, because if you presented it incorrectly people would get very nervous and perhaps not want to advertise with you at all. You had to give confidence that everything was fine and this was just a little temporary thing. . . . In fact, our backs were against the wall; we were in deep trouble. I'd come back on Tuesday or Wednesday with checks in my briefcase that we'd deposit so we could make payroll for that Friday."

On one or two occasions, they were unable to make their full payroll. Mary Alice Williams remembers paying the staffers in the New York bureau from her own personal checkbook. And according to Hogan, "There were even a couple of times when people would forgo a paycheck, although that was only done on a voluntary basis. It was very rough times financially."

Meanwhile, as CNN II head Ted Kavanau recalls, "the fight got uglier, because they [Westinghouse] were picking up our *people,* too. It became very bitter. After the union struggle, many of the people at CNN who had been behind the union were picked up by SNC. . . . The war was draining us."

In the middle of all this, Turner was frantically running around trying to boost staff morale and keep the cash flowing anyway he could. One of the tricks he had learned back in the early days was to sell his company. Or at least, that's how it appeared: "He'd let guys come in and talk about buying the company," says Mike Gearon. "When you're losing money every year on paper [while building a company with tax-sheltered dollars], he had to do something to show that he was creating value. This was a way to show he was creating value." In other words, Turner would pretend to sell the company, then take the best offer to the bank and borrow against it.

That was exactly what happened when 3M made an offer for Turner Advertising. "3M thought they had bought it," Reese Schonfeld recalls, "but Ted walks the contract into a bank and says, 'They're willing to pay me thirty million; won't you at least up my loan to twenty-two million?' 3M will never forgive him." Eventually, Turner would convince Citicorp and Manufacturer's Hanover to lend TBS $50 million over three years, but to get it he had to pay a $3-million fee and tie up all of WTBS's accounts receivable for those three years. It was yet another huge risk, and the crushing load of all this debt started to weigh Turner down.

"Ted Turner's tired a lot lately, it seems," read one article from the period. "Winging toward New York at seven in the morning, he buries his head in his hands and moans, 'I'm going to collapse. The only question is, Will it be tomorrow, next week, or next year? I mean, I'm just so tired I can't make it.'"

But each time it seemed as if Turner might slowly collapse, he'd suddenly reappear, his batteries recharged, his mouth flapping, his arms waving, his live-wire body all juiced up again. "They want to beat me," he'd exclaim. "They'll have to kill me first!" He'd show up at yet another cable convention, attacking ABC as purporting to enhance cable with its Satellite News Channel but, he railed, "in the other hand is a dagger to put us away." He'd pass out buttons and T-shirts and pictures of himself in a cowboy hat with a guitar and a slogan: "I WAS CABLE WHEN CABLE WASN'T COOL."

And even though he was deep in debt, he continued to spend money on new projects. Ira Miskin had brought him the idea for Turner Educational Services, linking TV and education, and although the price tag was $1 million, Turner OK'd it in less than ten minutes. He also signed on for the *Portrait of America* documentary series, with its staggering $20-million budget.

Ted's gung-ho attitude throughout 1982 and 1983 is best illustrated in a story Reese Schonfeld tells about the weeks right before he left. The Falklands War was raging—a naval battle, Ted's kind of battle. Turner loved this war. CNN was originating in Buenos Aires and London with split screens—a breakthrough in war coverage—and still Ted asked Reese, "Can't we do more?"

"Yeah, Ted, we could do more," Reese said. "But it's going to add a million dollars a month."

"Go ahead and spend the money."

"Ted, you don't have the money," Reese replied.

Turner told him, "If I have to, I'll mortgage my house."

So Schonfeld marched down to Bill Bevins' office and announced, "Bill, we're going to spend a million a month more while the war goes on. Ted wants to do it. It's going to be great."

Bevins stared at him. "We don't have the money," he stated flatly.

"Ted says he'll mortgage his house."

Bevins looked at Schonfeld and shook his head. "That house is already mortgaged," he said.

Tight as finances were, Turner continued to press ahead, and some of his most successful moves were sneak attacks. Spying on SNC, Turner's team discovered that their sales staff was taking a Christmas vacation. So TBS came up with a blitz plan. "If you sign up for our two news networks over the next three weeks," they told the cable industry, "we'll pay you a dollar a sub[scriber] for the next three *years*." It was, according to Terry McGuirk, "a grand, grand play by us. That was a big move."

Meanwhile, Ted pulled another sneak attack in Washington. The government was again threatening to raise royalty fees for programs, perhaps as much as six times, and Turner had been lobbying individual senators hard, showing up in their offices to beg and cajole them not to pass the increase. He even fell flat on their floors, exclaiming that if it passed, "I'm dead!" The increase was strongly backed by broadcasters and producers, but on December 17, Turner succeeded in persuading a fellow Georgian, Republican Senator Mack Mattingly, to slip an amendment blocking the measure into an unrelated omnibus emergency-spending bill. Network and movie-industry lobbyists were caught flatfooted and grimly labeled Turner's sneak move "an end run." John Summers, executive vice president of the National Association of Broadcasters, said that the amendment was a case of "one guy with a lot of charisma just sweeping through the Senate and captivating people to vote his way."

In the early months of 1983, the cable-news battle between CNN II and SNC entered its final phase. "There was a damn good chance this war was going to drive us both out of business," says Terry McGuirk. "We were killing each other, barely hanging on. We figured it was gonna wind up with the both of us dead. We couldn't coexist. We knew we were within probably a year of going under, maybe less." ABC-Westinghouse publicly admitted to a $60-million loss—twice what they had originally planned to invest. According to Robert Wussler and others, the actual loss was closer to $100 million. ABC's stock began to falter. And as Turner had predicted for the split command, what followed was bickering. Meanwhile, Turner Broadcasting stock zoomed up by nearly 40 percent, as investors began to see which way the wind was blowing.

It was Turner's opportunity to seize the moment, and he did. Moving quickly, he offered ABC-Westinghouse $25 million to buy out their service and their 7.5 million viewers. And within days, ABC-Westinghouse had caved in and sold off SNC. A single entrepreneur had convinced two giant conglomerates with fifty times more economic muscle to back down. A joke began making the rounds that Westinghouse had financed CNN twice—once in 1980 with Channel 36 and once again in 1983 with SNC. "On a level playing field," Turner exulted, "ABC and Group W got their brains kicked out!"

"I don't know how close he was to throwing in the towel near the end of the CNN-SNC battle, but we were pretty broke, and pretty desperate at that point," recalls Ira Miskin. "I think more his bravado than his maneuvering got the suits at ABC and Westinghouse to flinch. Because, in truth, they had the pockets; they could have beaten us. They could have been the network. And I think he was very lucky that they believed that he would go until the last drop of blood."

In the end, then, it seems to have been Turner's bravado that made the difference. For who could look at the gleam in his eye and not wonder if this was not the wrong adversary to challenge in a game of chicken, if this was not the wild man who would just keep going and going until you both crashed head-on and exploded in flames?

"How much of it was luck?" asks Mike Gearon. "To what extent was this all destined to happen? When you go back and examine it, there had to be a certain amount of fortunate timing. But how much of it was just because of this guy's tremendous will? Ted's successful because he's so relentless. Ted's a guy that if you went to war, you'd want him on your side. Tomorrow I'd venture if you were to scratch the great military generals and redo World War II, he'd be up there with the all-time greats. Because he knows how to go to battle. And he knows how to win. Was Patton a great salesman? I don't know if he was a great salesman or not, but he darn sure was going to scare the shit out of you, and convince you that he was going to win."

In his second great battle, Ted Turner had won again. His victory, however, had come at a cost.

CHAPTER 17

THE COST, as always, was paid by his wife and children. While Turner was tallying impressive victories on business battlefields across the country, he was amassing casualties on the home front.

His family—his own kids—took to calling their father "Hurricane Turner," as he continued the pattern that he had begun in the early years of his marriage to Jane, staying away from home for weeks on end. He would miss one Christmas after another, first for the Sydney-Hobart ocean race and later in pursuit of various cable business deals. Then he would blow back into town and rattle the doors and windows.

Ever since the days he spent at Binden with his father and stepmother, Turner had loved plantation life, but once there it was never long before he began to get bored. Sometimes he'd round up the whole brood at dawn and take them off on interminable hiking excursions—forced marches across the enormous Hope Plantation—exhausting his children with his sheer energy and enthusiasm. Other times, he'd fly off the handle, shouting at Jane or Rhett or Teddy as if they were inept employees, then cuffing them on the head. "Dad was often away in those days," recalls Teddy. "When he was home, there was a lot of yelling and tension and getting smacked around. I never could decide which was worse—having him away a lot or having him home."

For Jane Turner, there was no question; it was the absences that stung even more than the occasional blow. Being yelled at was hard enough, but being ignored was worse. Alone with the children at the plantation, surrounded by thousands of acres of woods and marshland, she found the silences long and painful. Everything she took pride in—being a sweet, gentle, devoted wife and raising happy, well-rounded, good kids—none of these achievements seemed to have much value for her husband. He liked his wife, and he loved his kids after a fashion, but the bottom line was that none of them was terribly important to him. Work—and other women—were more exciting.

Nancy Roe, who in 1963 had helped bring Jane and Ted together, shakes her head and admits, "It may have been a big disservice introducing poor Janie to Ted. He played around all their married life, for goodness sake! Everybody knew it, including Jane. But she liked being Mrs. Ted Turner. She liked raising her children. . . . If she didn't put up with it, it was 'Get out.' And she wasn't ready for that. . . . Back then, if you had some problems with your marriage, you kept your mouth shut. You didn't go around talking about it. And Janie was quiet."

Still, all that sorrow had to be directed somewhere. And so, over the years, the occasional social drink for Jane Turner grew into two and three and more; out of control, her drinking deepened into full-scale alcoholism. The liquor seemed to blunt the sharp edges of her pain, fill up some of the emptiness. Alone at home or in a crowd at cocktail parties, Jane drank to make the unhappiness go away. And gradually the liquor came to free the tongue of even this most reticent and demure of well-brought-up Southern women. More than once, her drunken pleas for help startled guests and staffers at Turner Broadcasting parties. "So you think I'm crazy to stay married to Ted?" she asked a surprised local reporter at one such bash at Atlanta's Peachtree Hotel.

"I remember being in the ladies' room at Techwood once when Jane Turner came in," recalls former CNN staffer Fran Heaney. "She'd had a few cocktails, and she just went on talking and talking about Ted. I was embarrassed, but it was obvious she wanted to talk to somebody. She just went on and on about his girlfriends, about how could he do this to her? She was so very, very sad. She looked like she was going to burst out crying."

"Jane was soft, sweet, vulnerable," says ex-CNN anchor Bill Zimmerman. "I'd see her separately from him—at a party that either he wouldn't be at or he'd be at the other end of the place.

She'd be feeling sorry for herself and drinking very heavily. She seemed desperate for anybody to talk to. She just reached out to everybody, and poured out her sobs to whomever would listen."

At CNN, according to one former staffer, "We used to say, 'Lock up your wives and daughters. Ted is on the loose.'" The female shape, in all its infinite variety, was endlessly fascinating to Turner. He'd make passes at women at ball games, on airplanes; he'd try to pick them up almost as a reflex. Although long-legged peroxide blondes seemed to be his personal favorite, he wasn't all that choosy. Flirting was second nature, and he might be captivated by almost anyone who happened to be sitting next to him.

But flirting was the least of it. Colleagues at cable conventions, players on his ball teams, a number have stories of walking down hotel corridors and seeing Ted standing in the open doorway of his hotel room wearing nothing more than a towel around his waist, while behind him one flashy blonde or another lounged about in a scanty robe that barely covered her charms.

Staffers and outside reporters couldn't believe how brazen he was, how public he was about all his various female companions. Turner became known for bringing his girlfriends to the ballpark and installing them in the press box or the stands, sometimes not far from where he was sitting with his wife and kids. "I think he'd seduce the first woman tenor in the choir if he got a chance," said one observer. "He and Elmer Gantry would have gotten along famously."

Choirwomen aside, Turner was even reckless enough to proposition the wives of his employees. Braves pitcher Dick Ruthven became so incensed when his boss tried to seduce his wife that he made a large scene, denouncing Turner publicly, demanding to be traded. He was. A Turner employee tells of traveling to San Diego with her husband, her boss, and several others for a convention and having Turner proposition her. "C'mon up with me to Los Angeles," he said. "I'll show you around. You can stay with me at my hotel in Santa Monica. It'll be great." He stopped only when her husband, who had stepped away for a minute, suddenly reappeared.

Ted Turner hated to be alone, and he loved women—or-namental, exciting, and thoroughly satisfying for as long as the conquest took. With a narcissist's need to be admired, he single-mindedly notched up sexual "victories" as if trying to offset the loneliness of his earlier years. As his first wife, Judy Nye, commented, his lovemaking made up in volume for what it lacked in quality. For him, an attractive woman was a necessary companion. Not that he paid too much attention to any one of them. Turner's attitude, recalls TBS president Robert Wussler, was "Women? Great, in their proper place." But a female who wanted to talk business or debate ideas or question his theories would be met by a narrowing of the eyes, a tilt of his head, and a lazy, careless insult: "I don't want to listen to you," he'd say. "I don't listen to women."

Perhaps his relationship with a professional journalist like CNN's New York anchor and bureau chief Mary Alice Williams is most indicative of his attitude toward women at that time. Williams, who would later become a friend (though not a girlfriend), still laughs when she recalls their first meeting: "Reese had hired me, and I flew down to Atlanta to meet Ted. It was clear from the second the door opened that he had been told only that he was having a meeting with the New York bureau chief. He was told nothing else. Ted leaped up from his desk when he saw me, as though red fire ants had bitten him. 'A woman!' he said. 'A woman! You didn't tell me it was a woman!' He stood behind his desk, screaming and ranting. I never even walked in the door; I stayed in the doorway, which is where you're supposed to stand in an earthquake."

A few months later, as CNN was preparing to go on the air, Robert Wussler threw a party at his Park Lane Hotel penthouse on Central Park South. Mary Alice Williams had in fact been hired, and she was there, and so was Ted Turner, who once again had had a few too many vodkas. Walking up to Williams with a sloppy lecherous grin on his face, he slipped his hand up her dress in full view of the startled party crowd and grabbed her crotch. According to witnesses, Williams didn't hesitate. Cocking her right fist, she cracked him in the jaw, and Ted Turner went down. Her six-inch punch dropped him to his knees. Leaping to her honor, fellow-newsman Tom Snyder pressed Turner's forehead into the floor until he apologized profusely. After that, Williams took to wearing her "nun suit" (a discreet black business suit with a high white collar) whenever she was around Ted, and began applying her "twenty-five-minute rule": Never stay in the same room with Ted for longer than twenty-five minutes, lest things get out of hand.

From the mid-1970s to the mid-1980s, Turner was known to employ various high-priced call girls. One TBS employee who attended the national cable convention in Las Vegas with him talks about how they were met at the airport by a voluptuous blonde in a stretch limousine, whom Ted had hired out of Los Angeles. Clad in a skintight jumpsuit, she ran her hands through her hair, wriggled her curvaceous form, and purred, "Call me Su-*zanne*." When they arrived at the hotel and Turner went up to the podium to give his speech, Suzanne promptly offered to service his employee as well. Whispering breathily in his ear, she ran her nails along the seam of his pants with a wink and a smile, and then handed him her business card.

If Ted Turner had a taste for prostitutes and pinup girls, he certainly wasn't unique in that. But what set Turner apart was his public posture as a family man who advocated family values. He not only advocated those values, he set himself up as their champion. Rare was the interview—radio, TV, or newspaper—in which Turner would fail to blast "sex and smut" and the "stupid sleazy sex" on networks like CBS, which he likened to "a cheap whorehouse." A self-proclaimed ally of Jerry Falwell and Donald Wildmon, Turner piously claimed, "I don't like off-color humor." He seized each opportunity to lambaste not just dirty jokes but anything that he believed set "terrible role models for children." He strictly forbade his own kids to watch R-rated movies.

For Turner, there seemed to be no connection between his private life and his public pronouncements: one was about fun, and that was for himself; the other was about public policy, and that was for everybody else. Almost in the same breath, he would point accusatory fingers at "sleaze artists" and then admit, "I love pictures of nude girls." Turner, in fact, was not just libidinous, he was incurably libidinous, and his proclivities ran toward the exuberant. "I photograph nudes myself," he confided to *Playboy*. Some sources claim that he also had a taste for videotaping himself in the sex act. He'd set up his camera, pop in a videocassette, and capture himself and his lovely partner of the moment in living color. Over the years, he came to assemble a collection of these blue movies. "If you ever want to see some great sex tapes," he once told an aide, pointing to a stack of cassettes, "I'm starring in these."

The contrast between his public pronouncements and his private life extended to other areas as well. He lectured his baseball team about drugs, telling them, "No more" (adding, for his Latin players, "No mas"). One newspaper account read:

> The Braves are still reeling from the drug arrest of Pascual Perez, a starting pitcher who is in jail in the Dominican Republic. Outfielder Claudell Washington has also admitted that he has used drugs, and relief pitcher Steve Bedrosian had conceded that he has experimented with cocaine.
>
> Turner admitted that he was mistaken a year ago when he declared his team drug-free. "I was absolutely [surprised]," Turner said. "I just assumed we didn't have a problem because they're such a good bunch of kids. . . ."

He told the *Savannah Morning News,* "I pleaded with those guys to stay off drugs."

But like his "good bunch of kids," Turner himself, according to several aides and acquaintances, did his share of experimenting with drugs. Partygoers from Hong Kong to Atlanta recall seeing him light up joints from the 1970s well into the 1980s.

Yet despite the double standard, there was a certain inescapable innocent charm to the private Ted Turner, a peculiarly Southern courtliness that—combined with his energy, his lean, rugged good looks, and the fact that he was rich—made him attractive to men and even more so to women. "Ted's the most interesting guy in the world," enthused former girlfriend J.J. Ebaugh. Another ex-girlfriend, Kathy Leach, concurred: "Our first real date, he flew me down to his plantation for a candlelit dinner. Then we went out for a walk in the moonlight. I'll never forget it."

Although scores of women would come and go, there were a few who (at different times) played a larger role in his life and charmed Turner as much as he charmed them. Through his sailing days, the Parisienne Martine Darragon was one such woman, and in his early CNN years Liz Wickersham was another. A model and would-be actress, Wickersham had been seeing Ted on and off for over three years when she appeared on the cover of the April 1981 issue of *Playboy* in a sleek plum-colored teddy, her pert breasts straining against the satin fabric, her dark hair (the original shade) cascading down to her shoulders, and a come-hither pout on her face. She showed off a lot of skin and a little negligee, which led the *Playboy* writers to deep-breathe, "Her teddy just goes to prove what we've always said: Less is more."

For Ted, who could resist anything but temptation, the voluptuous Liz became a frequent companion, accompanying him on those many occasions when his travels took him away from his family—to Los Angeles, Las Vegas, Greece, Cuba. He liked to show her off to his cable colleagues and to buddies like Fidel Castro. With her good looks, he thought, she belonged on TV, and she thought so, too. In short order, Wickersham was hosting a string of programs on WTBS and CNN: *The Lighter Side, Showbiz Today*, and *Good News*. Everyone soon knew she was Turner's companion, and there were the inevitable resentments, but Liz was sweet and vivacious enough to win over many of those who worked with her. Unfortunately, pretty as she was, she could never be termed a natural on camera, at least with clothes on, and every show she was in flopped.

Along with the excitement, the international travel, and the glamour of being Ted Turner's girlfriend, Liz also knew what it was like when he got into one of his uncontrollable tempers and his mood lurched suddenly toward the black end of the scale. She confided that once on a trip to Greece, Turner had gone ballistic and booted her in the shins like a field-goal kicker trying for a 50-yarder.

No doubt equally depressing for Liz was the realization that although Ted was cheating on his wife to be with her, he was also cheating on her to be with other women. CNN anchor Lois Hart tells the story of one Las Vegas convention where Turner showed up with Wickersham on his arm, proudly displaying her to his cable colleagues. But then, across the room, another woman caught his eye. "This one had to be a pro," Hart remembers, "all done up in leopard-skin pants, with enough makeup to cover a battleship. Ted takes one look at this babe and his eyes pop out of his head. Bye-bye, Liz! That's one of the things that made him so appealing. He had a capacity for work beyond anything I've ever seen, but he also had a capacity for play nobody on this earth has ever seen. You did have to feel a little sorry for Wickersham. She really had no idea what she'd gotten into."

Turner had come up with the idea for *Good News* around 1983, and he planned to use it as a showcase for Wickersham. He brought in high-priced outside talent, veteran CBS journalist Zeke Siegel, to help produce the program and especially to coach Wickersham in her shaky on-camera delivery. "I want you to work with Liz," he told Siegel. "Make her as good as she can be. Make her a star." Then he added confidentially, "If you can make her a star, make her stand on her own two feet, then I can get rid of her." A few weeks later, Siegel was working with Wickersham in her dressing room, going over a script, when she suddenly looked up at her coach. "Do me a favor," she begged. "Make me a star. Make me a star, so I can get rid of him."

In the end, even Zeke Siegel would not be able to transform the lovely Liz Wickersham into a star. But she would find her own kind of equilibrium with the roving Ted Turner. As much as he wandered, so would she. Unlike his wife Janie, Liz was not cut out of the old, long-suffering female mold. She was a modern, independent woman: "Her charm for Ted was that she wouldn't be faithful; she wouldn't be only his," Reese Schonfeld says. "She was involved with four men, as I understand it, at that time. A Hollywood guy, a congressman, a New York guy, and Ted. And she said, 'Ted, when you want to marry me—when you get a divorce and you want to marry me—then we can talk about it. Otherwise, here's who I see. You know who I see. I'll try to see you first, but sometimes I can't.' . . . That was the reason she stayed in his life for so long."

Meanwhile, Ted had begun another serious relationship with yet another blonde. He had run into Jeanette (J.J.) Ebaugh on the 12-meter sailing circuit, where she accompanied celebrated sailor Tom Blackaller, a rambunctious cavalier after Turner's own heart. When the 1980 America's Cup competition returned to Newport, and Blackaller and Turner found themselves sharing the same dock, Ted was quick to spot J.J. He wasn't the only one. Tall and tan and young and lovely, the tawny blonde caught the eye of many at Newport with her open, fresh-faced beauty and her lean, supple athleticism. "J.J. was the best-looking thing on the dock that summer," recalls Barbara Pyle, who was photographing the races for *Time*. More than that, Ebaugh also had a certain native intelligence. Her father was a professor of medicine, and she had grown up in the academic communities of Dartmouth, Boston University, and Stanford. She liked to read. But above all, there was a wild streak in her, a hunger for adrenaline and speed. She had spent her earliest years skiing and climbing in the White Mountains of New Hampshire; off the coast of Massachusetts, she learned to sail, and by the time she showed up in Newport she had become a competitive race-car driver and a commercial pilot. Ebaugh was twenty-three; Turner was forty-one. As Barbara Pyle snapped them on the dock, she thought they would look good together—the old rascal charmer and this golden girl. Here was someone who also believed in motion, who could keep up with Turner stride for stride. And when J.J. smiled, it was all teeth.

As Ebaugh's relationship with Blackaller gradually wound down, Turner beckoned to her from the other side of the dock. That winter, the winter of 1980–81, he invited Ebaugh aboard the *Tenacious,* sailing out of St. Petersburg, Florida, for the Southern Ocean Racing Circuit. "Being passionately interested in sailing, it was a wonderful invitation," she recalls, "to be on *Tenacious* with one of the greatest helmsmen of all time." But the invitation was clearly more than platonic. "Him platonic!" she laughs. "With Ted Turner! You've got to be kidding."

Although he had a wife and a lover and assorted other women, Turner decided he needed to have Ebaugh around as well. In fact, Ebaugh would become his own personal pilot. So what if he didn't have a plane? It was like something out of a Beatles song—"Baby, You Can Drive My Car." Just like Ringo, Ted could sing that he had found a driver and that was a start.

Within a matter of months, Ebaugh was installed in Atlanta. She moved into Barbara Pyle's house, and all three became good friends. With Turner's financial help, Ebaugh bought herself a sleek black race car and picked out a plane—a used Merlin IIB six-seater—for him. Soon she was flying him to speeches and meetings, shuttling Ted and Jane and the kids down to the Hope Plantation for weekends, and bringing him back to Atlanta on Monday morning, just as Jimmy Brown used to chauffeur his father.

What a strange and uncomfortable situation that had to be for Jane, or any wife—being flown around by your husband's own private Pussy Galore! It was perhaps at airports most of all that Turner's life resembled a French bedroom farce. Reese Schonfeld, among others, recalls flying to a cable convention with Turner aboard a commercial jet. "We got on the plane in Boston; Ted kissed Janie goodbye. We flew down to Houston. And Liz Wickersham met him at the airport in Houston. It's just like Ted, kissing good-bye to Jane and hello to Liz."

Another staffer recalls a story that became legend around TBS. Ted and Janie had driven out to Atlanta's Hartsfield Airport. She was going to catch a plane down to the plantation, while he was heading off to New York. Ted dropped Janie at the gate, kissed her, and hurried over to the gate for the New York flight, where Liz was waiting for him.

But a storm was descending on Atlanta; the sky turned an ominous gray-green color, and Jane started to get nervous. At the last minute, she decided she had no intention of braving that weather in a tiny propeller plane. So she hurried across to the New York gate. She'd surprise Ted, and they'd go to New York together for a fun weekend. But when she got to the gate, there was Ted with Liz, who turned bright red. "Jane, what are you doing here?" she asked. "What are *you* doing here?" Jane fired back. "What the hell am *I* doing here?" Ted said to nobody in particular, and sighed. As things turned out, he flew up to New York with both women.

According to family friends, "Jane wasn't supposed to know what was going on." (Ted would laughingly say to friends, "You know, I've never lied to anyone but my wife.") But it would have taken someone deaf, dumb, and blind not to notice. Seeing him on TV smooching with Liz in the stands at a Hawks-Knicks game was bad enough, but getting this pilot J.J. rubbed in her face each week was worst of all. Even the gentle, soft-spoken Janie had to erupt: "One time," friends recount, "J.J. was going to bring them to their beach house, and Janie went berserk at the airport, just yelling and screaming at Ted, she was so pissed off."

Unfortunately for the second Mrs. Turner, she was battling a force of nature. Ted Turner was too erratic, too unruly, too powerful for her to handle. She had no idea how to control him; she had no leverage. So, much of her rage was directed inward. One evening, while at their beach home on St. Phillips Island, off the South Carolina coast, she began to drink and then headed out along the shore to visit friends. Unhappy and disoriented, she fell into a tidal creek and got swept into the marshes. The more she struggled, the less she seemed able to extricate herself, and the more her feet and legs were cut up on the shells and barnacles in the marsh. Fishermen found her the next morning clinging to branches, lucky to still be alive. She was hospitalized for shock and her infected cuts.

Other friends bore witness to similar instances of self-destructive behavior: "Teddy had invited us over to the house, the Hope Plantation, and when we got over there Jane was kind of distant," Billy Roe recalls. "She was banged up—all scratched up. And Ted said something about 'Janie's problem.' And the kids said, 'Well, Mom got lost in the woods last night.' Evidently, Janie had been drinking, and she got lost for about twelve hours out in the plantation, just stumbling around in the dark. She was all scratched up and she was on crutches. She was also extremely embarrassed about it."

If her friends were worried, even her husband's colleagues felt concerned about her. At one point, Reese Schonfeld offered Jane a job at CNN. He believed that if she had a career to call her own, some small portion of independence, she could begin to be more positive about herself. He felt sorry for her. Jane went home to talk the idea over with Ted, and Reese never heard anything more about it.

Old acquaintances like the Roes saw Janie's decline, and they were upset. "Nancy and I went down to Palm Beach Polo," Billy Roe recalled, "and there was Teddy being interviewed." Located in West Palm Beach, the Polo Club attracted the rich and famous. On this occasion, Prince Charles was on hand and so was Gregory Peck, but once again Ted Turner was monopolizing the cameras. When he spotted his old friends, he came over immediately, his usual exuberant self, and proclaimed, "You gotta see Janie. Janie's here. She'll die if she doesn't get to see you!" According to Nancy Roe, "We're following Teddy right up to their box. We go around a corner, and coming down the stairs is Jane. And Ted says, 'Janie, look, here are the Roes!' Janie looks, but she's so drunk she can't focus." Billy Roe adds, "It's two o'clock, and Janie didn't even know where she was." Nancy says, "She just mutters, 'I gotta go to the bathroom,' and she stumbled on past."

Janie had grown up in the 1950s believing that a wife puts her faith and trust in her husband and finds her identity there as well. With a husband like Ted Turner, by turns abusive and caring, devoted father and outrageous philanderer, she had to be wondering just what her identity was. She took some solace in her children. Ted, for his part, had neither the time nor the inclination to think too deeply on all these issues. His own father had shown him how to be a businessman, how to build a company, but Ed and Florence had provided seriously flawed examples of how to build a successful marriage. "Ted was a horrible husband," Mike Gearon says. "But

I just don't think he knew the rules of a marriage. I don't think Ted knew anything more about marriage than he did about world affairs or anything else. I think in a way he's almost like the guy you take out of the jungle who's been raised by wolves. There's almost a sense about him that there's a wild man back there who doesn't know all the fine points, who isn't aware that you should be afraid of this, or subtle here or coy there. . . . I just don't think he knew how to be married. I don't think he knew anything about it."

Beyond all the mistakes he would make, intentional and unintentional, Ted Turner was fundamentally more interested in the excitement of movement and change than in roots, more involved in seduction and conquests than in relationships. Romancing, conquering, abandoning—that was a pattern he repeated in encounters with everyone, from the Braves' Gary Matthews and CNN's Reese Schonfeld to his sexual partners. Friends would marvel at the way they'd be forced to baby-sit his dates at parties, while he went off to pick up yet another woman. When one old friend talked to him about love, Turner, a man with one wife, one ex-wife, two mistresses, and countless one-day companions, said, "I wouldn't know. I've never been in love."

Whether or not he was motivated by some vestige of love, or just an abiding sense of loyalty, Ted did finally, in the early 1980s, agree to see a marriage counselor with his wife. They went to an Atlanta psychiatrist, Dr. Frank Pittman, a clinical assistant professor at the Emory University School of Medicine and an adjunct associate professor at Georgia State University. Dr. Pittman prided himself on his fashionable roster of rich and successful clients, and his specialty, extramarital affairs, was particularly apropos.

Bright and somewhat unconventional, Frank Pittman had established a reputation for being able to handle high-profile patients, a reputation he burnished by seeking to make himself high-profile as well. Over the years, he would spend long hours with the Turners. In that time, he published three books, and while of course no client names are mentioned, certain details suggest just how rich a case study his new patients were for the doctor:

> Many women are attracted to philanderers, even when they
> should know better. These men can be charming, wonder-
> fully comfortable and adept at sex, and can seem to be

winners in the world. . . . If the philanderer is successful
enough in life and amasses enough power, the wife may feel
she has a pretty good deal. She may get only a fraction of
his sexuality, but she gets all the legal rights and prestige
that would come to a real wife in a real marriage. The price
she pays in anxiety, loneliness, and humiliation may make
the rewards look meager, though.

Pittman's prognosis for such patients is clear: "I am not optimistic that long-standing, dedicated philanderers (the sort who
started in the first few years of marriage . . .) will change easily. . . . Philandering is addictive behavior, and, like all addictive
behavior, is difficult to change without great honesty and willingness to put yourself under someone else's control."

In fact, Pittman could do little for the relationship of Ted and
Janie. But he was more successful in getting Ted—the man who
had once said, without irony, "I don't wish I had more time to
reflect. I'm involved in such an intensive series of negotiations and
business deals, none of my mental powers can be spent reflecting"—to finally stop and look at himself.

Pittman helped Turner identify two crucial legacies he had received from his father. The first was biological. Ted's bursts of
energy and elation, his grand schemes, large sexual appetite, mood
swings from hyperactivity and excitement to despondency, and
preoccupation with suicide—all these were classic symptoms of a
mild form of a manic-depressive disorder known as cyclothymia.
And manic-depressive disorder is known to center in families; many
researchers believe it is inherited.

Manic-depressive disorder is not uncommon. In fact, it is now
generally believed that many great leaders and artists suffered such
symptoms, among them Winston Churchill, Ernest Hemingway,
John Berryman, and Alfred Hitchcock. Looking at the voluble,
erratic Ed Turner and his equally wild son Ted, it's hard not to
surmise that the father passed on his manic genes to his boy. But
in Ted's time, unlike Ed's, there was a treatment for such mood
swings—lithium, which smooths out the peaks and the valleys,
guarding against both the mania and the depression.

By the mid-1980s, Ted Turner was taking lithium under Dr.
Pittman's supervision, and according to those closest to him it made
a significant difference. On the negative side, TBS colleagues like

Martin Lafferty say they noted "a dramatic decrease in energy." But according to J.J. Ebaugh, the change was entirely for the better. "Before, it was pretty scary to be around the guy sometimes, because you never knew what in the world was going to happen next," she says. "If he was about to fly off the handle, you just never knew. That's why the whole world was on pins and needles around him. But with lithium he became very even-tempered. Ted's just one of those miracle cases. I mean, lithium is great stuff, but in Ted's particular case, lithium is a miracle."

The second legacy was behavioral and psychological, and it concerned the long shadow Ed had cast over Ted's life, by beating him and loving him, by molding him and pressuring him with his dreams about the future of the business. Ed had given Ted whatever it was in his power to give (a business, a multimillion-dollar fortune, a gift for oratory, and all his affection), and some not-such-good gifts as well. Ted was the child of an alcoholic, a philanderer, and a suicide. Inevitably, the dark inheritance took its toll.

For all the emotional turmoil of his early years, Ted, whether consciously or not, was following the blueprint of his father's child-rearing technique, down to some of the smallest details. Like Ed, he wanted his children to understand the value of a dollar. Just as Ted had mowed the grass around Turner billboards on blistering summer days, so too would his children. "We all worked, even the little kids," his son Teddy remembers. "I had to come home from school and pick weeds for an hour every day—no ifs, ands, or buts." While on vacation at the plantation, they would be sent out to clear the duckweed and the finger-slicing palmetto brush, in temperatures that soared into the triple digits, earning a dollar an hour for general labor and two dollars an hour for pulling out the duckweed. Despite this minimal wage, they, too, might be expected to pay rent when they got older.

As Ed had, Ted challenged his son Teddy to stay sober through college with an inflation-adjusted bet of $15,000. Unlike Ted, Teddy actually won the bet. ("It's the only thing I ever beat my father at," he said.) And like Ed, Ted also insisted that Teddy go to boarding school at McCallie, whether he liked it or not. When Rhett and Beau were of age, he wanted them to go, too. Janie and Ted took up an argument that must have sounded amazingly like a replay of Ed and Florence thirty years earlier.

"I don't understand why they have to go off to school, when

they could have just as fine an education here in Atlanta," Jane insisted, in one of a series of heated exchanges. Reese Schonfeld's wife Pat had even helped her pick out a promising local school for the boys.

"Why? To get them away from their mother, that's why," said Ted. "Boys shouldn't be around their mother too long, it makes them into girls."

"That's the most ridiculous thing I ever heard, Ted," Janie shot back.

"But it's true."

As it turned out, only Teddy attended McCallie. It was one of the rare instances in which Jane would, at least temporarily, prevail. Eventually, all three boys would follow their father's wishes and attend South Carolina's all-male military institute The Citadel, known to its critics as a stronghold of racism, sexism, and militarism. As described in the school handbook, however, The Citadel was "a fortress of duty, a sentinel of responsibility, a bastion of antiquity, a towering bulwark of rigid discipline." It was just the sort of place Ted wanted for them—a school where their fellow cadets were, for the most part, from conservative families in small towns in the Carolinas. Ted didn't want things to be too comfortable for his boys. They needed discipline plus a bit of insecurity. "I wanted it to be harder for my sons than other kids," Turner says. And it most definitely was harder for them at The Citadel, which, at least at the beginning, all three hated.

Like Ed, another way he'd make it hard for them would be at the end of a belt, venting his own brand of tough love even on the littlest members of team Turner. "I remember once when we were driving down to the beach," says Teddy. "For a change, Dad was in a good mood, and he made a comment about how unusual it was that he hadn't had to give any of us kids a whipping in quite a while. He said somebody was gonna have to get whipped. We thought he was joking. But I remember none of us breathed a word. Finally it was Jennie—the youngest, mind you—who volunteered. Dad stopped the car right there, put her on the hood and whipped the daylights out of her. My stepmom went ballistic."

Dysfunctional? "Basically, it was a nightmare," says Turner's youngest son, Beau. "Dad was out of control, and Mom didn't know how to deal with all of us. We each found a way to hide. Teddy hid out in the basement reading about sailing, and I grew my hair and used to blow up mailboxes. The interesting thing to me now is that all that tension made us kids incredibly close. We sort of protected each other against our parents. Dad wanted us to be individuals and tough, and we were."

It was Ted's friend Peter Dames who first took him to see *Citizen Kane* (dragging him off to a nearby theater in the middle of a workday), and according to colleagues like Robert Wussler and Gerry Hogan and neighbors like newspaper publisher Peter Manigault, Turner watched the film many times after that, caught up in the classic cautionary tale of the media magnate who gained the world but alienated all those who ever loved him. When he consulted Pittman and confronted the pain of a desperately unhappy wife, trembling kids, lost friends, and the threat of loneliness, he must have realized at some point that his most difficult battle might very well be to avoid just such a fate as Kane's.

"Here's a guy who had everything," says his son Teddy, "a big business, a big family, couldn't be more successful—and he was all alone. Who was there? Janie? His kids? His Mom? His friends? What friends?"

CHAPTER 18

AT 9:04 A.M. PACIFIC DAYLIGHT TIME, on October 26, 1984, a small crowd of less than a hundred people gathered in a makeshift studio at the old Fox Theater in Hollywood. The room was festooned with balloons—green and yellow, red and pink—but the early morning crowd was more subdued than festive, and the smattering of reporters on hand looked downright skeptical. The TV monitors all around the room displayed a logo of a dark blue star and a golden name—Cable Music Channel. Up on the jerry-built stage under an enormous banner stood Ted Turner, preparing to launch yet another new business venture, this one conceived as a full-fledged competitor of MTV Music Television. He seemed a little rumpled in his dark green suede sportcoat, a little fatigued, but he summoned up his energy and jabbed down at the large red album-size button on his right, exclaiming "Here we go! Take that, MTV!"

And so the Cable Music Channel was launched, going out to a viewing audience that Turner estimated at 2.5 million. It seemed a shrewd business move. MTV was the only mainstream music channel on TV; with its rock-and-roll fare, it catered solely to teens and young adults. Turner promised less talk, no VJs, fewer commercials, and a wider range of music—from Stevie Wonder to Willie Nelson, from the Beatles to Sergio Mendes and Frank Sinatra.

As he circulated around the room, chatting up the reporters from various trade journals, Turner was not as out of place in the world of pop music as one might think: his Turner Broadcasting System included several rock-music stations, and his WTBS ran a six-hour program of video music clips titled *Night Tracks*. He had a few "old radio tricks" of his own to try.

But behind the scenes there was another agenda. By attacking MTV, Turner was putting himself in the good graces of his old friends at the cable companies around the country, who felt that the burgeoning music network was getting a little too big for its britches. MTV intended to raise its rates, and cable operators like John Malone of Tele-Communications, Inc., needed a bargaining tool to beat the network back into line.

So, in the words of one cable president, "Turner went into the music business on behalf of the cable operators. . . . He was a stalking horse for them." Robert Wussler concurs: "CMC was basically started as a favor to the cable operators." While it's unclear exactly how formalized this arrangement was, the benefits to both sides were obvious: Turner was offering the cable operators an alternative music channel free of charge (versus MTV's price of ten to fifteen cents per viewer); they, in return, could offer him increased viewership for his other networks, like CNN and Headline News, as well as a fighting chance to get yet another network off the ground. (According to a little-known TBS prospectus filed with the Securities and Exchange Commission, "The company antici-pated that Cable Music Channel will lose in the range of $5 to $10 million in its first year of operation and may continue at that level thereafter.")

With exquisite timing, Turner's announcement of CMC not only upstaged MTV's launch of their new adult music channel VH-1, it also appeared on the very same day that MTV was slated to sell over five million new shares to the public, more than one-third of its stock. In fact, the MTV stock would plummet some $3 a share with the CMC announcement, for an estimated loss of $15 million. "It is a typically competitive Turner ploy, to announce something like this at a crucial time for his major competitor," said one Wall Street analyst.

According to Turner himself, the launching of CMC had little to do with business; CMC was a moral crusade. Interviewed by *Rolling Stone* magazine, Turner declared: "I was really disturbed with some of the clips they [MTV] were running. You can take a bunch of young people and you can turn them into Boy Scouts or into Hitler Youth, depending on what you teach them, and MTV's definitely a bad influence. My wife used the word 'satanic' to describe it."

Turner clearly didn't like MTV; he and Janie didn't want their teenage kids watching all that raucous, diabolical hard-rock stuff. Objecting to "the violence and sadism of MTV," Turner singled out the video clips of heavy-metal groups like Twisted Sister for special scorn. He attacked the video for the song "We're Not Gonna Take It," in which a boy throws his father out the window after his father tells him to turn down the volume on his radio. "I figured we'd clean it up," said Turner. "Something that would not be damaging for young people to watch."

Ted Turner was out to save the youth of America from MTV. It was unclear if young people really wanted to be saved, but Turner was sure he was the man to do it. The subtle irony behind Ted's moralizing came in a footnote from Scott Sassa, vice president and general manager for Cable Music Channel, who admitted that about 80 percent of CMC's selections would in fact duplicate those of MTV. But Turner wouldn't be denied; he was going to rescue the children of the United States from themselves.

Such crusades, in all their grandiose glory, were in fact nothing more than standard operating procedure chez Turner. "I want to be the hero of my country," he once said, without a trace of a smile. "I want to get it back to the principles that made us good. Television has led us, in the last twenty-five years, down the path of destruction. I intend to turn it around before it is too late." Thinking big was Ted Turner's preferred mode. Now in his mid-forties, and growing in media sophistication, he was still the kid who dreamed of being Julius Caesar. "I'll tell you what I want to do," he told one stunned reporter. "I want to set the all-time greatest personal-achievement record, greater than Alexander Graham Bell or Thomas Edison, Napoleon or Alexander the Great. And I'm in a great position to do it, too."

His meetings with Fidel Castro and other leaders had convinced him that he had been "just ambling through life." After starting CNN, he began to realize in the early 1980s that "I needed to find out what was happening in the world. You know, what was really happening."

It would come as no surprise to anyone who knew him well

that the Ted Turner of the 1980s was harboring presidential ambitions. He had been thinking about the idea of running the country
since the early 1970s, at least. Catching a flight out to Los Angeles
with his then accountant Irwin Mazo, Ted had confided to him his
four great ambitions. Two of them dealt with getting to the top of
the TV business. "Number three," he said, "I'm gonna be the
richest man in the country. And fourth, I'm gonna be the president
of the United States."

"Ted, you have no political base," Mazo exclaimed. "How are
you going to be president?"

Turner shifted in his seat and smiled. "I've got the boob tube.
When this country collapses, I'm going on the boob tube. And
that's how I'm gonna be elected president."

By 1977, Turner had publicly declared that he could win the
presidency "easier than any man except Jimmy Carter." And by
the early 1980s, after being contacted by members of the Georgia
Republican party, he was thinking about running for office so seriously that there was even talk among CNN staffers of bumper
stickers about to be printed up. With surprising frequency, as he
and his top aides gathered for a drink and a chat at the end of the
day in his office, Turner would lean back in his chair, his feet up
on the desk, and announce, "Well, I guess I'm gonna have to run."
He set his sights on the 1984 election. "The country needs me," he
told advisors. "It can't wait four years."

Unfazed by his lack of political experience, Ted Turner was
contemplating becoming a Ross Perot years before Ross Perot
emerged on the political scene. But despite his ambitions, Turner
had no clear-cut political agenda. He wasn't a Democrat; but he
wasn't exactly a Republican, either. He was a friend of the John
Birch Society's most ardent members, the American Eagles
("They're all my friends," he told Reese Schonfeld. "I've been very
close to them . . . for years"), but he was also a buddy of Fidel
Castro. He was a right-winger who had grafted on a few leftist
notions. Although fundamentally conservative in his upbringing as
the child of a staunch Southern traditionalist, he had over the years
been exposed to more and more liberal ideas. The basic truth about
this would-be presidential candidate, however, was that when it
came to politics he didn't have any; when it came to a political
philosophy, he had rarely bothered to sit down and work one out.
"I think that Ted was almost a virgin in terms of an awful lot of

things," says his friend Mike Gearon. "That's why he's so susceptible to new ideas. Fundamentally, I don't think he cared a whole lot about things that didn't have an application to what he was doing at the moment. While very opinionated, Ted's always a pragmatist. . . . He was apolitical, if anything."

Never the policy wonk or the urban analyst, Turner's political statements from the early years were mostly outrageous one-liners. There was the time he derided both the MX missile program and the unemployment program by joking that out-of-work blacks should carry missiles on their backs from silo to silo, "just as the Egyptians carried rocks during the building of the pyramids"; it was a quip that succeeded in alienating just about everybody, from left-wing supporters of African Americans to right-wing champions of the arms race. Other times he joked about "kikes" and "Eye-talians"; "Imagine the Eye-talians at war," he told a Washington audience. "I mean, what a joke! They didn't belong in the last war. They were sorry they were in it. They were glad to get out of it. They'd rather be involved in crime and just making some wine and having a good time."

Friends like Gearon and old associates like Mazo emphatically deny that Turner has any prejudices. In fact, both claim he's among the least prejudiced people they know: "Anti-Semitic or antiblack he is not," according to Mazo. "Ted isn't anti-anything. That just isn't the way he thinks. . . . People who worked for him could be Greek, Turk, black, white. . . ." They point to his hiring and championing of Jews like Reese Schonfeld and blacks like Bill Lucas—the first black general manager in baseball—and Henry Aaron. Even the *Atlanta Constitution* would observe, "Shallow and naïve, maybe, but Turner's no racist." If so, then the problem lay with his mouth, which seemed to operate in a different time zone from his mind. Actually, Ted Turner was an equal-opportunity offender, alienating people right across the ethnic spectrum with his legendary lack of tact.

Clearly, this brash marauder didn't know how to operate in the sensitive political sphere. Although he understood that his media companies gave him a certain power, just as William Randolph Hearst's newspapers had done for him, he didn't know how to translate that into political clout. "Ted has no understanding of that kind of power," Reese Schonfeld says. "He understands personal power, but not the kind of institutional power . . . that publishers and great editors used to have, where candidates would come begging for endorsements in the old days of American journalism. He has a sense of what he wants to happen, but he doesn't know what you have to do to make it happen. . . . In twelve years, CNN hasn't even had a dogcatcher fired." Mike Gearon concurs: "Ted's not that kind of guy. He doesn't realize the power he has. He's not like Henry Luce or Bill Paley, who knew the levers they had. . . . He doesn't even begin to think about it."

If Turner had no clear-cut political agenda, what he did have was a kind of inchoate yearning, a nebulous idea that the country needed help and that he was the man to deliver it. With a curious combination of Perot's rolled-up-sleeve pragmatism and his own wide-eyed idealism, Turner wanted to fix America, to make it all better and everyone friends: he wanted to teach the world to sing in perfect harmony. Under headlines like "TURNER HOPING TO SAVE THE WORLD" and "TURNER SEEKS BROTHER-HOOD," Ted would opine, straight-faced, that he desired to be "the Jiminy Cricket of the United States—the country's con-science." Drugs, poverty, crime, education—Turner had no po-sition papers on any of these. What he did have was the admirable and thoroughly childlike desire to ride in on a white horse and save the day. "The culmination of his life would be if our country gets into such a crisis that there is an outcry that Ted take over and save us all," says Gerry Hogan. "He carries that dream around every day."

It was Turner's old teenage missionary fantasy, to "go to some little Timbuktu and try and get the people to help me build a church, and, you know, convert them . . . to save their souls, help them save their souls." Of course, in his early years, as much as he might have wished and prayed, he had been unable to save those he loved most: unable to save his sister, unable to save his parents' marriage—most terrible of all, unable to save his father from depres-sion and suicide. Over the decades, as if trying to even the score, he would throw himself into "saving" things—crippled birds, dying businesses, companies at death's door, like Turner Adver-tising and WTCG, the Braves and the Hawks. He would save America from "sleazy" TV; he would save kids from "satanic" music videos. Ted Turner was going to save them all.

Only one thing could stop him. "Someday, somebody will put

a bullet in me," he would often say. "I would like to stay around for a while, but I really do believe that I'll be assassinated."

By 1984, Ted had to acknowledge his liabilities as a presidential candidate. With his raucous, freewheeling style, folks didn't take him seriously. He didn't just have a few skeletons in the closet, he had an entire cemetery. But Turner was never one to dwell on the negative for long. Besides, at about that same time, another idea was bubbling in his mind.

So what if he couldn't win the presidency? He would take another route—do an end run around elective office. He was building a communications empire; he intended it to be the largest media conglomerate on the face of the earth. Now he started to feel that it would not only make him colossally rich but it would give him an Archimidean lever to move the planet. Although he was probably unaware of the famous jurist Learned Hand's observation back in 1951 that "the hand that rules the press, the radio, the screen and the far-spread magazine rules the country," Turner sensed that truth instinctively. With television, he would make things happen without ever needing to be elected. He could end global suffering and make the world a better place. At a visceral level, Turner was combining two of his old dreams—to make more money than anyone else and to have everyone love him in the process. Ted was going to bring his message to the people: "We're really a pretty terrific species. We're the top drawer. And we can be so kind and loving and helpful to each other. We just have to learn to live together like brothers."

Things were going to be OK. Currently they were terrible, but soon they'd be looking up. What this country needed, Turner believed, were some happy programs on TV. Why were there no documentaries about the Boy Scouts, or the other institutions that made America great? Why did TV always carp and criticize about the latest scandals? He created a series called *Portrait of America*. Ira Miskin, the producer in charge, remembers: "The mandate was, America has all kinds of problems, and we're not necessarily interested in the problems of America in this series: we're interested in the positives, the strengths of America."

Then there were the two other Turner creations, *Nice People* and *Good News*. Zeke Siegel, a longtime CBS journalist brought in to oversee *Good News*, had his doubts. He and his wife were sitting with Turner and Liz Wickersham in Harrison's, and even before the vodkas arrived, Turner had mounted his latest hobbyhorse. "But, Ted," said Siegel, "there's no such thing as good news." Turner leaned forward intently. "Yes, there is," he insisted. "For example: Christ died on the cross. That's bad news. But then he rose again. That's the good news." He settled back in his chair with a satisfied smile. "You see?" QED.

It wasn't just America that needed his help. There was this whole cold-war thing going on. Ronald Reagan was calling the Soviet Union "the evil empire," and American and Soviet missiles were locked into a cycle of mutually assured destruction. As CNN expanded into seventeen and then twenty-seven countries around the world, including Japan and the USSR, Turner began to think globally. Any given week might find him in Boston or Greece, Washington or the Soviet Union. And one thought kept coming back to him: All these missiles all over the globe, all the fear, it's terrible. It interfered with business, and it interfered with people's lives. What the world needed was . . . PEACE!

It was a classic bit of Turner-think, simultaneously breathtaking in its bottom-line truth and its absolutely reductive simplicity. Like some Georgian Zen master, Ted Turner's shallowness was his profundity and vice versa. He had no time to analyze first-strike options or debate the ramifications of international border-patrol policy. He cut right to the chase with an executive summary: missiles are bad; we're all on this planet together; let's just get along. From a chest-thumping America-Firster, as well as a military-school graduate who insisted that all his boys also go to military school, he had made a startling leap, one that surprised and offended other conservative Southerners, who had counted him among their allies.

But Turner, the avid gun owner and hunter, had no truck with good old boys who elbowed him at parties and declared, "Let's nuke the commie bastards!" In fact, on a visit to Moscow, Turner had attended a gathering that included leading Soviet party bureaucrats. Chatting with one top Soviet official, he suddenly stuck his index finger as if it were a gun up to the startled man's skull. "Now," he said, "stick your finger up to my head. *There*. Now, can we talk to each other like this?"

"No," said the wide-eyed official.

"Of course not," said Turner, warming to his subject. "We've got to stop pointing these at each other. You love your children, right? So do we." Then stretching out his arms, he gave the man a big bear hug, and the smiling official hugged him back. "They loved him there," Charles Bonan, vice president for Turner Program Services in Europe, says. "He transcended propaganda and rhetoric. They couldn't react to him in any way but with their emotions."

It was in the summer of 1984, as the Olympics were unrolling at the L.A. Coliseum, that Ted Turner strolled into the office of his top aide, Robert Wussler. He took one look at the TV set, jabbed his index finger at it, and declared, "You know, Wussler, that's wrong. That's absolutely wrong." Wussler thought his boss was criticizing him for having the TV set on during the workday. He began to think up excuses ("I always have the TV set on; it's the way I was trained at CBS"). He nervously cleared his throat and ventured, "What do you mean?"

"The Russians aren't there!" exclaimed Turner. "The Russians should be there. And we should have been at their games in 1980. Goddamn it, Wussler, that's wrong!" Turner paced back and forth. "Where are the next Olympics?"

"Seoul, Korea."

"Let's buy the rights. How much do the rights cost? Let's buy the rights and make sure that everybody comes. Or better yet, I want you to go to Moscow. That's it—I want you to go to Moscow. I want to buy the rights to the whole world for the Olympics with the Russians."

Wussler hemmed and hawed about the difficulties, about the Soviet position, about the subtleties of dealing with the International Olympic Committee, which oversees everything Olympic. He tried to explain how unfavorably the IOC might view such a move. Turner was unfazed. And ten days later, Wussler found himself on a plane for Moscow. "I knew it wasn't possible, but I went ahead," recalls Wussler. "I did what he wanted to do."

As it turned out, the Soviets had an excellent relationship with the new IOC head, Juan Antonio Samaranch, who just happened to be the former Spanish ambassador to the USSR, and they turned Wussler down flat. But they did have some other ideas. They were, for instance, planning their own alternative games in the USSR, to be held in only a matter of weeks. Athletes from the Eastern bloc and other communist nations all over the world would be invited. How would TBS like to broadcast these games?

Turner loved the idea. He gathered his top aides in his office to discuss the Soviet games. One after another, they raised objections: the Soviets weren't well organized; there wasn't enough time; who could say what caliber of athlete would actually show up; and finally, wouldn't this be a terrible PR move, to broadcast the Soviet games just after they had boycotted our Olympics?

"All right, fine," Turner said, "that's enough discussion. Let's put it to a vote." He went around the office: Bevins, Hogan—every one of his aides except Wussler—voted no.

"OK," Turner said. "I'm still gonna do it." Perhaps noting the sour expressions on the faces around him, Turner paused. "I tell you what." He pulled out the Krugerrand he carried in his pocket for luck. "I'll flip the Koogie. Heads we do it, tails we don't."

Turner put the gold coin on his thumb and flipped it high into the air. Glittering, it arced up toward the office lights and came down tails.

"All right," said Ted, shaking his head. "Best two out of three."

In the end, Turner would be dissuaded by his aides. But there was another proposal that Wussler had brought back with him from Moscow. One Soviet official, while nixing the Olympics, had declared: "The idea of doing something with your country to honor the last time we were all together in Montreal in 1976, now *that's* something we would like to do." An eager Turner listened intently as Wussler described what the Russians had in mind. "They don't like the Olympic deal," he told him, "but they'd like to do something in '86, commemorating the tenth anniversary."

"Whaddya mean?" asked Turner.

"Well, you know, maybe a multisport event."

Tuner's voice thickened in amazement. "They'd like to do that with *us?*"

"Yeah."

"Boy!" Turner was stunned. This wasn't just broadcasting the Olympics; this was his very own Olympics. And he'd bring the United States and the USSR together even before the IOC could. This wasn't big; this was enormous. "Boy!" he repeated. Jumping up, he started doing somersaults—literally doing somersaults!—up and down the hallway, tumbling head over heels through the executive offices in his enthusiasm.

Shortly thereafter, Bob Wussler began his shuttle diplomacy between Atlanta and Moscow, almost twenty trips back and forth over the next twelve months. A year later, in August of 1985, Turner Broadcasting would announce the formation of a sporting extravaganza, afterward to be dubbed by Ted, with his flair for the obvious, "the Goodwill Games."

If in the early and mid-1980s Ted was slowly learning to become an advocate of universal brotherhood and one world, there was another fundamental sociopolitical belief he had grown up with and long held almost without thinking—environmentalism. His was not the fashionable ecology of the Green movement or the Hollywood hug-a-tree celebrities. Rather, it was the environmentalism of the outdoorsman, the hunter, and the fisherman. He was a *Field & Stream* ecologist, a *Rod & Ammo* ecologist. His concern came from deep in his childhood, back in the days when he collected and cared for animals (when they died, he tried to preserve them by learning taxidermy). It grew out of the time he had spent wandering through the lovely old oaks at Binden, walking down to the marshes, hearing the wind off the Atlantic whispering through the tall reeds. It was the love of those damp mornings he spent side by side with his father, each with a shotgun, waiting to spot the first fowl winging across the sky. It was the memory of long hot South Carolina afternoons, setting off by himself along the shore after escaping from the house and the quarrels between his parents, watching the waves soothingly lap the sand.

Turner began becoming actively involved in national conservation groups, like Ducks Unlimited, in the mid-1970s, attending their banquets and charity auctions to raise money to save wetlands. And after buying the Hope Plantation in South Carolina, where being outdoors was all one did every day, his enthusiasm for the land seemed to grow even stronger. He began to donate trips to the 5,000-acre plantation as part of the charity auctions. "I remember visiting the plantation just after he bought it," Jim Kennedy, head of Cox Enterprises and a fellow member of Ducks Unlimited, says. "We were driving around in an open jeep, over the old rice fields, just going like hell. And Ted, he was so exuberant he ran out of words. 'Isn't this great?!' he kept saying. 'Isn't this great?!'"

Coupled with Ted's love of the land was his bleak, doomsday feeling that, beautiful as it was, the earth might all turn to dust tomorrow. Whether this notion was born of the sudden shocks and desertions of his youth or had grown out of conversations with his favorite professor at Brown, the disaster theorist John Workman, Turner at thirty-five and forty-five would ramble on about the subject to anyone who'd listen, conjuring up, in the words of one writer, "nightmarish evocations of a ruined earth populated by *Homo sapiens* gone to seed."

Everything he read confirmed his fears. By the late 1970s, according to friends like J.J. Ebaugh, Turner had read such books as *The Limits to Growth* and other reports from the Club of Rome, an international think tank, in which the burgeoning human population was seen as devouring more and more of the planet. If there was one talent Ted Turner had, it was his ability to think ahead, and as he looked toward the future of the earth, he didn't like what he saw.

In June of 1980, the same month that Turner was launching CNN and sailing in his third America's Cup trials in Newport, a slim white volume called *The Global 2000 Report to the President* was published. It was a federal report issued in response to President Jimmy Carter's directive to several government agencies to study "the probable changes in the world's population, natural resources, and environment through the end of the century." Although there were appendix volumes that ran on for hundreds of pages, the main book (even with charts, graphs, and footnotes) numbered barely fifty pages.

It wasn't necessary to read past the opening summary to get the message, and a somber message it was: The world's population would grow from 4 billion in 1975 to 6.35 billion by the year 2000. By the end of the subsequent century, it would approach 30 billion, a figure that would "correspond closely to . . . the maximum carrying capacity of the entire earth." Meanwhile, "regional water shortages will become more severe. . . . Significant losses of world forests will continue over the next 20 years. . . . Serious deterioration of agricultural soils will occur worldwide. . . . Extinctions of plant and animal species will increase dramatically. Hundreds of thousands of species—perhaps as many as 20 percent of all species on earth—will be irretrievably lost as their habitats vanish."

One didn't have to be a Nostradamus to read the writing on this wall. And it wasn't an extremist group or even the Sierra Club laying out the facts. It was the United States government stating that "if present trends continue, the world in 2000 will be more crowded, more polluted, less stable ecologically, and more vulnerable to disruption than the world we live in now. Serious stresses involving population, resources, and environment are clearly visible ahead. . . . Barring revolutionary advances in technology, life for most people on earth will be more precarious in 2000 than it is now—unless the nations of the world act decisively to alter current trends."

According to some of Turner's closest friends, *The Global 2000 Report* hit him like a ton of nonrecyclable bricks. "It put his darkest fears about the environment in one neat little book," says one longtime buddy. All the more powerful for its pithiness, the *Global 2000* study seemed to sum up Turner's concerns, and to present the situation as more potentially disastrous than even he had imagined. "That report was really, I think, the cornerstone for his philosophical views," Mike Gearon says. "Ted soaked it up. He just grabbed it." Over the coming years, Turner would hand out copies of the report, gift-wrapped, to his friends and colleagues for the holidays.

For Turner, a series of ideas began to coalesce. Things looked bad . . . this horrifying report . . . the planet's gonna die and humanity's gonna be wiped out. Here was a major world problem to tackle, the grandest of all missions—to Save the Earth. It was a rescue mission worthy of Citizen Turner.

By the mid-1980s, what had been a vague desire to be useful, to be a Jiminy Cricket or a heroic martyr, had become a little better defined. Now when he talked to the media, which was often, he didn't just talk about business and sailing, as he had in the 1970s, or business and some nebulous notion of being the country's conscience, as the 1970s became the 1980s. Now he talked about the environment and nuclear weapons and (of course) business. Once again, he did it in the same bold saber strokes, short on detail and long on drama, running on and on like a Cassandra or some onenote Ancient Mariner who buttonholes you at a party and won't stop: "Population growth is analogous to a plague of locusts. What we have on this earth today is a plague of people. Nature did not intend for there to be as many people here as there are. [Medical science] eliminated the diseases that were nature's way of controlling numbers and keeping everyone in balance. And like a bunch of termites, we're just going across the world, cutting down trees, changing the ecology, siphoning coal and oil out of the earth's surface and burning it in massive quantities and polluting the atmosphere and messin' up the ozone layer. . . .

"And then on top of that, now we've gone into nuclear weapons and the capability of fighting our wars; the whole issue of warfare is to improve and increase firepower. Now we've increased firepower to the point where we can blow up the entire world. We ought to be in a daily communication with the Russians about disarmament. We ought to be working like beavers on alternate energy.

"We got to the moon; we should learn to live on this beautiful planet we've got here until the sun burns out, instead of ending it. We have two alternatives, humanity does: we can either make this world into a garden of Eden or we can go fight the battle of Armageddon. I'd like to see it be the former."

As long-winded as Turner was, the great thing about him, for environmental advocates, was that he was loud. Ted Turner cared passionately about the land—there wasn't a shred of hypocrisy here—and he wasn't afraid to shout what he felt. Jimmy Carter, near the end of his term in office, appointed his friend to the President's Council for Energy Efficiency, a group formed "to help encourage citizen participation in the national effort to achieve greater energy efficiency . . . at home, on the road, and on the farm." The honorary committee also included Mary Tyler Moore, Kirk Douglas, Stephen Stills, and Leontyne Price, all caring celebrities known for their ability to project.

In his own life, Turner took the lessons of ecology to heart. He cracked down on litter and smoking at TBS and CNN; he quit smoking cigars and eventually chewing tobacco himself, and in all his homes—even Avalon, the $6-million estate he would buy in 1985 outside of Tallahassee—he would not only insist on recycling but continue to ban air-conditioning, letting his family and guests drip with perspiration through the hot Southern nights rather than install the energy-guzzling, Freon-consuming machines he couldn't stand. For a multimillionaire, Ted had a surprisingly spartan mind-set; at home on the plantation, he grew his own vegetables and thought more like a live-off-the-land survivalist than one of the richest men in America, which is what he was fast becoming. And he went out of his way to learn from those who could teach him more about ecology.

In July of 1982, Turner packed up his two younger boys, Rhett and Beau, and flew down to South America to join Captain Jacques Cousteau aboard his research vessel *Calypso,* as it made its way along the Amazon River. For a week, they traveled with Cousteau as he filmed a documentary, enduring the constant rain and the sweltering, humid heat of the dense green rain forest. By day the forest seemed to slumber, but at night it came alive with the chirp of frogs and insects, the calls of birds and howler monkeys, the sound of water dripping from the leaves. It would have been hard not to be impressed by the sheer staggering volume of life along the river, the streams of ants and termites, snakes and rodents, hummingbirds and tapirs, monkeys and antelopes. One acre of this Amazonian forest alone could contain fifteen hundred varieties of butterflies (compared to seven hundred butterfly species in all of North America).

Turner was impressed by the scene and inspired by his long conversations with the French captain. One day, after Cousteau had finished filming, he and Turner were talking on board the *Calypso,* and the conversation came around, as it did several times that week, to the cold war and the environment. Cousteau thought that man himself was in danger of becoming an endangered species. Turner asked, "Do you think there's hope for us, Captain Cousteau?" The captain responded with a Gallic shrug. "Well, Ted," he said, "even if I knew we were going to lose, I'd still work as hard as I could to save us." He looked at Turner. "What else can you do if you're a man of conscience?"

Turner came away convinced on several levels, and, as always, his pocketbook followed his heart. He had already signed up Cousteau for TBS. Now he decided to pour in the money: "I gave him four million dollars for his work this year. We'll get four hours of programming out of it. Of course, I'm losing my shirt on it. That's double the budget of network programs. But at least I'm keeping Cousteau operating. He's on my team."

The hiring of Cousteau was only part of Turner's green initiative: he would also, over the coming years, purchase documentaries from the Audubon Society, and in 1985 he fulfilled a "boyhood dream," when, after four years of negotiations, he succeeded in bringing *National Geographic*'s magazine show *Explorer* to WTBS as well. In Turner's new approach, he was now using his TV networks to further his agenda of social change.

Perhaps most significant for WTBS, and for Ted Turner personally, was his meeting in July of 1980 with *Time* photographer Barbara Pyle. She not only had a feisty personality that intrigued him but she had also just read the *Global 2000* report, and was every bit as concerned as he was. Pyle had grown up a farm girl in Oklahoma, where her parents had founded the first Audubon Society chapter in the state, so she came by her environmentalism as naturally as Turner did. When he discovered that the young brunette was not only fun to flirt with but also shared his deepest fears, something clicked for him. "Ted had always been a closet environmentalist," Pyle says. "He'd just never had anyone to listen to his concerns. I became his sounding board."

At the Newport bars, Pyle would listen to him well into the night. It was following one of these evenings that Turner had a brainstorm. "You're going to come down to Atlanta and work for me," he said. The gleam was in his eyes as he laid out his idea. Pyle would work for WTBS and make documentaries on subjects like energy, overpopulation, and nuclear devastation. "But I don't know anything about making documentaries," exclaimed Pyle. That didn't faze Ted. "Teach yourself," he told the photographer.

During their long rambling conversations that summer, Pyle took extensive notes on the programs that Ted felt needed to be done. By summer's end, she had a job description, and it was basically the *Global 2000* report. Just as Pyle's lack of experience in filmmaking didn't stop Turner, neither did the obvious programming danger: that films on doomsday topics like global warming and nuclear winter were never big ratings winners, and that on WTBS especially—where the typical viewer was a lower-middle-class, undereducated, middle-aged male whom insiders called "Joe Six-Pack"—insightful environmental fare seemed an odd fit. Where, for example, would a somber program on the hole in the ozone layer go—before *The Andy Griffith Show* or after *Gilligan's Island*? Although the only thing these programs had in common was that they were all Ted, that was enough. Barbara Pyle was installed in Atlanta as the TBS executive in charge of all environmental issues.

"If there was a person who educated Ted to the environment, it was Barbara and not the other way around," says Pyle's former boss, Ira Miskin. "Despite what Barbara may say and despite what Ted believes, Barbara Pyle is the environmental conscience of Turner Broadcasting. Ted embraced those ideas, but Barbara has been talking about saving the environment since she could talk." According to Miskin, Pyle and Turner grew to have a special bond: "They're like two crazy people. They're good friends. But they're both speeding freight trains at a hundred and fifty miles an hour all the time, no stopping, unrelenting. It's like you've got a brother and a sister who are exactly alike, and they have the same personality, and they can't be in a room at the same time for too long because they explode. That's what it is. It's a nutty brother-and-sister act."

Barbara Pyle would also be a catalyst for her roommate, Ted's new girlfriend, J.J. Ebaugh. When one or the other wasn't flying off to some far-flung locale, they would lounge around the house and talk of Pyle's new job, her occupation and preoccupation. Ebaugh read the *Global 2000* report, as well as Jonathan Schell's *The Fate of the Earth* and Marilyn Ferguson's *The Aquarian Conspiracy,* and she too began to worry about acid rain and toxic waste, and the fact that so many people had so little understanding of these problems. "Suddenly, my life was directed," Ebaugh recalls. "I realized the technology was in a holding pattern waiting to be used—the satellite." And the man she was flying for had the satellites. By 1982, she had started to write a book on how TV could nurture global cooperation on environmental issues. The book was never finished, but it did eventually lead to an abbreviated gig for Ebaugh on a TBS show called *Planet Live,* which was described in press-release language as "an ecological *Entertainment Tonight.*"

Along with environmentalism, Ted's J.J. was also passing through a New Age phase—a crystals-and-gurus period of self-discovery and self-optimization. Ebaugh, a young woman with a proclivity for philosophizing, had taken a wrong turn and gotten lost in space. The hardheaded Turner was, for the most part, not buying. "She tried real hard to get him into that New Age stuff," says one friend, "but Ted didn't want to. It didn't catch on with him."

Still, it was clear that Ebaugh and the man she called her

"partner" shared more than a bed. Their ecological interest was a strong link: "That was the first step of a relationship," she says, "a commitment to the environment that we had in common." And both would come to be strong believers in "verge theory"—that the world was on the verge of something huge and unpleasant. "The next ten years are gonna be very difficult," said Turner. "Probably the most difficult that man has ever faced. . . . The planet is collapsing . . . the ecosystem is collapsing under the sheer weight of five billion people."

But Ebaugh, ever the optimist ("I can't tolerate negativity"), made it her mission to appeal to the cheerleader in Ted. Together, she believed, T.T. and J.J. could alleviate the world's ills: "As systems start to collapse," she said, "Ted will be a major source of leadership—a problem solver . . . both planetary and small."

Throughout 1984, both J.J. and Barbara Pyle had been bending Turner's ear. Ted and Barbara would chat around the office, Ted and J.J. would talk in the minutes before bed. Together the two women were encouraging Turner to create a foundation, a "Better World" society. Both Pyle and Ebaugh had their reasons. Pyle had come to realize that WTBS had neither the money nor the time to produce environmental programs on the scale she and Ted envisaged. She was pushing hard for a solution.

Meanwhile, J.J., according to friends, had begun to worry about Ted. He had often claimed that he would be struck down by an assassin's bullet. He had even dreamed up a last line of repartee, a specially tailored *bon mot* if he ever did come face-to-face with his assassin: "Thanks for not coming sooner."

Now, however, J.J. believed there was real cause for alarm: Turner had been receiving death threats, notes left at the office and on his car: "You dirty rotten bastard," said one, "we're gonna get you." Read another: "Make sure your insurance is up to date." Staffers felt they were just gags, but J.J. began to feel certain that she and Ted were being followed. Was it paranoia, a harmless crank, or a real danger? Ebaugh was scared. She started to take defensive driving courses. And while she took measures to protect herself and Ted, she also wanted to protect their legacy—to create a mechanism for the documentaries to continue, to institutionalize their notion of improving the world through TV, to keep spreading their message even if something should happen to Ted. And so with

Ebaugh's persuasive arguments and Pyle creating the framework, the Better World Society was born.

In late 1984, Ted approached Tom Belford, head of a Washington consulting firm that advised the National Audubon Society, among other organizations, and asked him to develop a foundation. Turner put up about half a million dollars of his own money, and gave Belford the assignment of raising grants and matching funds to sponsor TV documentaries that would never get made any other way, films on "the so-called larger agenda issues," in Belford's words—overpopulation, global warming, and all the other crises that make for dull television.

In late 1984, Belford gathered together in Washington such environmental advocates as Russell Peterson, president emeritus of the National Audubon Society, Lester Brown, president of the Worldwatch Institute, and Jean-Michel Cousteau, son of Jacques, to realize the dream of Barbara, Ted, and J.J. And in June of 1985, in the elegant University Club, just off Fifth Avenue in New York City, Turner unveiled the society, its heart-shaped globe logo, and its credo: "Harnessing the Power of Television to Make a Better World." Ted had recruited an all-star international board of directors, for globalism was a fundamental premise of a society that set out to tackle transnational problems. Around the board table were such leaders as former president Jimmy Carter, Soviet Politburo member Georgi Arbatov, Chinese ambassador Zhou Boping, the Aga Khan, and Nigeria's former head of state Olusegun Obasanjo. And so Turner's Better World Society was set in motion.

On Friday, November 30, 1984—little more than a month after it had first begun airing—the Cable Music Channel, Turner's bold foray into the world of pop music to save the hearts and minds of America's youth, shut its doors and sold off its assets to archrival MTV for $1 million, plus $500,000 of advertising buys on CNN and WTBS. A chagrined Turner announced, "We are very disappointed, but feel that the discontinuance of service now and the sales arrangement with MTV, Inc., are in the best interests of the company." It had taken all of five weeks for Ted Turner to pull the plug on this particular moral crusade.

Why? Although CMC officials had declared that they were launching the channel with 2.5 million subscribers, word had soon begun to leak out that the actual figure was lower—much lower: only three hundred and fifty thousand viewers. Turner's team had inflated the figure nearly 800 percent. Even the shamefaced Headline News had to report: "There is no explanation as to how the first subscriber figure was determined." Meanwhile, MTV was cranking along, viewed in almost twenty-four million homes.

Ironically, MTV's response to the threat of CMC was just like that of Turner and CNN to the Satellite News Channel. They had arrived first on the field; they would hold their turf. Fighting back hard, they had prevented CMC from airing many videos by aggressively signing "exclusivity" deals that barred performers from appearing anywhere else, and they rushed forward the launch of a second network, VH-1, that would appeal directly to CMC's older audience. (As it turned out, on January 1, 1985, VH-1 was beamed out on the same satellite feed on *Satcom 3R* once reserved for CMC.)

For all his idealism, Turner had once again proved himself a pragmatist when it came to his wallet. No one was watching CMC. *The Electronic Media* had called the network "unwanted, unneeded." And the cable operators had proved to be not so solidly in Turner's corner after all, especially when the bogus viewership numbers were revealed and it became apparent that there was little demand for Turner's network. So Ted decided that the battle against the "satanic" forces of MTV would have to be waged by someone else.

CMC was Ted Turner's first major business fiasco, and for him it was a defeat on the moral battleground as well. In the end, Turner lost $2.2 million on his Cable Music Channel after only thirty-one days. Yet even this storm cloud had a certain silver lining. "He lost some money on it," says one cable president, "but he proved a certain amount of loyalty to his customers." According to Robert Wussler, "It was a two-million loss, but it was really *de minimus*. He had convincingly showed his business associates how valuable he could be. Cable operators like Trygve Myhren (chairman of Time's ATC) got back to him, and said, 'You did us a great favor. MTV knocked their rates down. We won't forget you for that.'"

CHAPTER 19

By EARLY 1985, the forty-six-year-old Ted Turner had rounded an important corner. For the first time, his expanded TBS was actually showing a profit. It wasn't much of a profit, just a little over $10 million for calendar year 1984. But because of the way Turner had developed his business, that number was much larger than it looked. Ed Turner had taught his son to build companies by building value—plowing the profits back into the firm. That's the way Ed did it, and Ted was doing it the same way. By constantly spending and expanding, the company showed a loss on the books and could grow nearly tax-free. And now, although CNN was still losing $15 million a year, Superstation WTBS was turning into a great cash cow. And even CNN itself had begun to show every sign of going into the black by next year.

In fact, many of Turner's most optimistic predictions—the same ones that had gotten him into trouble on the *Donahue* show—were now coming true. WTBS was available in thirty-four million U.S. households (nearly 85 percent of the homes with cable) and CNN was appearing in thirty-two million households, a number Turner wouldn't have imagined earlier even in his wildest dreams. It looked now as if Ted Turner—controlling as he did 80.3 percent of Turner Broadcasting—owned 80.3 percent of a winner.

With the survival of his company assured, Turner began to focus on his next goal. He believed that over the coming years U.S. broadcasting would be consolidated. The future, he felt, would belong to the giants. "I don't think that this is the time to be small," Ted would say. "I think that, long-term, there are just going to be a few major companies in the programming and distribution business." Clearly, he intended to be one of the fortunate few.

It was time for his next roll of the dice. The economic situation seemed to call for it, and Turner's own character practically demanded it. Ted Turner's great strength was that he was always prepared to reinvent himself. He changed from billboards to TV, from UHF to cable, and he even changed the starting airtimes of TBS programs to five minutes after the hour.

"That was pure Ted," says Gerry Hogan, laughing. "We called it 'Turner Time'. . . . Ted decided that if viewers were on a different commercial-break cycle—5:05, 5:35—they would be less likely to leave. And alternatively, if they were changing channels with commercials on, they might stop at our program. I'll tell you, there were many long, passionate arguments about the whole thing. . . . Most people generally thought it was a goofy thing to do. But no one could come up with a valid reason why it shouldn't be done." And like other Turner changes, this one, too, bore unexpected fruit—separate listings in *TV Guide,* where his programs appeared not at the end of a long list of on-the-hour entries but all alone at five minutes after the hour.

By early 1985, then, Turner had some major changes in mind. First, he decided to lead CNN downtown to an unlikely new home. Like many urban centers in the early 1980s, downtown Atlanta had become a sad collection of ramshackle buildings, winos on the street, and a few glassy skyscrapers valiantly jutting into the sky as if trying to distance themselves from the squalor below. Smart city residents stayed away after dark, and many businessmen stayed away altogether. But Turner decided that downtown was where he would center his company, and he proceeded to purchase the Omni, a huge white elephant of a complex that had steadily lost money for its developer, Atlanta realtor Tom Cousins. At the time, the complex was 75-percent unrented—"a nightmare," according to TBS senior vice president of administration Paul Beckham. Turner's friends told him, "Ted, everybody's lost their ass there. And you don't buy real estate when you're so deep in debt." But in January of 1985, Turner declared that he would pay $64 million for what he called "that old amusement-park space"—rescuing the

Omni's combination hotel/shopping-center/office complex from bankruptcy. According to Turner, the officers of the Canadian Imperial Bank "just about gave me a hug."

As Turner wrapped up the details of the real-estate deal, he had begun to contemplate another move, something still more eye-opening. It was an all-out assault on one of the great bastions of American broadcasting. In April of 1985, Turner stunned the media world—stunned all America—by announcing a hostile takeover bid for the television giant CBS.

Surprisingly, for all the brickbats he had hurled at the networks as "traitors" and "scum," Turner had privately been trying to merge with any one of them since 1981. That was, at least in part, why he had hired Robert Wussler, an ex-CBS president, as his number-two man; that was why he had held meetings with NBC, ABC, and John Kluge of Metromedia all through the early 1980s. And it was why, in September of 1981, with the future of CNN and all of TBS at stake from the threat of the Satellite News Channel, Ted Turner met with some of the top executives of CBS in a meeting they described as "ultrasecret."

CBS had commissioned a study to see if it could establish its own twenty-four-hour news service, and though it had a fine news-gathering operation already in place, the company decided the costs would be prohibitive. Meanwhile, Turner had started offering CNN service to the CBS affiliates, trying to undercut the network. CBS network officials began to joke that "it might be a hell of a lot cheaper just to buy CNN." So one day in September 1981, after a phone conversation between Wussler and his former colleague, CBS News president Bill Leonard, Wussler walked into Turner's office to suggest a meeting.

"CBS?" said Turner, his eyes wide in amazement. "Is it your idea or theirs?"

"Actually," said Wussler, "it's mutual. Bill and I were talking about it. They're willing to come to Atlanta and make it easy."

"Hey," Turner said, "that's great! Let's do it!"

A warm September morning found Bill Leonard and his boss, CBS Broadcasting president Gene Jankowski, on the company jet as it pulled into Hangar One at Hartsfield Airport. Turner and Wussler were there waiting for them. The four adjourned to a nearby Marriott Hotel room and, over club sandwiches and iced teas, proceeded to map out alternative visions of the future for two of the world's largest news organizations.

It was an odd meeting, this secret rendezvous between two sides who were publicly bad-mouthing, even suing each other, while privately looking for common ground. The two CBS executives were dressed in their corporate suits, but Ted had come in jeans, an open-necked sports shirt, and Top-Siders without socks. He started the conversation, and never seemed to stop. "How are you guys?" he asked. "How's old Paley?" Leonard and Jankowski were a little put off. "Uh—fine," they said. Ted skipped right to the punch line. "I'll sell you CNN," he said abruptly.

Leonard and Jankowski sat silent, stunned.

"How much of it do you want to buy?" Turner asked.

"Fifty-one percent or more," said Jankowski.

"You want control? You don't buy control of Ted Turner's companies. Forty-nine percent or less."

"No, honestly, Ted," said Leonard. "You know CBS. We just wouldn't be interested in less than fifty-one percent."

Turner started working on his chewing tobacco, chewing and spitting out the brown juice into the water tumbler at his side. For the next two-and-a-half hours, he talked and chewed, chewed and talked as he strode around the hotel room. Forget about buying CNN, he said. Let's talk about CBS merging with TBS. "We got the greatest news organization in the world," he said. "We're non-union, and we're gonna open bureaus all around the world. I can help you. We can team up, and we can both cut costs. We don't both need bureaus in every place."

The CBS executives were cool to Turner's pitch. TBS and CNN were laughably small compared with their communications empire. The idea of synergy had little appeal to them if they didn't have control. They told Turner they'd report back to CBS, Inc. CEO Thomas Wyman, and left it at that.

"You guys," said Turner, "you CBS guys are something. Someday I'm gonna own you. You bet I am. Remember I told you so."

Everyone chuckled.

"Sure I can't sell you something? I can't come away from here without selling you *something*. How about my wife? Lovely lady."

The four drove back over to Hangar One, and Ted went aboard the CBS plane to check out how the competition was flying. "Hey, some terrific plane!" said Turner, who had just acquired his own pilot and jet. "Wanna sell it to me?" He looked around. "What's the difference? I'll own it anyway, one of these days."

About a month later, Turner and Wussler flew up to New York, and met with Thomas Wyman at his suite in the Essex House hotel, on Central Park South. The second meeting was even less successful than the first. After the initial pleasantries, the two men almost immediately started jockeying for position. Perhaps Wyman didn't take to the style of the tobacco-chewing Southerner; perhaps Turner felt put off by the smooth talker from New York's business elite. Whatever the reason, the conversation soon degenerated into a school-yard argument about who had the bigger sandbox.

"I'm here to buy you," Wyman said.

"No," said Turner, "I'll buy you."

"That's ridiculous," said Wyman. "I'm buying you."

"No," Turner insisted, "I'll buy you."

Gene Jankowski and Bob Wussler saw that tempers were getting frayed and hustled everyone downstairs, out into the fresh air. Turner and Wyman walked side by side along Central Park South and down Sixth Avenue toward CBS's Black Rock headquarters, where they said their good-byes. "We'll get back in touch with you," Wyman promised. But of course they never did.

In typical fashion, Turner soon blew the lid on the "secret" talks. "CBS came down to Atlanta trying to acquire CNN," he boasted to the Washington Press Club. "I didn't let them make an offer. I was too busy telling them I was going to bust them." The next day, a miffed Wyman shot back, "We were *invited* to come to Atlanta by Mr. Turner. He said he wanted to talk about some sort of collaboration. . . . But the suggestion that we were dying to make an offer for CNN is outrageous and untrue."

There would be other meetings with each of the networks throughout the early 1980s, especially in February of 1983, as Ted sought to merge or team up with anyone who could give him money, stock, a large distribution network, and control of their operation. As late as February of 1985, Turner was on the phone to CBS officials proposing yet another "business combination." And while these encounters didn't degenerate into the playground name-calling of the early CBS meeting, they didn't get very far either, given the scope of Turner's demands.

But as the 1980s unfolded, the full impact of the Ronald Reagan presidency was starting to be felt, and changes were on the horizon—changes in the way business was being done. Deregulation meant that government was allowing free rein to many corporate dealings (and limiting antitrust suits). The brakes were off. Meanwhile, business itself had become chic, and the formerly boring class of corporate executives had taken on a new glamour, a new image—"Masters of the Universe."

In this great, glitzy game, the highest art form was mergers and acquisitions—the multibillion-dollar corporate takeover. Raiders and junk bonds, greenmail, poison pills, and shark repellents—these were the catchwords of the day. In the first two weeks of April 1985 alone, headlines reported that Lorimar wanted to acquire Multimedia, Inc., for a billion dollars; Mesa Petroleum pursued Unocal; Triangle Industries, a $65-million company, proposed to acquire National Can for $428 million; and Golden Nugget, a $230-million company, set out to buy the Hilton Hotels for $1.8 billion. Everywhere, minnows were sharpening their teeth and going after whales.

In March of 1985, Capital Cities acquired ABC for $3.5 billion, and when the Reagan government did nothing to block the takeover, it seemed clear that even in the heavily regulated communications industry there would be no holds barred. According to Bill Bevins, Turner's CFO, the ABC takeover "crystallized where the various regulatory agencies stood, and that the timing was propitious" for media takeovers.

This was exactly the signal Ted Turner had been waiting for. As he looked around the media world, knowing that he needed to expand, he decided that the company that seemed especially ripe for the taking was his old nemesis, CBS. According to Wussler and Hogan, Turner saw CBS paradoxically as both "the Tiffany network" and the network that, because of management blunders, exemplified "all that was wrong with television." When Turner publicly vilified the networks, he often singled out CBS, and yet, as a youngster, says Wussler, "Ted's father imbued in him the spirit of quality. The result was that, like his father, he saw CBS as terrific

TV, *National Geographic* as the best magazine, MGM as having terrific musical movies. These things stuck with him."

Turner wasn't the only one interested in the network William Paley had created. On Wall Street, the perception was clear that CBS was a company "in play." The *Wall Street Journal* quoted an analyst who said that "CBS is on the ropes," and declared that "on Wall Street, CBS shares have been transformed from a solid but unsexy institutional staple into a favorite of investors who make their living speculating on takeover proposals." Ivan Boesky announced that he had bought 8.7 percent of CBS's stock. Clearly, he had a feeling that something was about to happen.

The rumors had started in January of 1985. That was when a group of North Carolinians, under the wing of archconservative North Carolina senator Jesse Helms, had filed papers with the SEC declaring their intention to buy enough CBS stock to change the company's policies and end its alleged "liberal bias in news reporting and editorial policies." The group called itself "Fairness in Media," and had sent out a letter to one million conservatives, urging them to "buy CBS stock and become Dan Rather's boss." By February 20, Turner had come to Washington to meet with Jesse Helms in his Senate office, and to talk about joining forces in the "Fairness" run for CBS. They chatted for an hour, and Turner called the campaign "a great idea" from his "good friend" Helms. "I made it very clear," he would later say, "I mean it's no secret that I've wanted to acquire or be acquired by a network. . . . It goes back a number of years. I knew Helms wasn't going to leave the Senate to run a network if he was successful. And I suggested that I do have qualifications to run a network, and that was what my interest in meeting him was."

As Turner and Helms chatted, the senator dialed up Tom Ellis, acting director of Fairness in Media. Over the speakerphone in Helms' office, the three men discussed how they might join forces. Turner emphasized that he would have to remain in control of the news operation, but added that he'd be making changes in the entertainment programming that they'd probably like: "less violence, more pro-family, pro-American-type programming, and less sex and violence and stupidity."

The next morning, Turner talked with FCC commissioner James Quello about the implications of a network takeover, and during the following two weeks, both he and his Washington attorney Charles Ferris (a former FCC head) were seen around the capital, talking to various legislators and commissioners about the mechanics of a CBS bid, paving the way for what was to come. Turner also chatted with both Georgia senators, Sam Nunn and Wyche Fowler, about the wisdom of his joining the Fairness in Media campaign. They shook their heads: "You've got such good friends up here in Washington," Fowler said. "It wouldn't be smart to align yourself with one [political] side or the other." Taking their advice, Turner met with the Fairness in Media leaders only one more time. Their interest was cooling as much as his, because while Turner shared some of their values, it was obvious that what he was looking for, above all, was a business opportunity.

But if nothing came of the Fairness in Media meetings, it was clear that Turner and others (Time, Disney, New York realtors Larry and Zachary Fisher, and General Electric—which would later buy NBC) were still hot on the trail of CBS. And Turner himself was caucusing with MCI and former treasury secretary William Simon, trying to raise $100 million for his own CBS bid. From the end of February to mid-April, the rumors continued to swirl around Wall Street, and CBS stock shot up from $81 to $118 a share, as nearly two-thirds of the company's outstanding shares changed hands in a flurry of excitement.

In public, CBS laughed off the threat of a takeover, especially from a redneck rogue like Ted Turner, who operated on such a comparatively small scale. After all, the newspapers reported that he had a mere $50 million in available cash and an available credit line of only $190 million—hardly enough to tackle a multibillion-dollar company like CBS. Internally, Thomas Wyman shot off a memo to his department heads—a tart, tongue-in-cheek dismissal of his former sparring partner Turner. While acknowledging that the increased activity in CBS stock had "attracted heavy press coverage and speculation," he went on to say, "As we have indicated, we have no interest in any kind of association with Ted Turner. To the best of our knowledge there is no substance to reports that he plans a network takeover attempt. It would appear that his financial resources are already fully engaged." As CBS News head

In a rare pensive mood with the launch of CNN only days away, Ted Turner in May 1980 gazing out the French doors at his Hope Plantation home in South Carolina.
*(Beau Cutts/*Atlanta Journal and Constitution*)*

June 1, 1980, at the Atlanta inauguration of CNN, the first 24-hour all-news television network. "I guess we are on the air," Ted Turner said and then added, "I hope so."
*(Joe Benton/*Atlanta Journal and Constitution*)*

Turner family annual Christmas portrait in the mid-eighties with (from left) Beau, Rhett, Janie, Ted, Jennie, Laura, and Teddy Jr.

Hitting the large button, Turner launches his ill-fated Cable Music Channel in October 1984 at the station's Los Angeles headquarters. (AP/Wide World)

Chris Curle and Don Farmer, anchors of CNN's *Take Two* (a daily news and newsmaker program), interviewing their boss. (Courtesy Chris Curle)

On the cover:

SEXUAL HARASSMENT: THE NEW CORPORATE NIGHTMARE • PEOPLE WHO SEE UFOs

...SAY SCOTT, YESTERDAY'S HERO

ATLANTA

NOVEMBER 1989 $1.95

The Woman Who Tamed Ted Turner

Ted Turner and J.J. Ebaugh

His pilot, J.J. Ebaugh, for whom Ted left his second wife and a marriage of twenty-four years.

(Reprinted with permission from Atlanta Magazine)

On St. Phillips Island, S.C. (1985), to discuss African environmental issues, Ted Turner is seated between his second wife, Janie, and girlfriend J.J. Ebaugh. Seated on bottom step are (from left): CNN anchor Mary Anne Loughlin, executive vice president of WTBS Robert Wussler, TBS's executive in charge of environmental affairs Barbara Pyle, financial consultant Mike Mitchell, and conservationist Rick Lomba. Behind Loughlin is Lester Brown, head of the Worldwatch Institute. Behind Turner is CNN executive producer Bob Furnad.

With TBS's superhero Captain Planet, both battling to save the environment.

(Copyright © Bill Swersey/Gamma Liaison)

THE POWER IS YOURS!

Turner holding forth at the White House on the occasion of the signing of the Panama Canal Treaties in September 1977 with (from left) his wife, Janie, President Jimmy Carter, Rosalynn Carter, former President Gerald Ford, and Lady Bird Johnson.
(Courtesy Jimmy Carter Library)

Ted Turner talking with President Ronald Reagan at a candlelit White House dinner. *(Courtesy Ronald Reagan Library)*

Turner and his second in command, Robert Wussler, meeting in 1986 with Soviet President Mikhail Gorbachev and aides in the Kremlin on occasion of the first Goodwill Games in Moscow.

Enjoying a joke with Russian President Boris Yeltsin at the opening ceremonies of the Goodwill Games in St. Petersburg, July 1994.
(Photo by Alexander Zemlianichenko. AP/Wide World)

The new Board of Directors of TBS in 1988, following the bailout of the company by a cable consortium led by TCI and Time. In front row, from left: Michael Fuchs, Hank Aaron, Rubye Lucas, Ted Turner, John Malone, Bill Bartholomay, Gerry Hogan, Stewart Blair. In rear row, from left: Terry McGuirk, Tim Neaher, Gene Schneider, Joe Collins, Jim Gray, Robert Wussler.

Ted and his former Bible teacher, John Strang, with McCallie students on the occasion of his January 14, 1993, visit to Chattanooga. In 1994 he gave a gift of $25 million to the prep school.
(Courtesy The McCallie School)

With his date, actress Jane Fonda, Ted Turner appears at the 62nd Academy Awards in Los Angeles on March 26, 1990.
(Reuters/Bettmann)

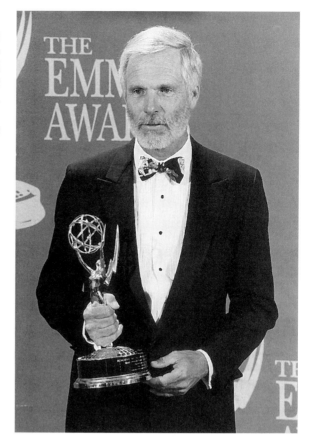

A bearded Ted Turner receiving the Governor Award at the Academy of Television Arts and Sciences Emmy Awards telecast in Pasadena, California, August 1992.
(Reuters/Bettmann)

Ted and Jane on their wedding day—December 21, 1991—at Avalon, his plantation in Capps, Florida. With them at the far left is her brother, actor Peter Fonda, her son, Troy, and behind him Henry Fonda's widow, Shirlee Fonda.
(AP/Wide World)

Ted and Jane at home on their Flying D Ranch in Montana with trophies on the wall and a telescope at the window. Turner says, "I'm spending more time in Montana than I should, and a lot less than I'd like to."
(Doug Loneman/Bozeman Daily Chronicle)

Ted Turner looking toward the future and his grandson, John Seydel III, Laura's child.
(Gamma Liaison)

Ed Joyce would later comment, it made one wonder where Wyman was getting his information.

Meanwhile, behind the scenes, the top CBS brass had decided to get tough. If they were under attack, perhaps from several directions, they would fight back on all fronts and hard. Senior CBS executives were quoted as saying that they'd make any foe "sorry he ever got up in the morning." In March and April, they filed lawsuits against both Fairness in Media and Ivan Boesky—clever courtroom maneuvers designed to knock their opponents out of action. And Thomas Wyman himself called the heads of investment-banking houses throughout New York City, looking for support and dropping not-so-subtle hints. At Shearson Lehman Brothers, where executives had been holding talks with Turner, he was especially blunt. He warned James Robinson III, who oversaw Shearson as head of American Express, that to represent any would-be takeover artist would constitute conflict of interest, as Shearson had worked for CBS in the past. (In fact, Shearson had represented CBS only on one small matter, but CBS was playing hardball.) All over New York, law firms got the message loud and clear that anyone who did business with a CBS foe could kiss good-bye any future work from the blue-chip broadcasting client.

Wyman and his team felt confident that they had established themselves as tough, pugnacious foes. Of all the possible threats to CBS, from Helms to Boesky, it was clear from Wyman's memo that Turner's was taken the least seriously. He was easy to dismiss. Turner was like the little kid who always whined about wanting to play with the big boys; he was from the boondocks of Georgia, not New York or Washington, and he ran the Chicken Noodle Network. His reach constantly exceeded his grasp. Having seen proof of this back in 1981, Wyman wasn't worried about Turner.

In late March, CBS's Ed Joyce traveled to Moscow, where he met with Vladimir Lomeiko and Valentin Kamenev, the chief and deputy chief of the press department at the Soviet Foreign Ministry. The Russians knew all about Ted Turner, having met him in Moscow, and were eager to hear the latest gossip about his CBS takeover attempt. Joyce shrugged off Turner as "overly ambitious, without adequate resources." But as he was leaving the Foreign Ministry, Kamenev, who'd spent eight years in Washington at the Soviet Embassy, stopped him at the door.

"Don't underestimate this Turner," he said, smiling. "Remember, the Bolsheviks were a small group, and they took over an entire country."

On the morning of April 18, 1985, in the red-carpeted ballroom at the back of New York's Park Lane Hotel, more than two hundred reporters and photographers jammed together under the massive cut-glass chandeliers to hear what the man up at the podium had come to say. Surveying the surging crowd and the flashing strobe lights—"like a scene from *Day of the Locust,*" an aide would later say—even Ted Turner himself seemed slightly flustered by all the commotion he had stirred up. "OK," he said, "wow," as if the full enormity of what he was about to do was only now sinking in. The flashes kept popping, and in the mirrored room the light bounced crazily back and forth off the glass and the ornate gold wallpaper.

Unfolding his prepared statement, Turner began to read: "We have been very interested in joining forces with one of the three networks, because of our desire to be number one in the business." Then he laid out his bid for CBS. TBS would offer $5.41 billion to take over the media giant, paying the shareholders $175 a share for all of CBS's 31 million common shares. The deal would be financed with no cash at all, just a combination of stocks and junk bonds. "I'll never call them junk bonds," Turner told his audience, grinning. "That's what the doomsayers call them. . . . I like to call them *high-yield.*"

Turner was trying to do with CBS what he had done with WTBS and the Braves—to borrow against the company's own assets to buy it. In this case, according to David Good, of E. F. Hutton & Company, who had put the deal together, the proposal was "the first leveraged buyout directly to shareholders." In other words, the CBS stockholders would have to give up their real stock, worth about $110 a share, for a Turner IOU of $175, which might or might not pan out. Turner had been unable to find support among the normal New York banking and law circles and had turned to the fledgling mergers department at Hutton and the West Coast legal firm of Latham & Watkins ("We couldn't get a law firm," recalls Robert Wussler. "Every big New York law firm said, 'We have to deal with CBS'"), but they had finally come up with a creative bid. "The only thing Turner didn't throw into the deal were the bamboo steamers and the Ginzu knives," one analyst said. "It's all paper, nothing but paper—there's no money," said another.

For sheer impudence, the CBS-takeover proposal was breath-taking. TBS, a company with revenues of $282 million, was attempting to swallow a huge multimedia conglomerate that had revenues of $4.9 billion. It was like a little kid with a lemonade stand trying to buy out the neighborhood supermarket. Reporters called the bid "novel," "his longest shot yet," "the most audacious takeover bid ever." No one had ever accused Ted Turner of lacking chutzpah.

But on Wall Street, the mood was skeptical. Most analysts dismissed the offer as frivolous. One even called it "a joke." That day, CBS stock dropped $3.50 a share, in what the *Wall Street Journal* termed "an extraordinary repudiation" of Ted Turner by the market. One analyst commented that "the probability is zero that Ted Turner will end up with CBS." And even Turner's own advertisers were less than supportive. "They were shocked," Gerry Hogan remembers. "They couldn't believe it. We were outsiders. Cable was still a very small deal, and its future was somewhat questionable. There's this strong old-boy network in advertising, and they just couldn't believe we would even consider buying CBS. They said we were crazy."

Other groups responded more or less as expected. At Fairness in Media, cofounder James Cain declared his organization "delighted" by this move from an "old friend of Jesse Helms." At CBS, the reaction was a harrumph of outrage. Mike Wallace termed Turner an "interloper," and CBS affiliates declared the proposed buyout "totally repugnant . . . a disaster for the American people to have the CBS network come into his hands." Internally at Turner Broadcasting, employees regarded the whole thing as a grand adventure, except for those who worried and shook their heads, declaring, "This is too much of a stretch." The uproar was just beginning.

One night toward the end of June, as New York high society gathered at the Metropolitan Museum of Art to honor oilman Armand Hammer, one particularly distinguished navy blue-suited guest stared across the flag-draped Arms and Armor Hall at the tall, slim figure standing beside one of the displays of armor and stopped in midsentence. "That's not Ted Turner," he said. "It can't be."

As it turns out, it was. There among the champagne and hors d'oeuvres was the brash Atlanta cable king himself, showing up at the height of the CBS war, in the backyard of his enemy, to honor his friend Hammer. "I don't know what all this ranting and raving is about," he announced to the guests who came over to surround him. "What am I but a free-enterprise guy? The same way as Bill Paley when he started CBS." Turner shifted self-consciously from one foot to the other. "Paley was an outsider when he started, too."

If Turner initially seemed a little ill at ease among the cream of New York's business and social elite, as the night wore on he was soon entertaining and befuddling the partygoers. He threw his arms out theatrically; he kissed the hands of socialites and shadowboxed with tycoons. And always his pale blue eyes swept the room, constantly looking ahead to the next guest. At his side, pretty and perhaps just a little overly made up, stood Liz Wickersham. "I'm from Texas," she warbled. "From Orange, a little town you never heard of. . . ."

But Turner was dominating the conversation that night. "As I've grown older," he declared to anyone who'd listen, "I've become less desperate. I've decided to be more statesmanlike. I'm going to be a gentleman."

Liz Wickersham started to chuckle. "If so," she asked, "how will I know you?"

Ted pointedly ignored her, his gaze fixed on the party crowd. "Gee, will you look at that!" he exclaimed. "That's Walter Cronkite." He launched into the crowd to confront the white-haired elder statesman of journalism. Pulling up in front of Cronkite, he declared, "You and I would get on like a house on fire!" Then in a move that went back to his old Gary Matthews baseball days, he promised Cronkite a million dollars a year if he would go back on the air. "When I own CBS, . . ." Turner rambled on, as if it were only a matter of time.

The evening held one last surprise. At the postreception dinner at the swanky "21" club, as Turner arrived carrying a briefcase full of his papers from the CBS bid, he was informed that Joseph Flom, the celebrated mergers-and-acquisitions lawyer working for CBS, was in attendance. Turner insisted on being introduced to the man who was leading the charge against him. And as they were introduced, Turner, perhaps thinking that mere words were inadequate for the occasion, promptly kissed Flom on the top of his head.

CBS, however, was not in the mood to pucker up. The company had offically responded to Turner's bid in what, given the restrained language of financial dealings, was a broadside assault. CBS said it was rejecting Turner's proposal as "grossly inadequate . . . financially imprudent . . . [a move that would] bankrupt the company." They attacked Turner on all fronts, first with the SEC, and when that didn't work, with the FCC. They stirred up their affiliates as well as community, labor, and civil rights groups from the NAACP to the United Church of Christ. Letters to the shareholders cited his old racist MX missile blunder and "a number of pejorative statements by Mr. Turner about various minority, religious, and ethnic groups." Over the previous several months, Thomas Wyman himself had declared that he didn't think Ted Turner was "moral enough" to run the network, that he lacked "the conscience."

The more forceful argument was an economic one. Wyman and CBS stated that the Turner purchase would put CBS on a "collision course" with financial disaster. Turner had shown a profit only because of tax credits, they declared. WTBS ran nothing but reruns and ball games. Turner didn't know how to operate a real network. He would lose the advertisers; the company would go into a "death spiral." There was a "high risk of financial ruin."

Yet as outrageous as Ted Turner's proposal was, he did in fact have a certain financial plan. He believed, as many would later come to agree, that multimedia companies were being fundamentally undervalued, and that the parts were worth more than the whole. He planned to raise $3 billion by selling off the subsidiaries of CBS he didn't want—the radio stations, the record business, the book-and-magazine publishing group.

In his proposed combination of CBS and TBS, there were also potential economies of scale: "You could gather the news for both CBS and CNN with one organization," Turner said. "That's one hundred million in savings right there." Then, too, according to Robert Wussler, you could spread other costs: "The idea was you could buy two runs of a sitcom for CBS and six more for TBS, two runs of a movie for CBS and two for TBS, or buy the rights of a sporting event with secondary rights for TBS."

Turner felt that there were valuable synergies possible between the two companies, one operating on cable, one on broadcast. And he felt that it wasn't just a question of savings: the moneymaking potential was huge. He wanted to grow his company. It needed to expand. Like a youngster who has outgrown a pair of pants, Turner Broadcasting was ready to step up to the next size. While Ted talked about taking over CBS for moral reasons, to remove the "sleaze" and "stupidity," the key issue for him was distribution—he needed the ability to reach more viewers. CBS had ten to twenty viewers for every one watching WTBS or CNN. "I needed distribution," he would later admit, "and CBS represented distribution."

But CBS itself was fighting back with everything it had. In Washington, the company continued to lobby the FCC to reject the Turner bid out of hand, just as it had thrown out Howard Hughes' attempted purchase of ABC back in 1968, when the rich recluse refused to appear before the commission and be judged on his fitness to run TV stations. On Capitol Hill, CBS got Thomas Eagleton and five other senators to introduce a bill requiring extensive evidentiary hearings before any network takeover. And in Albany, CBS persuaded the New York State Legislature to introduce a stiff anti-takeover measure that would make almost any hostile noncash takeover a virtual impossibility by requiring a two-thirds approval of stockholders. The bill passed the legislature almost unanimously. But it touched off a major ruckus, as pro- and anti-takeover law firms and investment houses realized the ramifications for their revenues and jumped into the fray, inundating Albany with telegrams. Governor Mario Cuomo agonized over whether or not to veto the bill, and decided to put off his decision until it would be irrelevant—and a lot less politically risky. (The measure would eventually fizzle away.)

CBS had meanwhile retained two powerful blue-blooded firms to represent the company, the Morgan Stanley investment group and the law firm of Cravath Swaine & Moore. Why so much ammunition, if the Turner threat was so laughably thin? That was a question even CBS staffers were asking. "There's a lot of money loose in the marketplace," CBS general counsel George Vradenburg told them. "If Turner can make any progress, others will ask to join him, and he'll improve his offer."

By July 3, CBS had decided to stomp out the pesky Turner bid once and for all. The company would swallow the so-called poison pill, taking on a huge amount of debt—nearly a billion dollars—

to buy back 21 percent of its stock. This would be such a heavy burden that Turner would find it impossibly expensive to buy CBS. It was a bold and somewhat self-defeating move, like a young woman deliberately scarring herself to drive away an unwanted suitor. But it was nonetheless effective.

Ted Turner immediately realized that his takeover hopes were looking bleak. He attacked the CBS strategy at the National Press Club: "Poison pill," he scoffed. "The last guy I know who took a poison pill was Hermann Göring." But now the roles had switched. Now Turner was the one asking the FCC to block his opponent's plan. Now he was asking the FCC to hurry hearings, which the FCC refused to do. Sending his lawyers to a federal judge in Atlanta, he challenged the CBS buyback.

On July 31, the newspapers announced that two rulings—one by the FCC, one by the federal judge—had allowed CBS to go ahead with its stock repurchase. "This kills Turner's existing bid," declared one analyst. "It's dead in the water." Turner himself vowed to press on, to appeal, to do whatever it took, but the next day the FCC declared that it was canceling all hearings on his takeover bid.

The commissioners' announcement that such hearings were "not appropriate at this time" was restrained, understated. But if you listened closely, you could hear the coffin lid closing on Turner's network dream.

In the aftermath of the CBS takeover attempt, new facts came to light. Five days before the April bid was announced, Turner had met with his board of directors to inform them that he intended to take over CBS. "Why are you doing this?" one board member asked worriedly. "Ted, you're going to be throwing our money away."

Turner then admitted—in private, and well before he had begun his CBS bid—that he himself estimated that it had only a 10 percent chance of coming to fruition. "I probably won't make it," he said. "It's a long shot, but I'm gonna take my shot." Then he added a crucial point. Even if the bid didn't work, he said, there would be other benefits: the publicity would make it easier for TBS to sell ads. "We would be perceived as a more major player on Madison Avenue," Ted declared. Turner was once again pursuing his old strategy: by choosing major-league opponents, one immediately places oneself on their level, even in defeat.

The directors weren't overwhelmed with his logic, but they went along. Said one director, Mike Gearon, "I didn't see that it had much business merit, or likelihood, but the company had just made a lot of money, and if Ted wanted to do this kind of crazy thing just to kick CBS in the ass—if he wanted to spend twenty or thirty million doing it—well, he had made twenty or thirty million at that point, and if he wanted to do it, he could do it."

In the end, TBS lost $18.6 million in legal and other fees on the CBS takeover attempt, a not insignificant sum. Still, the loss had its compensations. For all the outrageousness of his attempt, the CBS fight had in fact established Turner in the minds of journalists and the American public as a "player." Just as the America's Cup victory had brought him to national prominence, just as the launch of CNN had marked a step up from notoriety to a certain kind of distinction, the CBS battle was a milestone—one that, despite all the derisive comments from the media establishment, established Turner as a business leader.

As for CBS, the company's victory was Pyrrhic. "In the end," says Gearon, "it was worth every cent Ted spent. It really left CBS crippled in many ways." Burdened with some $954 million in debt, Thomas Wyman opened CBS's doors to a "white knight," financier and philanthropist Laurence A. Tisch, of the Loews Corporation, who bought up 12 percent of the company in three weeks in July. CBS undertook a massive cost-cutting drive and, ironically, ended up doing many of the same things Turner had proposed, and even more: it eliminated 10 percent of its national news department staff and sold off St. Louis station KMOX, as well as the company's entire record and publishing divisions. And as for Thomas Wyman, Tisch, head of CBS, Inc., and Bill Paley proceeded to boot him out the door.

CHAPTER 20

ONLY A FEW DAYS after CBS swallowed its poison pill, Kirk Kerkorian was on the phone with Ted Turner. A quiet, low-key financier, the scion of a family of Armenian rug merchants, Kerkorian had made a fortune in the airplane business and knew when the time was ripe to cut a deal. Turner was still smarting from his CBS loss, smarting so much that he wouldn't even admit that he had lost; he was still talking about proxy fights.

"Ted," Kerkorian said, "I admire what you've tried to accomplish with CBS. Let's see what we can do with MGM." Kerkorian had been the partial owner of the old Metro-Goldwyn-Mayer film studios since the early 1970s and had acquired United Artists in 1981. All through the early 1980s, while Turner was talking to CBS and the other networks about mergers and combinations, he was also talking to Kerkorian about creating a joint MGM/TBS.

About four times a year, they would meet in New York or Atlanta or Los Angeles to see if there was some way they could form a strategic partnership or a united company. On one September evening in 1982, Kerkorian and three of his senior executives came to Manhattan for an early dinner with Turner at the New York Yacht Club. Ted and Robert Wussler had just flown back from Scotland, and Turner was in fine form. He led the group through the yacht club for the full guided tour, pointing out displays, regaling them with stories of previous America's Cup winners, and waving his arms as he spun seamen's yarns. Walking by the various boat models, he reeled off the name of the captain of each, the name of its designer, and a synopsis of its record.

The group settled down at a round table in the darkened dining room, and began to talk strategy. If the food was mediocre, the conversation wasn't. Kerkorian, laid back though he was, had clearly enjoyed the yacht-club treatment, and Turner, as always, was full of ideas: "Could we carve out two hours each night at TBS and just run MGM films? Could we carve out the weekends for MGM?"

While no immediate plan would come out of this dinner, Turner was clearly captivated by the notion of teaming up with MGM. In his father's corporate pantheon, MGM (like CBS) had stood for quality.

So when Kerkorian spoke again with Turner on the phone that July day in 1985, he knew he was talking to an interested buyer. And with the CBS bid a failure, Kerkorian felt that the time was propitious. Recent interest in MGM from the wealthy Al Fayed family of Egypt and Charles Knapp, the flamboyant "Red Baron" of the savings-and-loan world, would serve only to pique Turner's appetite.

On paper, MGM was no winner. In fact, in the previous nine-month period, MGM/UA had posted a $66.2-million loss. As wily a deal maker as Kerkorian was, his movie-industry track record had been very poor. In the 1970s, just as the film business was becoming highly profitable, MGM was scaling back production, and through the early 1980s the company had accumulated a string of flops. All in all, financial problems dogged the studio.

On July 25, Kerkorian told Turner that he could have MGM/UA for $1.5 billion. But he added, "I don't want to sit here forever and not know whether I have a deal or not." In fact, he didn't even want to wait two weeks. Applying the screws, Kerkorian declared that he'd give Turner until August 6 to make a decision. After that, he was going to put MGM/UA up for auction.

With only two weeks to make a decision on the biggest deal of his life, Turner sent TBS into a maelstrom of activity. Forty lawyers and accountants, under Bill Bevins, were dispatched to MGM headquarters in Hollywood to examine the company's financial records and to try to sort through the maze of contracts that promised rerun rights of the old MGM movies to thousands of different TV stations around the world.

While that was going on, Turner called a directors' meeting to sell the MGM deal to his board. It is indicative of the magnitude of the risk of the proposed purchase that even this essentially rubber-stamp group of Turner's friends and employees had serious reservations; they wondered whether their chairman was overreaching, feared that he was being led astray. "The directors' meeting where MGM came up was like a baited trap," one board member recalls. "Turner couldn't avoid it. He had to go for that goddamn deal because he had just walked away from losing CBS. And so here was MGM—something that in theory we could afford. At least the way Bevins explained it. There was a lot of fast talk about what we would do—you do this, you sell off that—fast talk. Bevins was marvelous at fast talk."

Turner himself had plenty to say that day in Atlanta, and he spoke with passion. He didn't like TBS's strategic position. To be a long-term factor in this business, he said, you either had to grow in viewership (the CBS deal) or in the programs you owned (the MGM deal). And TBS wasn't strong enough in either. "It's essential to have this additional programming to make TBS competitive," he said. "We gotta do this to survive. We *need* to do this."

He described just how valuable the MGM film library would be to WTBS, his core business. "Movie fees are rising," he said. "Everytime we sign a new contract [to rent films], it's more, more, more." Profits would keep shrinking. But imagine actually *owning* all of those classic films. "How can you go broke buying the Rembrandts of the programming business, when you are a programmer?" he asked.

Turner was dying to step up to the big leagues. "It's a business where the big are getting bigger and the small are disappearing," he'd later say. "I want to be one of the survivors. This makes us more of a major player. It gets people's attention. I mean, we *are* in show biz. . . . Without MGM, there's a question mark with advisors about our long-term viability. They used to throw that up to me. Early on, there was a certain group of people in the New York establishment—the same sort of thinking that gave me such a hard time at the New York Yacht Club. You know: 'He's from Georgia, we don't need him around here. Here's a quarter, go shine my shoes.' You know: Step 'n' Fetchit."

As he wrapped up the board meeting, Turner was once again rallying the troops. "We *need* to do this," he said. "Let's get on it."

And gingerly, anxiously, the board went along. Board member Mike Gearon recalls, "Ted realized how important it was to have these additional programs. But once he got into it—brother! He had barely made up his mind that he was going to get ahold of these guys, when he'd sink his teeth in, throw his parachute off, and jump out of the plane. And somebody better catch him before he hit the ground."

On August 6, 1985, Ted Turner signed a purchase agreement with Kirk Kerkorian to buy MGM/UA for the asking price of $1.5 billion, or $29 a share. The deal would include the MGM studio, the fabled MGM lot in Culver City, and, most important, a library of 3,500 MGM films, including 1,450 films from the old Warner and RKO studios—a selection of classics like *Gone with the Wind, The Maltese Falcon, The Wizard of Oz,* and *Citizen Kane.* (Not included at the time were either United Artists or the UA library—containing such films as the lucrative *Rocky* and James Bond series—which Turner would sell back to Kerkorian for $480 million as part of the deal.) Ted called the package "a tremendous business opportunity" and an "exceptional fit with our long-term business plan."

But others were less generous. On both coasts, in Hollywood and on Wall Street, at the star-studded Ma Maison and the fashionably sedate Four Seasons, loud guffaws could be heard. The way show-business executives figured it, Turner, the overly eager naïf, was going to have his pockets emptied on this one by the shrewd Kerkorian. The price of $1.5 billion was a good $200 million to $300 million too high. Industry bankers valued MGM at $22 a share; Bill Bevins himself put it at $24 or $25 a share; and as for Ted, he was perfectly willing to pay the full asking price of $29 a share without even negotiating. *Newsweek* reported that in Hollywood "Turner is almost universally regarded as 'a pigeon.'" One analyst declared that the movie studio was only worth that much if "Turner's found oil on the MGM backlot." It appeared that Turner's gambles were getting more and more outrageous, and this was the most outrageous one yet.

Some members of the TBS board were starting to get worried. Mike Gearon says, "I felt that all of a sudden Ted had gotten in over his head. We weren't even being properly informed. We had inadvertently given Ted too much authority—we had said he had the right to negotiate with a lot of latitude, but we were responsible for whatever the hell he did. And the price seemed to exceed the representations that had been made to us on the board. I thought there was a lot of directors' liability at that point; the company could be destroyed. Probably more than anybody, I was afraid and expressed myself, because I had some assets and I didn't want to lose them. And I saw myself getting wiped out by this guy."

Turner didn't disagree. According to Gearon, Ted was concerned, too. They were, after all, working from the same facts. "I just think that he was prepared to take the risk, and I probably wouldn't have been prepared. . . . It was a lot to lose."

If the board was anxious and Hollywood amused that Ted was paying the asking price without even haggling, it was not an unusual move for him. An eternal optimist, he always seems to imagine that what he's excited about and is purchasing may actually be worth the asking price. "Because he believes he sees hidden values," Reese Schonfeld says, "he pays more."

Schonfeld tells the story of how Turner had toured him around his new plantation, proudly declaring: "There's peat on that land. There's a peat bog. It's worth a fortune. They thought they were selling off a plantation, but I was buying a peat bog." Of course, Turner never did become a peat baron. And another real-estate deal—Pritchard Island—that Turner came within a hair's breadth of buying for the asking price of $25 million, sold ten days later on the courthouse steps for $12 million.

"Ted used to say that his father had told him, if he really wanted something, if he had to have it, it didn't matter what he had to pay," Gerry Hogan recalls. "In that context, Ted would spend huge amounts of money, more than were necessary . . . but on a day-to-day basis he was very tough with the dollar—very, very cost conscious. Unless he wanted a right-handed reliever—then it didn't matter."

Two days after the MGM deal was set in motion, as his board fretted and his aides scurried around feverishly trying to hammer out the details, Turner suddenly took off for Alaska with his family to fly-fish on a north-country river. The currents surging past his legs, the line singing through the morning sky, the fish rising to the lure, Turner concentrated on rainbow trout and Alaskan sockeye and let others worry about taking care of his business.

His lieutenants were hard at work around the globe. In Moscow, the fastidious Robert Wussler was making his way through the Soviet bureaucracy to arrange the Goodwill Games. In New York, the energetic Gerry Hogan was selling ads, trying to keep the dollars rolling into WTBS. And out in Los Angeles, at the Beverly Wilshire Hotel, the company's skinny, bespectacled chief financial officer, Bill Bevins, had plunged full-tilt into his new role as deal maker. A former accountant with a trace of his native East Tennessee accent, this intensely private Southerner had come to love the action of the Hollywood deal. And although he had had a heart attack back in the early 1980s, at the height of the battle with ABC/Westinghouse, when money was tightest, he was still smoking and working overtime to make the MGM buyout a reality.

At the core of the MGM deal was the king of the junk bond, Michael Milken, of Drexel Burnham Lambert. If Drexel was "highly confident" they could arrange the $1.5 billion in financing for Turner's eye-opening purchase, theirs was not a totally disinterested opinion: they had a huge stake in the proceedings. Because of the speed Kirk Kerkorian insisted upon, Turner had been forced to retain Milken, Kerkorian's own moneyman and the only one familiar enough with MGM's balance sheet. In other words, Michael Milken was working both sides of the street—getting paid both as MGM's banker and TBS's financier. It was a highly unusual setup and raised certain conflict-of-interest questions. Ironically, though, if it hadn't been for Milken's major stake in the sale, the buyout probably would never have happened, since Drexel Burnham was one of the few financial firms that had enough power to raise money for such a speculative transaction. In a way, the MGM/TBS deal—from the players to the financing—would come to be emblematic of the risky, high-flying go-go 1980s.

But first the buyout had to happen and, as *Variety* would later say, "the deal began unravelling before the ink was even dry." Between mid-August and late November, four new MGM films bombed, one after another. *Year of the Dragon* and *Code Name: Emerald* were bad, but *Marie* and *Fever Pitch* proved to be disasters, together losing almost $29 million of a $30-million investment. Bevins returned to Atlanta shaking his head and, face-to-face with his boss, declared, "Look, we've got a problem here." It would come to be a recurring refrain. Later he would comment, "When you suffer the kind of losses we suffered at the studio, the financing becomes a virtual nightmare." And there was an additional problem as well: Only two days before the August 6 agreement had been initialed, MGM/UA had signed a minor contract for pay-TV rights with a small East Coast distributor called Rainbow Services—a contract that had a killer clause, locking up all basic cable rights. Neither Turner nor Kerkorian had quite realized it at the time, but it would take months of lawsuits and countersuits to sort out.

Even before all these snags, Drexel had been having trouble rounding up buyers for Turner junk bonds to finance the MGM deal. But after the motion-picture flops, it seemed as if investors were avoiding the bonds just as assiduously as moviegoers were avoiding MGM films.

Drexel and Turner employees had been working long, desperate hours, but on October 1 and into October 2, they labored all through the night, bargaining and refining. Finally, they emerged with what one PaineWebber analyst, with a sigh of relief, described as "a brilliant piece of financing." The price of MGM was to remain the same, $1.5 billion, but now Turner would come up with the $29 a share by offering one share of Turner preferred stock (worth $4) plus $25 in cash. It seemed that the deal was finally on track.

Yet despite these revised terms and the avaricious appetites of the junk-bond-buying public of the mid-1980s, the studio was still not drawing investors as the weeks went by. So Ted Turner began to look around to see what he could sell off to raise money—anything to bring in the dollars. "There seems to be an atmosphere of urgency, if not panic" at TBS, the *Wall Street Journal* reported. Turner, never a recluse, now began to talk to anyone who could give him a desperately needed infusion of cash. Time, ABC, Gannett, Viacom—all these and others met with him over the next year, all circling around looking for bargains, all trying to bite off pieces of TBS and MGM. Viacom was interested in buying half of the MGM studio but kept dropping its offer, from $225 million to $175 million ("trying to squeeze blood out of a turnip," in the words of one Drexel partner), until the deal foundered. NBC was offering to buy half of CNN for about $285 million. Turner thought the news network was worth at least $100 million more, and besides, he didn't want to give up control of CNN to anyone. Silvio Berlusconi, Time's Nick Nicholas Jr., and HBO's Michael Fuchs also talked to Ted, as did Australian media magnate Rupert Murdoch.

If there was any individual on the media scene whose ambitions paralleled Ted Turner's in scope, it was probably Murdoch. In the same year that Turner had gone after CBS and MGM, Murdoch and his partner Martin Davis had acquired 20th Century Fox and a mini-network of six Metromedia stations, and he did so more efficiently and cheaply than Turner was doing in any of his proposed deals. Murdoch and Turner first met early in 1986, when the Australian tried to hire Gerry Hogan. Hogan decided not to leave Ted, but introduced the two men, and they met again three or four times over the following months.

On one of these occasions, Murdoch traveled to Turner's office in Atlanta. Adjourning downstairs for dinner at Bugatti (the Italian restaurant in the Omni Hotel, which TBS had just bought), Turner, Murdoch, Wussler, Bevins, and Barry Diller (then head of Fox) clustered around a table under huge, flashy pictures of classic cars, and tried to find a way it made sense to blend Fox and MGM/TBS. Quiet and understated, Murdoch kept his focus on the numbers. "What if we sell off this piece of MGM?" he'd ask. "What if we . . . ?" Ted, meanwhile, was talking about the big picture, the big ideas. "We could be the biggest thing in Hollywood," he'd say. "We could build more networks. If we put your library and our library together. . . ."

Cordial and intensely deal-oriented, the two men nonetheless attacked the same problem from very different angles, for though they got along well, their instincts and temperaments were quite different. Gerry Hogan recalls, "Ted was this wild, idea guy. Rupert is a tremendous financial guy; he understood numbers and the ways to get things done financially far, far better than Ted. Ted doesn't have many strengths in that area at all. He relied on Bevins to make those things happen. What Ted had was vision—how powerful the combination could be, how important."

But on this particular evening, the conversation would fall apart because of a third participant: Barry Diller. According to Hogan, the chemistry between Diller and Turner was worse than bad; it was terrible: "I don't think Diller wanted any part of Turner. I think the merger would have been a direct assault on Diller's position. If Turner and Murdoch had merged . . . it would have cut into Diller's area."

That night after the meeting, Hogan flew back to New York with Murdoch, who was still focused on trying to sort out the finances: "Murdoch was intently serious in trying to make the numbers work, but he was having a real tough time, because both companies had so much debt that it was as if two punch-drunk fighters were trying to hold each other up. In the end, it just couldn't be done."

The deal that came closest to happening was with Allen Neuharth and his Gannett Company, the owner of newspapers across the country, including *USA Today*. While specific dollar amounts were never agreed on, the idea was to do a straight merger of the two firms. Neuharth was eager to merge with Turner Broadcasting. He would be able to expand into cable TV and film, his newspapers and CNN could share resources, and Ted would get the massive infusion of cash he needed. "That deal was the road to Easy Street," says Robert Wussler. Nervous members of the board urged Ted to take it. "Ted," one said, "you can walk off with eight hundred million dollars in Gannett stock. Maybe that's the safe course to take." The merger showed every sign of happening. In fact, when Neuharth and his aides flew down to Atlanta and were shown around CNN by Turner, he introduced them to staffers and his wife Janie as "my bosses."

But then Turner started to have second thoughts. "He got nervous," Wussler remembers. "He said he wasn't sure." Half of the problem came from Ted's philosophical objections to the Gannett businesses. "Ted is antiprint," Wussler says. "He doesn't believe in the future of publications. He said he thought that newspapers would be dead in ten years. Ted's ecological sense was against newsprint, something you read each day and throw away. He always talked about that." In part, then, the eco-Turner sank the lucrative Gannett deal. But in part he had simply been following his standard operating procedure, the methods he had developed over the years. He had flirted that way with prospective partners for his billboard company in the mid-1960s and for CNN in the early 1980s, allowing them to think that they had bought the company, and then taking their offering prices to his bankers to demonstrate his firm's rising value. Whatever the reasons, Ted Turner fundamentally believed that the company was *his,* and he had no intentions of sharing it with anyone.

By late 1985 and early 1986, however, Turner's position was beginning to look more desperate. Not only was he short of money, but

NBC—once a CNN suitor—now seemed about to start a cable-news service of its own in direct competition with CNN. Reese Schonfeld had resurfaced at the head of this new cable-news venture (under NBC News president Larry Grossman), and so an NBC internal memo reflected personal experience when it warned that "Ted Turner will go to the farthest limits possible to bar entry of a direct competitor, especially if one of the three networks is behind it."

The NBC News cable venture was well on its way and serious talks were in progress with the potential customers, the cable systems, when Turner telephoned John Malone, head of TCI, perhaps the most powerful of all these cable operators. Turner was calling in a favor—the favor he had done the cable owners when he started his ill-fated Cable Music Channel at their urging. He was also calling to explain Ted Turner's facts of life: "By putting the Satellite News Channel in business, the cable operators nearly drove us out of business," Turner recalled. "They came within an inch." Now he was asking Malone and the rest of the cable industry to back him, a cable stalwart, against this network interloper. He said, "You put them in business and I'm gone. I would have to sell out. I wouldn't just disappear; I would sell the company. Because there's no way [I can survive], with MGM hanging over my head and all that additional debt." This time, Malone and the cable industry stood behind Turner. By refusing to buy NBC's cable news, Malone and his peers had put the venture out of business by February of 1986. As powerful a force as NBC was, as serious a programming option as they offered, the cablecasters had decided that it was time to rally around their pal Ted. His networks had worked for them so far, and if they had to make a choice, they would stick with a proven commodity from a longtime ally.

It was lucky they did, because out on the West Coast the MGM deal was looking worse than ever. By now, two more movies had been released, *9½ Weeks* and *Dream Lover,* and both had also failed miserably at the box office. Turner tried to put a brave face on it: "Mr. Kerkorian is no dummy," he said, tipping his hat to the shrewd businessman. "He knew what he wanted to sell, and what he sold was the troubled part of the company. . . . But I have always bought troubled things. Normally, things aren't for sale if they're in great shape. Right?" Still, as Bill Bevins would later admit, they hadn't realized just how bad things were: "In twenty-twenty hindsight, the fact that the studio was in free fall was not all that clear at the time."

Bevins was locked in yet another series of meetings in Los Angeles to renegotiate—yet again—the financing of this increasingly risky deal, as the junk-bond buyers continued to stay away in droves. Turner was experiencing "certain difficulties in completing the transaction," reported the papers. But Bevins had the bit between his teeth, and while he and his boss had had their share of conflicts in the past (Bevins had, on occasion, labeled Turner "a son of a bitch" behind his back), still he was just as eager as Ted to make the MGM purchase work. He loved Hollywood, loved doing a big-time deal, and now that it came down to crunching numbers, he was in his element. In fact, he was working so intensely and smoking so much that some of his colleagues, recalling his heart condition, literally began to fear for his life. But as Bevins and Drexel continued to play with the numbers, now dropping the cash component down to $20 a share, the one thing the CFO wanted to do was keep Turner back in Atlanta, away from Michael Milken and the negotiating table.

"Bevins played that extremely intelligently," recalls Robert Wussler. "He kept Ted and Milken apart. They were two gigantic egos, and you don't put two gigantic egos in the same room except when you really have to." Only occasionally would the two meet, in brief moments before the market had opened at 5:18 A.M. West Coast time. Milken prided himself on being scheduled to the minute; so they would talk for twelve minutes or seventeen minutes in Drexel's Wilshire Boulevard office or in the Polo Lounge at the Beverly Hills Hotel.

In late March of 1986, Turner flew out to Los Angeles for his most important meeting yet with Michael Milken. After eight months of frantic negotiations and renegotiations, eight months of grueling fund-raising and hard-core junk-bond sales, the Drexel team had finally found enough high rollers—or bottom feeders—for the TBS notes to finance the MGM purchase. The deal would go through. Turner, relieved and ecstatic, asked Milken, Bevins, and staffers from TBS and Drexel to join him at the conference table. And then, in a notion he must have picked up from his girlfriend J.J.'s New Age inclinations, he asked all these hard-nosed businessmen, in their rolled-up shirtsleeves and their loosened ties, to clasp hands around the table. Looking sheepish, the financiers complied, and as they joined hands, Turner led them in directing their positive energies toward the outcome of the deal.

Milken himself was bemused by the whole scene, but he wasn't about to forget where the bottom line lay. According to Robert Wussler, he announced to Turner and his men, "Oh, by the way, our fee is now one hundred forty million."

"Wait a minute," snapped a member of the Turner group. "We agreed to eighty million."

To which Milken replied, "Yeah. I changed it to one hundred forty million."

There was nothing more to be said, as Drexel controlled all the money. Besides, the deal had been a backbreaker.

On March 25, 1986, Turner publicly announced that he had completed the purchase of MGM, selling UA back to Kirk Kerkorian, and once again the industry reaction was amusement, perhaps even more now than before. "It's one of the nuttiest deals of all time," said one industry analyst, laughing. Another joked, "Ted Turner came to town fully clothed and left in a barrel." Even the sober *Wall Street Journal* declared the new TBS/MGM "one of the most debt-ridden companies of its time." In defense, Bevins staunchly noted that TBS had always been heavily leveraged.

But the figures did look scary. Together, the combined TBS and MGM were pulling in only $567 million every nine months, yet they were obliged to pay $600 million within the next *six* months or face rapidly escalating interest rates. Even worse, the way the deal was set up, if Turner failed to reduce his debt within those six months, he would have to start paying Kerkorian in TBS stock. Thus with each payment, Turner's control of his company would dwindle. "I think Kerkorian thought he'd end up owning TBS," says Mike Gearon. "He thought he was going to get Turner." Eventually, Kerkorian might even have another chance to sell MGM all over again.

The clock was ticking now, and with each tick, Turner came closer to losing everything he owned, everything he had worked to build. "Drexel has put a gun to Turner's head," one Wall Street banker said. "I think he's in terrible financial difficulty," said another. "Unless he has some plan no one knows about, so creative no one has ever thought of it, he can't do it. The whole empire could come crashing down around his ears."

In public, Turner put on a brave face, actually bragging about how much he owed—nearly $2 billion, in all. "That's more than [the debt of] some Third World countries," he boasted to a group of business leaders in Davos, Switzerland. "I'm pretty proud of that. Today, it's not how much you earn, but how much you owe." On another occasion, he would declare, "Two billion dollars I owe. Actually, it's closer to one-point-nine, but I like the sound of two billion better. . . . That's a million dollars a day in interest. . . . No individual in history has ever owed more. . . . Here, look at my picture in today's newspaper. Do I look worried?"

But at home with his friends and family, Ted would admit that for all his devil-may-care pronouncements, he *was* worried. One night, with his mother, Florence, and some neighbors gathered at his other South Carolina plantation Kinloch, he stood up in the middle of dinner and started pacing. As everyone else ate, he strode up and down the length of the long dining-room table. The scene was peaceful in this backcountry lodge—a large log fire crackling in the next room, lovely waterfowl prints on the walls of the wood-paneled dining room. But Turner's mind was anything but relaxed. "Goddamn it!" he said. "I've really done it this time." Turner shook his head, awed at his predicament. "I may have really done it. Maybe I shouldn't have gone into MGM. . . . But that library is great." His mother, sitting tall and aristocratic in her chair, smiled reassuringly at her son.

Turner paced some more, then, stopping suddenly, threw a hand up into the air. "How the hell am I going to pay that two billion dollars?" he asked.

"Teddy!" his mother exclaimed, the smile fading. "Did you say two *billion* dollars?"

"Sure," said Turner. "I told you it was gonna cost two billion dollars to get MGM . . . if I can find the damn money."

"Oh my," she whispered. "I thought you said two *million*."

As Turner traveled out to the MGM lot in Culver City, he clearly felt the pressure of the passing days. And walking through the main gate and under the famed entrance arch, with its huge roaring MGM lion, he had one overwhelming problem on his mind: How the hell was he going to hang on to this place and keep himself in the movie business?

His father had foundered at a similar moment when he tried to step into the big leagues. These next months would be a pivotal time. The strain that Turner was under caused him to erupt from time to time: One afternoon, he was talking to Nick Nicholas and other Time, Inc., officials as they made their way through the MGM parking lot. As Turner walked, he sketched out his plans for some of the films in the MGM film library. "Well, Ted," they pointed out casually, "you know we've leased a number of those films."

It turned out that HBO had already signed deals for several MGM movies at a very advantageous rate. "They're locked up," they said. Turner stared wildly from one executive to another, suddenly realizing that he hadn't really even investigated these details. "Goddamn it!" he exploded, throwing up his arms. Then turning toward the nearest car in the parking lot, he started violently kicking its tires, booting the inflated rubber with all his might.

In a calmer moment, Turner would admit, "I've never done anything like this before. It's like sailboat racing in a hurricane. It's like being in an airplane in a storm. You buckle your seatbelt."

Once again, Turner was talking to almost anyone who would listen. He tried unsuccessfully to convince Steven Spielberg to come in and run MGM. He talked about mergers and combinations with a series of companies. The Murdoch and Gannett conversations were ongoing, although nothing would come of them. And now two film companies, the Israeli-owned Cannon Group and Lorimar Telepictures, were discussing taking parts of MGM off his hands. Turner had wanted to hold on to the famous MGM studio lot and make his own motion pictures. In fact, according to aides, he was mesmerized by Hollywood and deeply wanted to be a part of it. But the one thing he absolutely had to have was the MGM film library. If push came to shove, everything else was expendable, and as the days ticked by and his debt grew ever larger, push was definitely coming to shove.

In early June of 1986, as the cable industry traveled west for a national convention, Turner invited a select group of his business friends out to the MGM lot—John Malone, Nick Nicholas, and Michael Fuchs, among others. He took them on a tour of the studio and invited them back to the Irving Thalberg Building for cocktails. That's when the conversation turned serious. What did Ted intend to do? How could he get out of this predicament?

"It still remains one of the most incredible business meetings I ever had," recalls one of the participants, "because everyone was in there sort of seeing if they could get a piece of the action." Several parallel conversations were going on at once, as these high-powered executives clustered around the office—discussions of all the different options and combinations Ted could pursue with TBS and MGM. Michael Fuchs, who was then feuding with Cannon's owners, Menacham Golan and Yorum Globus, was saying to Turner, "Ted, don't deal with those Cannon guys. I hate those guys." Meanwhile, Turner was scribbling furiously on the back of an envelope, crunching numbers, trying to see if there was any way he could hang on to the studio. Watching him, one of the executives remembers, "I was thinking about the way our company would do it—nine million accountants, ten thousand lawyers. Here was Ted, so seat-of-the-pants, doing this deal on the back of an envelope."

As the executives quizzed him about MGM's financial situation, it became clear how much Turner *didn't* know. "What's the future home-video value of those films?" they asked. "What are the contracts like with the distribution companies?" Wussler recalls, "It was like 'Ted goes to film school.'" After an hour or two, it became obvious to everyone there that Ted didn't have the answers. And so they told him, as gently as possible, "This is not a business you are prepared to be in at this point." And, says Wussler, "they were correct. We were too shallow, too overextended."

That night, when Malone, Nicholas, Fuchs, and the others left, Turner still hadn't made a final decision. But within a day or two, by June 6, he had agreed to perhaps the only decent option he had: to sell almost everything back to Kirk Kerkorian—the MGM studio, the video business, even the MGM lion logo—and all for $300 million, substantially less than Turner had paid for it just three months earlier. The studio lot and the film laboratory went to Lorimar, for $190 million. That left Turner with only the library, for which he still owed well over a billion dollars. "Kerkorian had taken him bad," says one of the cable executives, who had just visited with Turner. "It really looked like he had taken him to the cleaners."

But what upset Turner the most, as he struggled with his decision in the main suite of the Thalberg Building that night, was the fact that he would never get to make his own movies, never bring out his own *Gone with the Wind* or *Singin' in the Rain,* his own *Philadelphia Story* or *Ben-Hur*. There he was, sitting behind the executive desk in the fabled MGM glamour factory—"More Stars Than There Are in the Heavens"—and the glamour would never be his. He had come so close to running a studio, but somehow it had all gotten away from him. There, on the historic MGM lot, where so many great movies had been filmed, there with the ghosts of Gable and Garbo and Garland looking over his shoulder, Ted Turner watched a piece of his dream slip away. And he put his head down on that great wooden desk, and he cried.

CHAPTER 21

THAT NIGHT, only a few hours after sorting through the most painful business reorganization of his life, Ted Turner showed up in the dining room of the Beverly Hills Hotel with J.J. Ebaugh on his arm. More than one cable kingpin who had been with Ted earlier that afternoon and witnessed him dismantling the MGM movie empire stopped by his table to offer a few words of condolence and encouragement to cheer him up. They came away amazed at his ebullient mood. Ted was, after all, about to suffer the worst financial setback of his career, handing MGM and a multimillion-dollar profit back to Kirk Kerkorian. But Turner had no time for regrets. "You don't look back," he drawled, flashing his hundred-watt smile. "You gotta look ahead."

For all Turner's optimism, ahead didn't look any better than behind that evening, as he sat in the subdued opulence of the Beverly Hills Hotel dining room. While he had staved off the immediate threat of Kerkorian taking over TBS, he still had a crushing debt and no way to pay it off. Turner had, in effect, shelled out $1.2 billion just to buy the MGM film library, a figure expert analysts agreed was a good $200 million to $300 million dollars too high. Even after selling off all the other pieces of MGM for $490 million, he was still $1.4 billion in the hole. That was enough red ink to destroy almost any business, even the healthiest. And TBS was looking anything but healthy. In a little over a year, the stock had plummeted from a high of $29 down to the midteens. Although

CBS had also come out of the Turner attack bloodied, its stock was down a mere 10 percent. TBS, in comparison, had plunged 47 percent—nearly half its value. The stockholders clearly weren't predicting good times ahead for Turner.

Ted had been working hard in the preceding months. In one seven-day period, it was reported, he had appeared on three national TV programs, taken eight flights, given four speeches, met with executives from Ford, MGM, Drexel, Volkswagen, and the Motion Picture Association of America, attended Hawks games and business parties, and on one day had gotten up at 4:15 A.M. to speak live via satellite to a television convention in France. In the next months, he would try to surpass that pace, traveling from California to Belgium, from Colorado to the Soviet Union—trying to build the value of his company and keep it afloat. In New York, he agreed to write his life story for Simon & Schuster for a hefty $1.8 million. In Brussels, he kicked off CNN Europe in spite of heavy resistance from the European community. (Responding to accusations that he was an "ugly American" cultural imperialist, an uncharacteristically subdued Turner replied, "I don't want to upset the apple cart or make anybody mad. I just want to be a banana, one of the bunch.") His friends warned him that the long hours would take their toll. "Right now," Jimmy Brown said, "Mr. Ted is doing the same thing his daddy did. He's working himself to death."

"I've begged him to take some time off," said Jane. "I've had a doctor who specializes in stress talk to him, but nothing helps."

With his purchase of the MGM library, Turner had undertaken the boldest expansion of his company in a career marked by risky moves. He had staked all he owned on this one roll of the dice, placed all his multicolored chips on the table. For the next months—the next years, in fact—it would be unclear if he had won or lost. His strategy made perfect sense to him, but to most outsiders it looked as if he was losing on several fronts. And so 1986 through 1988 would be difficult years for Turner—years of trauma and upheaval, years when optimistic talk in public was matched with private angst.

It was right in the middle of the MGM debt crisis that Turner, with impressive resilience and in great expectation, abruptly took off for Europe to attend the Goodwill Games. In Moscow, he was center stage for fifteen days (July 5–20, 1986) along with thirty-five hundred athletes from around the world. In collaboration with Soyuzsport and Gosteleradio (and thanks largely to Robert Wussler and his staff), Ted Turner had created his very own Olympic substitute. At a time when the cold war was still frigid, when President Ronald Reagan was still talking about "the evil empire," and when the previous two Olympics had failed to unite the world, the Goodwill Games brought together Eastern bloc and Western athletes to compete in eighteen sports—everything from track and field to tennis, from judo to yachting.

Turner arrived in Moscow beaming. Whatever his business worries back in the States, he knew he was pulling off something historic here. And he was more than willing to provide the rhetoric for the occasion, as if he were running for some international political office, reminding everyone of the gospel according to Ted. On the podium, he dusted off a handful of old-fashioned platitudes, proclaiming, "We can best achieve peace by letting the peoples of the world get to know each other better. Not only will the participants compete together in the spirit of good sportsmanship, but the audiences worldwide will see the harmony that can be fostered among nations." In private, the words came out differently, but wherever he went the depth of his passion, the haunted ferocity of his apocalyptic "verge" vision, was evident: "How much time we got, huh?" he asked, referring to the superpowers and their nuclear arsenals. "What we need is a Big Daddy to take us behind the woodshed and take a big board and hold us by the ankles and give it to us good."

If Russians had a soft spot in their hearts for maverick American capitalists like Armand Hammer and Cyrus Eaton, a character like Turner was sure to find favor there, and he did. Earthy and emotional—by turns moody and euphoric—he was someone Slavs quickly recognized as a soul mate. But no one was quite sure exactly how he'd get on with Mikhail Gorbachev.

On July 17, 1986, at eleven in the morning, Ted Turner was ushered inside the Kremlin into the stately ceremonial office of the general secretary. According to Robert Wussler, who also attended the meeting along with Georgi Arbatov, the two men were clearly intrigued by each other.

Gorbachev rose from behind his desk, and came forward with his hand outstretched to show the visitors to his conference table.

With his natty European suit and his thick teamster's neck, Gorbachev came across as an odd blend of blue-collar strength and white-collar sophistication, a combination that Turner clearly responded to. As always, Ted was not shy about plunging right in, and Gorbachev, who had been fully briefed about the entrepreneur's personality, showed no hesitation either.

As it turned out, the two men talked about the Goodwill Games, the ostensible reason for their meeting, for only ten or fifteen minutes, skimming over the organizational difficulties and the promise for the future. Then the conversation spun off into world politics. When Gorbachev spoke of his goals for the USSR and the world community, he struck a chord with Turner.

"I see myself as a citizen of the earth," Turner declared emphatically. "I don't want to see any nuclear weapons going off over your country or over my country. Those are short-term victories. Those victories only last twenty-four or forty-eight hours. If we bomb you, it's gonna hurt us. And if you bomb us, it's gonna hurt you." Then, as he often did, Turner brought up his family: "Look," he said. "I have kids, you have kids. What's their future gonna be like?"

If Turner was perhaps a little short on ideas for the SALT negotiations and ICBM reduction numbers, there was no denying the genuineness of his emotion, and Gorbachev was clearly moved by the strength of Turner's feelings. They ended up chatting for over two hours, their conversation ranging from the future of the USSR to the seven-hour *Portrait of the Soviet Union* documentary series that TBS was working on; the latter would provide some good PR for the USSR at a time—just two-and-a-half months after the Chernobyl nuclear disaster—when favorable American coverage was not so easy to come by. As they wrapped up the session, Turner invited Gorbachev to come to his plantation in South Carolina for a little relaxation and duck hunting. Turner left the general secretary's office and strode out of the Kremlin that day into a lovely, warm Moscow afternoon, thoroughly elated at how well his meeting had gone.

The games themselves were another story. These Un-Olympics had been dogged with logistical problems from the start. Many athletes from countries other than the United States and the USSR chose simply to stay home, including the highly regarded Cuban boxing team. Israeli athletes had not even been invited by the Soviet hosts. And some sportsmen were scared away by other factors, like the specter of Chernobyl. Danny Harris, an American 400-meter-hurdle champion, had no intention of attending: "I don't wanna breathe no radiation," he told reporters.

If the games lacked certain top athletes, they were also missing something even more important—an audience. As the cameras panned around the Moscow arenas, there were large numbers of vacant seats, and the cheers were sparse in the empty halls. The fans had failed to be excited by an ersatz Olympics. Worse still from a financial perspective, few viewers were watching the games on TV back home in America. The television audience was less than a third of what had been predicted—not the seven or eight million promised to advertisers but barely two million. The struggling last-place Braves could attract more people than that. TBS would have to give advertisers free airtime to make good their predictions. Turner would end up losing over $20 million on the Goodwill Games. "As a business venture," declared one newspaper, "it's as hard finding the good in Ted Turner's Goodwill Games as finding the Muscovites in the stands."

Despite the cost of the games, Turner did not feel like a loser. In less than a year, he had created an international event, producing 132 hours of programming and establishing what he hoped would be a sporting franchise for the coming decades. Even more important were the nonfinancial considerations. In later years, he would say that he truly believed that the Goodwill Games had served as a linchpin in bringing the two feuding nations together. "He felt that it was a major mark in history," says Robert Wussler. In a sense, Turner regarded the Goodwill Games as a loss leader for world stability.

A little more than a month before the games, Wussler had realized that their original financial projections were inflated. "You know," he told his boss, "you're going to lose a ton of money on this. Do you want me to cut back? Do you want to cut a couple of days, or do less on the TV production? We're still in a position to cut something. I mean, there's not a lot of fluff. We're doing this with Scotch tape and baling wire, but we can do it with less Scotch tape and baling wire." Turner looked at him and shook his head. "No," he said. "I don't want you to cut. I want this to look first-class."

In the USSR, as Turner walked the streets of Moscow and went from one event or press conference to the next, he was clearly thrilled. He was, after all, the cohost of a worldwide party. This was a bold stroke worthy of his childhood heroes. "I'm so happy with the way things are going, I'm having a hard time keeping from jumping out of my skin," he said. Working from 9:00 A.M. until midnight, Turner dashed from one event to the next, handing out medals, visiting locker rooms, giving speeches, answering journalists' questions, hosting athletes and bureaucrats at one reception and cable operators at another, sprinting off to make sure that the entire U.S. women's basketball team would appear in a group interview on WTBS. Turner was having the time of his life. "I don't think I ever saw him happier for a concentrated three-week period of time than at the Goodwill Games," says Wussler.

Indicative of how much these games meant to him and how jam-packed his days were is the fact that he had brought along almost every significant person in his life to share this moment with him—his wife, his five children, his two girlfriends, and seventy-five of his closest cable-operator pals, all flown there at Ted's invitation and expense. A classic snapshot from the games that was later passed around TBS shows Ted out front, with Janie, J.J., Liz Wickersham, and Barbara Pyle all behind him in a row—Ted and his harem.

For many of those attending, there would be other snapshots from the '86 games—snapshots and memories—and in each, Ted would be grinning. CNN anchor Don Farmer recalls an incident in which he and his wife, coanchor Chris Curle, accompanied Ted, Jane, and Ted's daughter Laura to a TBS reception, and the limo left them off at a park dedicated to Soviet industry. "Ted was feeling expansive," recalls Farmer, "and he walked up to this giant metal statue of a bull, honoring the meat industry or something. It was huge—thirty feet up, fifteen feet long—and it was anatomically correct. Ted started walking around this statue, and then he stopped and pointed up at it and said, 'That's what's wrong with America today.' Everyone looked at him quizzically. Ted nodded emphatically. 'We don't have any heroes. Can you name any current American heroes?' "

To Farmer, though, the massive statue had another resonance:

"It kind of reminded me of Ted—this great big figure, with these great big balls."

Another moment from the Goodwill Games remains in Ira Miskin's mind—the send-off meeting as his team went into the field to start filming *Portrait of the Soviet Union.* They met at Gosteleradio, in the hallway outside one of the main production studios. Turner was on his way to Tallinn for the sailboat races, but he wanted to meet Miskin and John Purdie, a former BBC journalist who had been hired as series producer, to give them his vision of what the series should be.

Seating themselves on a bench in the Gosteleradio corridor, Turner looked at Miskin and got right to the point: "Don't forget that we're doing a story about a nation of *winners.*"

Miskin stared at Turner as if he were nuts. "What are you talking about?" he said. "This place, on a good day—it's Detroit on a bad day! We're talking about a nation of brilliant history, great literature . . . and a really oppressive, abusive, shitty system. What are you talking about?"

Turner nodded. "We know all that," he said. "But we're talking about winners here!"

Purdie, the British producer, looked up at his new boss gloomily. Oh my God, he thought. I've just destroyed my career. What am I doing here? I'll never produce another film for any organization as long as I live. He tried a weak grin, and said nothing.

Turner glanced from one to the other and nodded emphatically. "A nation of winners!" he reminded them one last time, and hurried off with a smile.

A few days after the end of the Moscow games, Ted finally took some time off for a long-planned vacation. In typical Turner fashion, it was to be less holiday than expedition—no Disneyland stay or visit to Paris museums but a trek across Africa to Botswana for a big-game safari. Friends and staffers like Don Farmer and Chris Curle wondered how this environmentalist could reconcile himself to shooting animals, but Turner was an outdoorsman who had never seen a contradiction between loving animals and shooting them. Like Teddy Roosevelt, he could work to establish nature preserves but still enjoy bagging a lion. Man is a hunter, he told them. It's just a part of his nature. Besides, he said, it's a great way to bond with your sons.

Led by conservationist and documentary filmmaker Rick Lomba, Ted set off with Rhett and Beau to the arid plains of South Central Africa, a dusty grassland home to some of the last free-ranging herds on the planet. There, as the sun baked the warm African earth, they drove in a Land Rover across the savanna through the sweltering heat of the day, while the animals dozed and slept. In the cool of the late afternoons, elephants and rhinos would come down to the watering holes to drink. While the boys were just as happy taking pictures of springboks and zebras with a telephoto lens, Ted intended to bring down at least one lion. When one finally ambled into view, he took aim and fired. Later, he would joke about how frail and elderly the beast was, but he nonetheless had him stuffed, mounted, and prominently displayed in the TBS conference room.

Out on the great African plains where they camped, the days for the most part passed slowly. Until one night, Ted heard something from Lomba that threw him into a panic. J.J. had apparently not liked the way she had been treated in Moscow, forced to play second or third fiddle to the other women in his life. She wanted more from him than that. Growing increasingly unhappy, Ebaugh decided that she had had enough. She was taking up with another man, a California podiatrist.

"It was a friendship that crept up on me," Ebaugh recalls. "I never discussed it with Ted. He'd be jealous. But it was OK for him to go home to his wife." Like Liz Wickersham, Ebaugh had realized that the only way to cope with Ted Turner's wandering eye (and mind) was to maintain other relationships of her own. And while they might be less glamorous, at least they held out the promise of a future in which she was the star attraction. "There were never any ultimatums," she says. "It was just me wanting to live my own individual life. I just wanted to get on with my life."

Ted was stunned. He immediately cut short the safari, made his excuses to his boys, and set off from his remote camp in the bush to chase Ebaugh halfway around the world. For almost twenty-four hours he drove across the African plains to the nearest airport, caught the first flight out to Frankfurt, and from there flew back to Atlanta. Several days later, he finally caught up with

Ebaugh. But what J.J. had to say probably made Ted wish he hadn't. Her mind was made up—she was leaving him. Although Ebaugh had a spacy New Age side, she was no blond airhead. She knew what she wanted and she went after it—and right now she wanted to move on.

Turner had a hard time accepting what he was hearing. In fact, according to one friend, "he fell apart emotionally. He was running scared, in a panic. He had to get J.J. back. His world was dependent on getting J.J. back. Before, it didn't matter who he was with, just someone with a good body. But now she was giving him the treatment he had been giving women. It was a crushing thing for him. This woman had really gotten to him."

By all accounts, Turner now became obsessed with the idea of winning Ebaugh back. He called their mututal friends, like Barbara Pyle, pouring out his pain and his hopes. He went to talk to Dr. Frank Pittman again to see if he could help. He opened his heart to J.J., telling her he'd change if she'd just come back. And above all, he lobbied—lobbied constantly, lobbied anyone who could make a difference and win a reconciliation. As he had done often before in his life, he waged a campaign. He may have been emotionally shell-shocked, but he was still capable of attacking and he still intended to win. Ebaugh recalls that "he put on the most aggressive campaign to get me back that I have ever heard about or read about in my entire life." Talking to J.J., Ted was passionately intense, fiercely insistent. He listed the things he'd do for her, the commitment he'd make. He told her, "I can't live without you."

Gradually, after less than a month of this siege, J.J. relented and came back to Turner. He was ecstatic, but Ebaugh wasn't so sure. She thought their relationship would last only a few months. She put the odds of things working out at 100 to 1. Above all, she doubted that Ted would ever really leave his Janie, who had stood by him for over twenty years and raised all his children. Although Ted had been continually unfaithful to Jane, he had his own brand of loyalty to the marriage. "Ted had told Janie that he'd never divorce her," says J.J.'s elder sister Sandy. "He never planned to, even when Jeanette was pretty much his number one. My sister never, ever expected that Ted would become a one-woman man."

But now, with Ted pouring out his love for J.J., chasing her back and forth across the world, even Janie, that most devoted and long-suffering of wives, decided that she had had enough

humiliation. She and Ted agreed to a separation that fall, the pro-
logue to a divorce. And during the Labor Day weekend of 1986,
Ted found himself in the process of moving into two new homes,
two new lives: transferring into new executive offices at the Omni
downtown (now known as CNN Center) and moving with J.J. to
his house in Roswell, in the horse country just outside Atlanta.

Any time of moving—of changing one home and one life for
another—is a time of disruption. Just the simple act of dumping
one's past into a cardboard box is unsettling. But the aftershocks
Ted experienced as he moved were merely the precursors to a whole
series of quakes that would rumble through his life over the next
two years.

First, Ted was handed the final figures from his wanna-be
Olympics; he had lost exactly $26 million. Then, in the fall of 1986,
the small Roswell house where he and J.J. were living caught fire
and burned down when an old TV set shorted out and burst into
flames. (To replace the house, Ted set his old friend, architect Bunky
Helfrich, to work again, this time building a log cabin for Ted
and J.J.)

The bad luck continued. An IRS audit charged that Turner owed
$250,031—and TBS owed $1.3 million—in back taxes for under-
valuing the worth of the Braves and the Hawks (a charge Turner
would later successfully appeal). Worst of all, early in 1987, Turner's
son Teddy Jr., was hurt in a terrible accident. Working as a cameraman for CNN in the Moscow bureau, he had begun a television
career filming everything from the Goodwill Games to the war in
Afghanistan. Late one wintry night, after having a few drinks in a
Moscow bar, he hitched a ride home with a CBS cameraman. It
was around 2:00 A.M. when their car spun out of control on the icy
streets and slammed head-on into an iron pole. Every major bone
in Teddy's face was cracked. Most of his teeth were knocked out.
He had glass in his eye, and was throwing up blood. The surgeons
in the Moscow hospital operated, wiring his jaw. And the Turner
family waited anxiously.

Six days later, as soon as Teddy could be moved, his father sent
an air ambulance to fly him back to Atlanta for additional surgery
at Emory University. "I think the accident shook us [the family]
up deeply," says Teddy. "I also think that was a turning point in
my relationship with Dad. For him, I think he began to realize how

he wasn't going to live forever and we really didn't know each other that well."

On the business front, there were other scares, one on the heels of another. None was quite so dramatic as the near death of his eldest son, but collectively they were daunting. With each passing day, the MGM purchase continued to plunge Turner's company into more debt, and it was becoming clear that TBS alone would never be able to pay off what was owed even after having sold parts of MGM back to Kirk Kerkorian. More pieces of Turner Broadcasting would have to be sold—especially since several of Turner's merger ideas showed no signs of panning out.

Meanwhile, Turner's first great moneymaking plan for the MGM film library—colorizing the old black-and-white movies to make them more attractive to a contemporary audience— almost immediately encountered a firestorm of criticism. The pundits who had fretted about his CBS takeover bid now attacked Turner as a money-fixated Southern hick, a philistine. Film buffs in New York, Hollywood, and at a specially con-vened Washington hearing publicly vilified him for turning the shadowy film noir *The Maltese Falcon* into a film red, blue, and yellow. Its director, John Huston, incensed at the way his movie had been treated, declared the colorization "as great an impertinence as for someone to wash flesh tones over a da Vinci drawing." Woody Allen proclaimed the process "a criminal mutilation." Even the soft-spoken Jimmy Stewart described colorization as "cultural butchery."

Turner wondered what all the fuss was about. The black-and-white originals still existed, and people could watch them whenever they wanted. So why was he being attacked by these bicoastal snobs? "Women put on makeup, don't they?" he asked. "That's coloring, isn't it? Nothing wrong with that. Besides, when was the last time anyone took photos in black-and-white? I know, Ansel Adams—but he's dead, too." For Turner, the old MGM motto "Ars Gratia Artis" (art for art's sake) was not just a foreign language, it was a foreign concept. As far as he was concerned, art, movies, TV, music had always been commodities to trade. What were all these artsy-fartsy elitist types complaining about? Later years would prove Turner's instincts—his *commercial* instincts—correct, for by colorizing these old movies he made them more appealing to a mass audience and raised their market value by hundreds of thousands of dollars apiece.

In a time of upheavals and adjustments, the biggest change for Turner was the new partners he would take on, partners personal and professional. According to many of his friends and colleagues, J.J. Ebaugh had a significant impact on the life of the inveterate male chauvinist. One longtime colleague says, "J.J. was the first woman who was able to deal with him straight on, head-to-head—who didn't seem to be totally subordinate to Ted." In Bill Tush's opinion, "She changed him around. Ted was from the old school—men are men and women are women. J.J. changed all that. Everybody thought she would be the next Mrs. Turner."

Working from a position of strength, Ebaugh demanded attention and respect from Turner. She forced him to make a stab at monogamy. She even opened his eyes to women's issues. Strongly against abortion in 1980, Ted soon became solidly pro-choice. She also made him concentrate on eating healthier food—fewer steaks, hamburgers, and pizzas—and on exercising. He lost weight and worked to keep it off: "My lifestyle's changed a lot," Turner said. "I mean, when you leave your wife of twenty-three years and run off with a thirty-year-old woman, that changes things. I've been hopping a little more."

One of the main concessions Ebaugh won from Ted—a congenital workaholic—was that he would finally start taking some time off to be with her. He agreed to work shorter hours, and when he went on the road, she went with him. Although they traveled extensively—Boston, Colorado, Washington, England, France, Germany, Sweden, and Greece were among their destinations in one three-month period—Ted and J.J. also began to put down roots, acquiring some new homes of their own, including one cliff-hugging residence on the rugged shoreline at Big Sur, California, a rustic lodge once owned by actor Ryan O'Neal. Together they enjoyed the countrified pleasures of flannels and jeans and home-cooked meals, walking the land and fly-fishing.

What Ebaugh insisted on was partnership, and it wasn't a natural concept for a leader like Turner. In fact, it was as hard for him as keeping his mouth shut. "He determined all the schedules," said J.J., "and he tried to second-guess me. We have eight homes and lots of things that needed to be handled. I wasn't given the responsibility I should have. He has his—the business. He wanted all this and some of mine, too. We had to do it his way. The couches . . . the chairs. Good Christ! He'd involve himself in every little thing, even the rugs and pictures. He knew in advance how he wanted things, and it wasn't open to discussion. Even if my ideas were better."

Turner had always been a one-man band, according to Ebaugh, and that was exactly what she wanted to change. After all, he had never had partners in his life. He had employees, he had sailing crews, he had wives and children, but all took their orders from him. Now here was J.J., demanding equality and talking about the trendy New Age ideals of her friend and mentor, the "evolutionary scholar" Riane Eisler, and such paradigms of Eisler's as "the partnership model" and "the dominator model." While Ted had desperately wanted J.J. back, living with her required some real growth and development, and he wasn't always certain he liked dancing to this new tune. "A reasonable compromise of give-and-take was a requirement," says Ebaugh. "It was an outstanding, probing battle."

Over the next months, there's no question that Ted Turner did make an effort to change. He went back to Frank Pittman and other therapists and tried to modify his character and improve his relationship with J.J. "I've changed quite a bit," he would acknowledge. "You can't have a very close relationship with someone without both of you being altered." Turner claimed that he was now attempting to listen, "to wait until someone was through rather than interrupting them, and then think about what they said before I prepared an answer. I learned to give and take better than I had previously." But as energetically as he tried, he seemed to have a congenital inability to listen, an irrepressible urge to run things his way. Almost as if his body were telling him that he wasn't really partner material, he began to develop an ailment: he was becoming hard of hearing.

If Turner was trying hard to cope with a single partner on the home front, he was about to be saddled with dozens of new partners professionally as well. By early 1987, it was clear that his financial situation—even after the studio sellback—was dire. There was no way that TBS could pay off its debts. The company was spending more than one-third of its revenues just to cover the interest on what it owed. Ted would have to do the one thing his father had always warned him against: he would have to sell off parts of his company. The only question was to whom.

"I counseled Ted not to do the MGM deal, since I knew he couldn't afford it," says John Malone, chairman of TCI. "I was terribly afraid he would have to liquidate." This was the specter hanging over Ted—an electronic garage sale. He had been talking to Fox, to NBC, to Gannett, and others. He was even exploring the idea of selling the entire CNN Center complex to Japanese real-estate developers for $175 million, then leasing it back. He needed to raise cash, and fast.

But although he now had a larger debt than some Central American countries, Turner was not exactly destitute. And he had a trump card: he understood how much his networks were worth to the various local cable companies across the country—the Sammons and the TCIs, the Paragons, Coxes, and Jones Intercables. Without his programs—without CNN and WTBS—they had only HBO, ESPN, and a handful of rarely watched channels to offer their viewers. Without Turner, the fledgling cable services would seem much less attractive. "We had enormous value to the cable industry," Gerry Hogan points out. "The worst thing that could happen to them would be that we'd sell something to Murdoch or NBC or someone else. So we had some leverage."

One cable CEO concurs: "I think the cable industry understood how important his brands were. They didn't want to lose CNN or TBS to an outsider. They realized how crucial it was to their very survival to keep Turner in business. . . . So I don't think it took the cable industry long to get off its ass, and Malone led the charge."

The steely-eyed John Malone, once the owner of a small, local Denver-based cable company, was now one of the emerging titans of the cable industry. While Ted was making headlines with his noisy crusades, Malone had quietly become perhaps *the* most powerful force in the business, as the head of Tele-Communications, Inc. He owned more of the local cable services than anyone else, and he wasn't shy about flexing his muscle, as both MTV and the proposed NBC cable-news staffers could attest. Vice President Al Gore, when he was a senator, labeled Malone the leader of "the cable Cosa Nostra."

Malone and Turner were in frequent communication, often chatting by phone at least once or twice a week. Malone was shrewd enough to recognize early on that several companies were starting to circle around TBS, and that Turner might have to sell CNN (or part of CNN) to an outsider like Rupert Murdoch or even a powerful insider like HBO (owned by Time, Inc.), either of which would make Malone's life more difficult when he had to negotiate over rates or industry alliances.

In the first few weeks of 1987, rumors began about a consortium of cable operators, led by Malone, buying a major piece of Turner's company. On January 13, cable leaders from many of the big companies, like ATC, Cox, Continental, United, and Warner, met to discuss the details. Turner's financial picture was looking grim. He had just lost $180 million in the MGM deal, and his stock had slid from 29¼ to 17⅛. Big losses were predicted, in Turner's words, for "the foreseeable future." But the cable executives made up their minds quickly. By mutual agreement with Turner, they offered to bail him out by buying roughly a third of his company. "If Turner's at risk, we're at risk," said a cable CEO. "We're not ready to face the prospect of someone who might favor a competing technology [like one of the broadcast networks] getting their hands on CNN." In short, his best customers would become his business partners.

The next six months were spent working out the details of the sale. After all, it was hard to cram so many major egos into one deal. Gerry Hogan remembers that the biggest question was "the balance of power—how could you fit the entire cable industry into this deal so that they all felt they were a part of it, and not change the balance of power, which at the time was pretty equal between TCI and Time? And also, how could we at Turner live with those guys? How could they be our customers and yet be our shareholders?"

On June 3, 1987, the agreement was announced: A group of thirty-one cable companies had bought a 37 percent stake of Turner Broadcasting for $562.5 million. The cable companies had the right of first refusal if Turner ever sold his company. They were to receive seven of the fifteen seats on a newly configured TBS board. Most significantly, when it came to approving the budget or any of seventeen other corporate functions, including any expenditure of $2 million or more, Turner was obliged to get a "supermajority" of the board—an affirmative vote from twelve of the fifteen directors.

Although Turner had been saved financially, it was at a tremendous cost. In this new TBS, his stake had shrunk from 81 percent to 51 percent—still a majority, but barely. (A reconfiguration of the stock would later give him about 65 percent of the voting shares.) Now he was compelled to share his company with some of the tigers of the cable world—Malone of TCI, Michael Fuchs of HBO, James Gray of Warner Cable, Trygve Myhren of Time's American Television & Communications Corporation. In short, he had given up control of his empire. Bill Bevins conceded, "It's clear from the transaction as it's structured, and I think it's an economic fact of life, that Ted no longer runs the company." As of June 1987, Ted Turner's life work, the company he had created, was going to be run by committee. And the characteristic that had made Turner Broadcasting so remarkable—Ted's seat-of-the-pants operating style in which, for better or worse, he could act instantly on any whim—that was all going to change. In a year that had included his son's accident and his house burning to the ground, Ted's loss of control of his company must have felt like the last straw.

Of course, that wasn't the public spin that Turner was putting on the buyout: "I don't consider it a rescue at all," he said shortly after the deal was completed, his characteristic nonchalance firmly in place. "We are not desperate in any way shape or form. . . . We had oodles of alternatives, and we picked the one that made the most strategic sense." As for the limitations on his control of the company, he declared, "I do have some negative covenants, but the negative covenants that I have with the cable operators are not significantly different than they are with the banks. . . . I'd like to think that we're like Brer Rabbit. When he got caught in the briar patch, he was hoping he'd get thrown into that briar patch."

But most of the players in the cable business recognized how awkward Turner's position was. He had been compelled to give up the major decision-making power for his company. And while Turner might hint that he had planned it all, the cable CEOs on his board knew better: "That was not Ted pulling off a coup—that was Ted being rescued," said one. "At the time, the MGM deal was a mistake. I mean, if you say I just made a deal that required me to give away half my company, when that wasn't the intent of the deal. . . . Well, it might work out in the future, but at the time

it was clearly a miscalculation. They totally miscalculated the deal."
Now Ted would pay the price of that miscalculation.

On the day before the first meeting with his new board, Turner
was nervous—very nervous, according to his friends and staff. As
he paced around his office, he was heard to mutter several times,
"I've lost control. . . . I've lost control."

For all of his professional life, his board had been composed of
old friends and employees, like Hank Aaron, Mike Gearon, and
Bill Bartholomay—basically a group of rubber-stamp directors to
meet legal requirements. He had always liked chatting with them,
according to Robert Wussler. He had liked seeing them nod yes.
As for his staff, they also nodded their approval, and when they
didn't, Ted usually ignored them.

He enjoyed playing the visionary. In the company drama, he
cast even his top lieutenants as passive straight men. From Paul
Beckham to Robert Wussler, from Gerry Hogan to Martin Lafferty,
all recall their role at staff meetings as primarily an audience for
Ted—a group of guys sitting in Turner's office as he restlessly paced
around behind the chairs, past the desk, gesticulating as he went.
In one meeting, Wussler began to keep count as his boss circled the
office, and he logged Turner at a record seventy-three circuits.
Sometimes during a meeting, Ted would wander into the outer
office of his secretary; sometimes he'd go into the bathroom next
door. Even while peeing, he'd keep right on talking at the top of
his lungs. The men in suits would remain seated throughout. They
were the role players; he was the star attraction. It was like assisting
at the levees of Louis XIV at Versailles as he reeled off instructions
from his bed.

Ted's managerial style had always been autocratic. The best
work at his company got done (according to aides) when he had
staked out a direction but then went off, leaving the staff alone to
take care of business. While he was on the scene, however, he clearly
ran the show with a dictatorial hand. Like Dagwood's Mr. Dithers,
he had a habit of firing and then rehiring his top executives. It's
said he sacked his veteran secretary Dee Woods at least a half-dozen
times. And anyone who grew too fast, who challenged his hege-
mony, was looked on with disapproval. "Ted is in charge. He
likes to be in charge," says Paul Beckham. "If his strong suit is
motivation, his weak suit is his management—his tendency to

divide and conquer. The division of his company makes people have to come to him." According to several staffers, Ted never really liked the idea of having a successor, of grooming someone who would eventually run TBS. Perhaps that was why he hadn't sought strong managers outside the company and often undercut those who were growing too powerful within it. Later that year, in fact, Bill Bevins decided to quit. According to inside sources, Bevins "wanted to be the successor to Ted," an ambition that would never come to pass.

According to one employee, Turner treated all his staffers equally, "like janitors—which is great if you're a janitor, and not great if you're a vice president." Or as one director would later say, "It's like slaves on the plantation—they get more and more authority, but they're still slaves."

In short, TBS had always been Ted Turner's company. But now, with the change of the board, Turner's family-style business was becoming much more corporate. Despite his name on the logo, the company wasn't really Turner's anymore. And all day long, as he prepared for his first new directors' meeting, that thought was gnawing at him. When the evening rolled around, he welcomed most of the board members to an informal gathering at Bugatti, the restaurant at the back of his CNN Center. In the large room with its oversize leather armchairs and private, screened banquettes, he greeted his old friends and business associates for the first time as partners. Despite his uneasiness, he tried to keep the mood light-hearted: "This is our first meeting," he said. "I've waited all these years for it. It's sort of like the first time I had sex. I waited eighteen years. I waited fifteen minutes for the second."

But the next day, neither he nor his partners were laughing as they gingerly settled into their new roles. In the large conference room looking out over the spacious atrium of CNN Center, they sat down around the long oval table. And Ted, surprisingly ill at ease in this new, more formalized role, called the gathering to order.

From the beginning, it was clear that Turner and his board of directors had different agendas. While the savvy group of cable operators respected Ted for his achievements, and hoped to make money with him, they were clearly concerned about his shoot-from-the-hip style, his Save-the-Earth philosophy. One of his new board members, Michael Fuchs, had publicly declared that the directors didn't want Ted to "take millions of this company's dollars and squander them away on an ideology that's out there in the wild somewhere."

Insiders described these first meetings as "combative," as the powerful figures around the table sought to find an equilibrium. "Remember," says Wussler, "they were all cable operators. Ted's not a cable operator—he doesn't own any cable systems. They all represented companies with cable systems. Herein lies a basic conflict."

In the coming months and years, that conflict would erupt several times as Turner attempted to expand his company and was repeatedly blocked by his board. In 1988 and again in 1989, Turner set his sights on the Financial News Network, which was a perfect fit with CNN, since it provided in-depth business reporting that complemented CNN's fare. With his CNN news staff already in place, he could run FNN at a significant savings. The Financial News Network was on its last legs financially, and the price was right—about $100 million. Moreover, Turner's old rival NBC was threatening to acquire FNN. "If we don't buy it," Turner told his board, "NBC is going to. And then they're going to be competition for us. We have a very low-cost news operation here, and it's not going to be low-cost anymore. . . . We've got to put them out of business, 'cause if they're successful, they're liable to come and grow over into our business." NBC had tried before; perhaps they might try again.

Before the next board meeting, the outside directors asked for a brief delay, and sat down by themselves to hash over Turner's first major request. They met privately because they didn't want to have their discussion in front of Ted's inside directors, who were mostly employees. And among themselves, before the full board had even assembled, they decided to check Turner. "For a number of different reasons, we just didn't think it was the right thing at the right time," remembers one of the new directors. "We thought it was too aggressive."

Clearly, the cable owners did not relish having any single person controlling too many channels, even if they had a voice in his affairs. The more channels Turner acquired, the more difficult he became to control, and the more he could potentially dictate terms to them, his customers. They didn't want Turner growing too big. (Certainly, they owned a piece of his company, but only a piece, and they had the profitability of their own firms to worry about.)

Equally important, the board members, representing the entire cable business, now brought to TBS meetings a series of larger concerns, industry-wide concerns. "I can remember personally feeling, along with others, that this FNN deal would have come under a lot of scrutiny," said a director. "Ted is a very competitive guy and clearly part of his motivation was to make sure no one else got it. . . . But for us, there were a lot of legal and political sensitivities. If too many channels were owned by too many cable operators [such as the heavyweights on Ted's board], it could raise eyebrows in Washington. . . . Then, too, with FNN you'd be taking over something where someone [like the former owners] could claim, 'I couldn't succeed because the cable industry was blocking me. All of a sudden, the cable industry [as represented on Ted's board] owns it, and *boom!* it's successful.'"

Just as Ted was prevented from acquiring FNN, he was also blocked on other occasions. Eager to own a midsize motion-picture company, he attempted in subsequent years to take over Orion and then MGM once again from Kerkorian, but members of the board repeatedly checked him in quiet behind-the-scenes conversations. "Ted once told me that he had planned to go after one of the studios, and he could have done it. But someone from the Time group on the board told him not to," recalls a board member. "I'm sure Ted would tell you that he'd have been in the studio business were it not for the board."

By late 1987, reports of Turner's clashes with his board began appearing in the newspapers. Ted tried to put the best face on it, using the formula he had learned from J.J.: "When you've got partners, you've got to spend some time listening. You can't be talking all the time." But he was clearly upset. In fact, he ordered his oval conference table replaced with a round one, in the hope of eliminating oppositional thinking. As for the cable operators on his board, the *Wall Street Journal* announced that they wanted Turner out and NBC in. (In this scenario, it was not clear exactly how the spoils would be divided, since both NBC and Time wanted to control CNN, and there was reportedly "intense suspicion and rivalry" between them and Malone's TCI.) But either way, the article stated that Turner "seems ready to bow out," and quoted one executive as saying that Ted was "positioning himself for the best offer he can get." The source went on to give a "scientific guess" that the cable operators and NBC would "buy Turner out for $840 million."

That guess was none too scientific. Turner was clearly frustrated, but he had no intention of bowing out. What was in fact happening, behind the scenes, was that the directors were jockeying for power, trying to increase their leverage over Turner and to reshape TBS in the mold of a more traditional business. Like a chastening parent, the corporate board was trying to tame the always volatile Ted. In the weeks before a private January 15, 1988, directors' meeting, the board began to pressure Turner to restructure his organization, proposing that he hand over the reins and relinquish his daily operational role to a more traditional CEO. John Malone declared, "The question is: Is Ted a one-man band? Ted says 'no' but he hasn't introduced the board to his people yet."

In fact, the board members had met many of Ted's "people," but what they really wanted (especially in the aftermath of Bevin's departure) was for Ted to step aside, to give up daily control to a strong number-two man. With Ted continuing his usual gallivanting all over the world, the directors were getting edgy. They knew how creative Turner could be and didn't want to stifle him. They felt sure he was the right person to build a company; they just weren't sure he was the right person to run it. The board wanted "a point guy the employees and investment companies could trust to call the shots on a day-to-day basis," said one director. "There had to be a restructuring. As it stands now, you have an unstable situation that's breeding an environment that isn't productive." Turner was clearly facing a showdown.

As it turned out, the board meeting came and went, and Turner successfully resisted all attempts to make him share his power. He refused to restructure his company. He would not relinquish his control. And the board had to retreat, grumbling; "We realized," a board member says, with a sigh, "that he wasn't going to live with that very easily, so we got off it."

Following the clash, Turner suddenly appeared in public interviews for the first time in months to declare that although he and the board had "nudged and shoved each other a little bit"—although he realized that the MGM deal was "too big for us" and that it had put TBS "between a rock and a hard place"—he was now "wildly enthusiastic" about the future of his company. "The only way I'm going out of here is feet first," he proclaimed. If it sounded a little

like whistling past a graveyard, Turner could be excused. He had
bailed out his company, but it was still unclear exactly what price
he had paid. John Malone reportedly bragged about his control over
Turner, claiming that Ted couldn't make a major move without
his approval. Squeezing his thumb against the table, Malone declared, "I've got him right where I want him."

On opening day of the 1988 baseball season, the Atlanta Braves
hosted the Chicago Cubs at Fulton County Stadium. Before the
first pitch of the season had been thrown, some Braves fans unfurled
a large banner and paraded it around the stadium. "WAIT TILL
NEXT YEAR," it declared. Those irreverent fans must have
known something, because the Braves promptly dropped the game
to the lackluster Cubs 10-9, and then went on to lose the next seven
in a row. In 1986, the Braves had come in last; in 1987, next to
last, but here they were on their way to a whole new low: the
Atlanta Braves had become the first team in a century to play their
first eight games at home and lose every one of them. It had been
that kind of year for Ted Turner.

Beyond the incidents and accidents, beyond the clashes with his
board, he had also fallen ill. In early 1987, Turner had been diagnosed as having Epstein-Barr, the so-called Yuppie virus. With all
of his travel, with all of his stress, he had been feeling run-down.
He had taken to spending hours on the fold-out sofa in his office.
His doctors told him that his problem was not just a lack of sleep
but that he was suffering from chronic fatigue syndrome, a poorly
understood immune-system dysfunction characterized by incapa-
citating fatigue, exhaustion, neurological problems, and a host of
other symptoms. They had traced his illness to a virus known as
cytomegalovirus—a herpes-family virus. The disease itself was
known as cytomegalic inclusion disease, or CMI, which Turner
laughingly called the country music disease. Little known and little
understood, this diagnosis of chronic fatigue syndrome would be
upheld over the following months by some doctors and rejected
by others, who claimed that Turner was merely suffering from
extensive jet lag. Either way, through 1987 and 1988, Turner was
clearly feeling ill.

The problems in his business and personal life were taking their
toll. Since separating from his wife, he had been wavering about
whether or not to get a divorce. J.J. was clearly in favor, and while

she had urged him to break off his ties and start a new life, she never quite believed he would do it. But for almost a year, from late 1987 through 1988, Ted and Janie and their attorneys tried to work out the details of a divorce. Originally, he had declared, "I'm divorcing her, and by God, I'm going to give her twenty-five million." Later he would state, "Ah, Janie would take three million," thinking that he was probably overpaying her for wear and tear; his friends tried to argue him out of such a low figure. At other moments, in the warm glow of nostalgia, Ted would look back on his life with Janie and tell Dr. Frank Pittman and others that he intended to give her $50 million.

In the summer of 1988, the divorce was finally agreed upon, and it was one of the largest settlements in Georgia history. Just as Ed had done with Florence, Ted was setting a financial record in leaving one life behind and moving on to the next. By selling off three million shares of stock to the cable board, Ted handed Jane an annuity of between $18 million and $20 million (which would eventually be worth an estimated $40 million) as well as the Kinloch estate in South Carolina. The settlement didn't approach the hundreds of millions of dollars that some friends thought she might have been entitled to, but Janie was happy just to be out of the marriage. "It was really the best thing for both of them," says Mike Gearon. "They get along better now than they did when they were married. Ted cares more about her."

But family friends have noted just how shell-shocked Janie was on leaving Ted. After twenty-four years of marriage to Turner, according to one friend, she emerged "like someone coming out of a prison camp in North Vietnam." With an extraordinary effort of will, she would eventually kick her alcohol addiction and start a new life. But to this day, Janie rarely mentions her ex-husband's name; she just refers to Ted as "him."

With Turner finally single again, the next months were supposed to be a glorious time for him and his girlfriend. Ebaugh gave the go-ahead to an interview for an *Atlanta* magazine article titled "The Woman Who Tamed Ted Turner," a piece that trumpeted their newfound relationship ("He has developed a lot of trust and confidence in me" said J.J.) and the changes it had made in Ted ("He is a new man; to many, a happier man"). Still, there were dark clouds looming. Although in the article Ebaugh seemed to suggest that Turner was ready to get married and start a new family ("He'll do whatever I want"), she would have done well to look at the writings of Dr. Pittman, who declared that a philanderer who left his wife for his lover stood small chance of making the new relationship work.

The Ted who had been caught up in the romance of the chase was not the same Ted who now found himself living day-to-day with a J.J. contemplating marriage. Ebaugh was pretty and intelligent, tough and sure of herself in a way that Janie had never been, but did he really want to get married again so soon? Perhaps the thrill of victory was wearing off; perhaps J.J.'s unending fascination with the outer limits of spirituality and "human consciousness" were finally starting to rub him the wrong way. Even a freethinker and earth-lover like Turner had to draw the line somewhere, and he drew it at crystals and pyramids, magicians and tarot cards. Then, too, there was a certain edge to J.J.'s character that was exciting but also scary.

Ebaugh had grown up in a family of adventurers. She had lost all three of her brothers to outdoor tragedies and, like her father, became even more daring as a result. In pursuit of the thrills of racing or flying, she didn't mind courting danger. She liked the adrenaline rush out near the brink; she liked how it made you feel more alive when you pushed close to death. Ted, by contrast, was perfectly ready to gamble a fortune when he had calculated the odds, but he was no thrill seeker. Strictly an early-to-bed, early-to-rise type, he didn't search for exhilaration by dangling off a mountain face. Mike Gearon recalls a phone call from Ted the day after J.J. had dragged him away to try ocean kayaking near Carmel, where the surf pounds down on the jagged California coastline. "It was pretty damn dangerous," Turner had to admit, with a note of surprise and trepidation in his voice.

And so, only a few months after the divorce that had been such a long time coming, just as J.J. and Ted were appearing on newsstands all over Atlanta (with J.J. announcing, "Our destiny is so inextricably mixed. This is totally inevitable"), the relationship of Turner and Ebaugh was, ironically, already starting to dwindle. Perhaps his down-to-earth personality and her flightiness were a bad mix, but the two had cared deeply about each other, and the end was sad when it came: "I can see why they were together," says a friend who knew them both. "[J.J. Ebaugh] was very loving, and he liked that, he needed that. And I think she taught him a lot of things about caring and relationships. I know they went through a lot. And I think that after a while, it was just maybe a little too much. Her out-there thoughts were just a little too much for him." The end of Ted's romance with J.J. was yet another heartbreak in one of the toughest thirty-month periods of Turner's life.

It came as no surprise, then, when Turner vetoed the idea of aides to turn his life into a TV movie. As for the autobiography that Simon & Schuster had commissioned, it slowly fizzled out after some two years of abortive starts. For each lively literary device or monologue that journalist and ghostwriter Joe Klein had tried to craft, Turner substituted a dull, straight-ahead memo style. Turner wanted to be taken seriously. He wasn't ready to reveal too much of himself. Besides, he just didn't like the way the book sounded. But according to those closest to the project, the real reason for Turner's discontent with his autobiography lay deeper: "It was a very, very rough time for Ted. The negotiations with Malone and the board were ongoing. Turner was afraid he might lose his company. He was nearing the end of the road with J.J. He wasn't really certain where he was going. That book—it wasn't going to have a happy ending."

CHAPTER 22

ON A BITTERLY COLD December night in 1989, spotlights blazed across the Atlanta skyline as the city celebrated the fiftieth anniversary of that seminal Southern event, the 1939 premiere of *Gone with the Wind*. Parades, look-alike contests, and a "re-premiere" at the Fabulous Fox Theater—the festivities would stretch into a week-long party, an orgy of publicity and promotion, courtesy of the film's new owner, Ted Turner. On this Thursday evening, December 14, all of Atlanta was talking about the Antebellum Ball.

As late-arriving limousines continued to pull up in front of the Georgia International Convention & Trade Center, inside more than fifteen hundred guests, most in period costumes, were waltzing the night away to the music of the Roy Bloch Orchestra. Hundreds of Rhetts and Scarletts, Confederate officers and Southern belles decked out in their finest 1860s regalia, strolled past a replica of Tara in the central banquet area, chatting gaily, their magnolia-blossom accents hanging softly in the air.

The entire evening had been designed as a replica of the fancy-dress Junior League Ball that accompanied the original 1939 Atlanta premiere, and ten of the movie's surviving actors (including Butterfly McQueen) had returned. They stood framed by massive Ionic columns of ice, overflowing tables of hors d'oeuvres with names like "Scarlett's Vegetable Terrine en Croûte." The Big Bethel Choir raised their voices joyously in song.

Ted Turner himself was one of the last to arrive. Dressed in a cream-colored frock coat, trousers, and cummerbund, Turner burst in the door with his statuesque new girlfriend on his arm. Interior designer Kathy Leach was resplendent in an emerald green satin hoopskirt trimmed with black lace. If they looked surprisingly like Rhett and Scarlett, both had worked hard to get the details right. Not only was Leach's dress a painstaking duplicate of the one Scarlett O'Hara had made from Tara's drapes, Leach had also spent hours at the beauty salon, getting Scarlett's curls just right. Coming through the doorway with flashbulbs going off all around them, the two made a dashing couple.

Turner wasted little time in striding up to the microphone on the dais. Smiling at the assembled crowd, he announced: "All I can say is, thank God they shot *Gone with the Wind* in color." Appreciative laughter rolled across the huge hall, and Turner warmed to his topic. He declared that if he were an alien from another planet, this film would be his textbook, his primer on human behavior: "Not what happened in Tiananmen Square or the dropping of the Iron Curtain, but *Gone with the Wind*."

Turner loved *Gone with the Wind*. He loved the romance, the drama, the epic sweep. He called it "a noble movie—the greatest movie ever made." He had seen it dozens of times; more than one guest visiting his plantation recalls being forced to sit through the film with him. Ted pictured himself as a Rhett Butler of the cable age—a brash, handsome, devil-may-care maverick. So, intriguingly, did the media. The *Atlanta Constitution* once ran an article comparing the two: "Both were tossed out of school. . . . Both are great sailors. . . . Both are full of political contradictions. . . . Both are outsiders in a landscape they dominate." Turner had, in fact, consciously gone to certain lengths to pattern himself on the character, admitting that he had fashioned his mustache after Rhett's. He had considered calling his first son Rhett, after a contraction of the intials for Robert Edward Turner, and actually did name his second boy Rhett (although Jane had drawn the line at naming their daughter Scarlett).

On this December evening, as Ted danced briefly with Kathy—a shuffling, senior-prom slow dance—and then stopped to eat and chat, the friends and colleagues with whom he talked could be excused if they felt that *Gone with the Wind* seemed now, more than ever, an apt Turner emblem.

With each passing year, Ted Turner's life appeared to have just a little more in common with the Mitchell saga of Southern tragedy and rebirth, with its big canvas, its melodramatic ups and downs, its sudden reversals of fate and fortune. His life had the roller-coaster drama of a Hollywood plot. In 1985 and 1986, he had been the laughingstock of both coasts over his failed CBS bid and his "successful" MGM deal. Like antebellum Atlanta, his company seemed doomed to go down in flames. Add to this his son's accident, his clashes with the board, the back-to-back disintegration of his marriage to Jane and his relationship with J.J.: "He's had a whole series of things that would have crushed a mere mortal," Jim Roddey said of his ex-boss at the time. "He's laying low, but I think he'll be back."

And here he was on this December evening in 1989, once again on top of the world—celebrating not just the golden anniversary of *Gone with the Wind* but one of the most dramatic turnarounds in American business. Ted Turner, the man who had "left town in a barrel," was now being called Captain Comeback. His stock had shot up, tripling, quadrupling, quintupling its value. Turner himself was now worth at least $1.76 billion, according to the *Forbes 400* list of the richest men on the planet. "He's on a roll," said the *Wall Street Journal*. "He's on a roll," said *Business Week*.

The reason for all this enthusiasm lay in the balance sheets: for the first time since before the initial MGM agreement in 1985, for the first time since it was buried under a mountain of debt, Turner's company was actually showing a small but tangible net profit ($5.5 million for the second quarter of 1989). "We're like the Allies after Normandy," crowed a triumphant Turner. "We have landed. We are there. The beachhead cannot be obliterated."

By mid-1989, Turner's empire looked impressive indeed. In two years, it had doubled in value, and was now worth $5 billion. The catalogue holdings included CNN and Headline News ($1.5 billion), reaching nearly fifty-one million homes in the United States and eighty-three nations abroad; Superstation WTBS ($1.5 billion), the single most watched basic cable channel in the country; Turner Entertainment ($1 billion), syndicating the MGM movies; the new network TNT ($650 million), just taking off; and the Braves, Hawks, and CNN Center real estate ($350 million). Ted was not only successful in terms of cable standards—with three of the six highest rated networks on basic cable—he was even more impressively successful in terms of broadcast standards. According to Turner Broadcasting annual reports, CNN and WTBS had *each* earned more than the ABC and CBS networks combined.

But if the media was describing Ted Turner's remarkable success as the turnaround of the year, if not the decade—if reporters talked of him as rising from the ashes like the New South after the Civil War—Ted Turner himself declared that the results merely proved what he had been saying all along. He was building a company; the MGM debt didn't mean a thing; he had planned it this way from the beginning. "It's obvious I was seeing things that other people didn't see," said an exuberant Turner, "just like Columbus discovered the world was round."

Why, in fact, had Ted's empire turned around? Was it, as he indicated, simply a case of the world finally catching up with his farsighted vision? Or was it the fabulous Ted Turner luck? Actually, while vision and good fortune both played a role, there were more mundane reasons for Ted Turner's renaissance. First, eighteen months earlier, in January 1988, as Turner stock hit a low of less than $10 (in the wake of the October 1987 market crash), he had instituted an investor-relations plan, the company's first organized attempt to talk to the Wall Street analysts who can make or break a company's reputation. In the middle of all the bad press, Turner sought to sow the seeds of reassurance, which would in fact bear fruit a year and a half later. Then, too, Turner was able to ride the overall growth of the cable industry. As interest in cable TV exploded, as more and more Americans turned from CBS, ABC, and NBC (whose prime-time viewership dropped from 90 percent of the television audience in 1979 to 67 percent in 1989) to the new upstart cable networks, Turner's channels were lifted and swept along on the tide.

But perhaps the main reason for Ted Turner's rebirth, ironically enough, was his board, the very directors with whom he had clashed repeatedly. For all their conflicts with him, these cable-system operators now had a clear financial stake in Ted Turner. And while they didn't want him to get too big or unruly, they nonetheless realized that for every dollar Turner made, they would get a share. Comprised of some of the best and brightest minds in the cable industry, this board was in a position to help Turner make those dollars in two ways: they could give him advice and insights not available to the average businessman, and they had industry clout.

Actually, they *were* the industry. If they wanted a new Turner channel to be shown to viewers across the country, it would get shown. Turner could bank on it.

One of the current directors describes this board as "a microcosm of the heavies in the cable business—I always get the feeling when I'm at a Turner board meeting that it's generally a board meeting about everything in cable." Even a former board member like Mike Gearon, who was replaced after the takeover, speaks with nothing but admiration for the new group: "What a board! The very thing that people thought was a liability. This isn't a board of yes-men. These are fucking tigers on that board. No one's sitting there with a hundred shares of stock because they belong to some civic organization. Every one of them is sharp as hell. This is *their* money on the line. This is *their* business. There are times when their best interests conflict with his, businesswise, but by and large the benefits far outweigh the shortcomings."

Of course, Ted was now reined in by the $2-million limit. The old swashbuckling style of TBS, where Turner would get a new idea and immediately charge after it, had changed totally. He was no longer steering a Lightning, responding instantly to the currents of wind and water. Now he was helming an ocean liner, and the steering was done by committee.

But there was also much more stability, much less chance of capsizing for Turnover Ted. The board brought to TBS a new sense of financial discipline and organization. "They gave the place structure," Robert Wussler says. "No one ever cared about the old board meetings. Now there are committees established—audit committees, compensation committees. Now there are reports to be made. It's a different kind of place."

Though clashes continued, by late 1989 and early 1990 Turner and his board had gotten past the initial shakedown period. Both had staked out certain turf, indicating the limits beyond which they would not budge (the board on FNN, for example, and Turner on any intrusion into the management personnel of his company, or any offer to buy him out). But having growled and flexed their muscles, both soon found common cause—the success of Turner Broadcasting—and each side was contributing to achieve that success. "The board," Gerry Hogan says, "was clearly operating with another agenda. Their agenda was, 'What's good for the cable industry.' It's proved, however, that what's good for the cable industry has been good for Turner and vice versa. Ted has really come into his own with their help."

The classic story of the successful entrepreneur is one of early triumph—a triumph that cannot be sustained. Entrepreneurs are typically sprinters, not marathoners. If they succeed at all, they start in a blaze of glory, then encounter trouble when they have to make the transition from creating a company to running it. Some businesses and businessmen hit this wall at the $1-million mark, others at $100 million to $250 million, but with surprising regularity entrepreneurs face the same dilemma: The skills of empire building are not the skills needed for empire administration.

With the combination of Turner and Malone, Fuchs, Levin, and the rest, Turner Broadcasting now had the best of both worlds—creative energy and discipline, entrepreneurial initiative and executive stability. Tied in as he was to the future of the cable industry, Ted Turner was almost guaranteed success. "I have to believe that this has worked out better than Ted ever thought it would," one of his directors speculates. "Quite honestly, I think this is a very good structure. I don't think you can have anyone with the kind of visionary skills and impulses of Ted and at the same time have a very careful, studied, thoughtful business operator. So what we've got is the good part of Ted still operating, and where the downside used to be, that's limited. When Ted made a mistake, it was a big mistake, because Ted likes to roll the dice, and Ted believes in his insights. Ted's capable of going for something and not worrying about the financial consequences. That's what visionaries do. But if you can combine a visionary with an operating person—well, that's Ted and the board."

The clearest example of teamwork between Ted and his directors came in the formation of TNT—Turner Network Television. Both Turner and John Malone had, for years, been contemplating the creation of a cable channel to rival the broadcast networks with big-time sports and entertainment—"must-see" programs to cut through the clutter of the fundamentally undistinguished medium of cable. The idea of this "next step" in cable had gone through several permutations. Turner had originally envisioned it as a broadcast network in the early 1980s—wooing the Hollywood studios for films and airing "special event" basketball games—while

Malone had contemplated an all-star cable network that pooled programs from all the different cable channels.

It was when Turner bought the MGM film library, with its nearly 3,500 titles, that he clearly had the makings of a new network, and when he combined those films with premiere NBA or NFL events, he had a network that people might pay extra to see. Here, Turner hoped, was that apotheosis, that grail of television executives—"appointment TV," programs like Sunday night football for which viewers would carve out a specific time in their lives. Although WTBS had become the most popular of all the cable networks, it had done so with *Andy Griffith Show* reruns, which even Turner, with his tin ear for quality, could tell did not add up to innovative programming. TNT, with its flashy, high-profile sporting events, classic films, and earnest TV movies, was his bid to acquire distinction.

By the fall of 1987, soon after the board was in place, Ted was talking eagerly of launching his new network, launching it even in the teeth of the October stock-market crash. The new board was less than encouraging. "Some of the directors want to see Turner on a more solid financial footing before undertaking something of this magnitude," said Tim Neher, president of Continental Cablevision and one of Turner's directors.

The board hemmed and hawed for months as they and Turner felt each other out early in their relationship. But as concerned as they were about giving Ted too much power in the industry by adding another network to his stable, they nonetheless could see that the whole point of his buying the MGM library was to start such a network. The financial logic was inescapable. "There was a clear sense that the cable operators wanted us to stick with what we already had," Gerry Hogan remembers. "But TNT had to be launched. That was really built into the deal with the partners— not as a deal point but as an understanding that the library allowed us to launch another network."

By mid-1988, the board gave its grudging support to this new channel, and on October 3, 1988, Turner Network Television was launched under the supervision of Gerry Hogan. Its first feature, of course, was *Gone with the Wind*. At the outset, TNT seemed none too revolutionary. Despite all its claims about spending millions ($250 million a year was the promise) for brand-new programming, the first week was merely a tepid film festival, one creaky old-timer after another, the most recent work completed in 1956. Over the first year, TNT would premiere only one new program a month, a halfhearted string of lukewarm cinematic wanna-bes, like Farrah Fawcett in *Double Exposure: The Story of Margaret Bourke-White,* Hal Holbrook in *Billy the Kid,* and *The Secret Life of Ian Fleming.*

On the financial side, however, TNT's debut was spectacular, the most successful launch in cable history. Where CNN had struggled to open with two million viewers, at a time when fewer homes had cable, TNT started with seventeen million. Within five months, it had shot up to twenty-seven million, and inside a year the pay service had fifty million subscribers. And the advertisers loved it, too.

The explosive rise of TNT can be attributed, in large part, to the board. After all, once Ted had convinced his board of directors, he had *de facto* sold nearly two-thirds of the cable industry. "Ted and the board coalesced on TNT," says one director, looking back with the rosy glow of hindsight. "The board was behind the channel, and helped get the channel distributed through cable. I mean, we supported it. We gave it the kind of support it needed through our cable systems. We also helped make decisions, like moving the NFL to TNT, which moved TNT from forty to fifty million subscribers. That was a board suggestion."

When Ted and the board worked hand in hand, it was a powerful alliance—Turner with the fresh ideas, the board with the old-fashioned muscle to make them come true. With this partnership in place, many past assumptions about Turner began to alter, like pieces in a kaleidoscope. Whereas the MGM deal had once looked like a classic case of overreaching that might sink Turner Broadcasting, it was now clear that, with the board's backing, the MGM library had become the core of both a syndication business *and* a successful new network—together worth a projected $1.6 billion by mid-1989. In only three years, Turner had essentially earned back all his money on the MGM purchase. The deal that had once made him look as if he were being taken to the cleaners now looked brilliant. Turner *had* paid too much, as even he would later admit, but the MGM library was simply worth more to him than to Kerkorian. And with the board behind him, it was worth still more. By cleverly repackaging the old movies—from the superlative *Adam's Rib* to the mediocre *The Devil at Four o'Clock*—as "classics," by colorizing them and cleaning up the old prints and re-presenting them as "all-time greats," he brought in a sizable new audience, often as large as the one for first-run movies. Turner's idea had always been to own, not rent, the programs he showed, whether it was a Braves game or an MGM film, and while he had overspent more than once, he would consistently recoup his money as long as the cable companies made a place for him on the crowded cable dial.

By the end of 1989, the financial reality of TBS was finally beginning to match all of Ted's hype. While it was often easy to dismiss Turner's overheated claims, he had in fact been proved right on several scores. With TNT, WTBS, CNN, and Headline News, he now controlled approximately one-third of all cable viewing. His company was worth over $5 billion. Even more important, this once high-risk venture now looked so financially secure for the future that banks were clamoring to lend Turner money. (Turner would comfortably be able to refinance his debt when eight banks, led by the august Chase Manhattan, offered him $1 billion in credit.) And with TNT, he had opened the door to a creative future as well, establishing a center for new, quality TV movies—although the network still continues to struggle to fulfill that promise.

But the grandest gamble that had paid off was CNN, which—looking back with twenty-twenty hindsight—now seemed almost inevitable. As Ted danced on that chilly December evening at his Antebellum Ball, celebrating all things Turner, his news network, which had started with long odds indeed—not just financially but in terms of credibility—was a clear winner, both monetarily and journalistically, as it approached its tenth anniversary. If its journalistic success could be attributed to Ted only because he had enough sense to keep his hands off (and let people like Reese Schonfeld and Mary Alice Williams get on with their jobs), CNN had nevertheless been his creation. He had backed the idea of an all-news network when others with experience in television news had laughed in his face, and in a decade it had gone from 1.8 million viewing households to 53.8 million. CNN was seen in eighty-nine nations around the world, from Buenos Aires to Budapest. It had an operating profit of over $85 million, which now dramatically figured as nearly three-quarters of Turner Broadcasting's total profit. In other words, the cash drain had become the cash cow.

Even more impressively, CNN now had a kind of international credibility that many diplomats lacked. Through its coverage of the Falklands War, the *Challenger* disaster, and the TWA hijacking, the Chicken Noodle Network had achieved a worldwide stamp of seriousness. True, the reporting could be haphazard, as when Moscow correspondent Steven Hurst mistakenly announced that Mikhail Gorbachev was considering stepping down, and sent international currency markets into shock. The network's pieces could still be sloppily conceived, written, and edited, and contextual analysis was still a weak point.

But more than any other American TV news entity, CNN had a global reach and a global mission. With four thousand employees, sixteen hundred full-time staffers in twenty-one bureaus around the planet from Nairobi to Moscow, CNN was building an instantaneous worldwide reach that would make it the video wire service of record wherever big events were happening. Margaret Thatcher, François Mitterrand, Mikhail Gorbachev, and Fidel Castro all received CNN in their offices and were, according to aides, faithful viewers; Saudi Arabia's King Fahd was said to watch compulsively all through the night. Even North Korea's supposedly isolated Kim Il Sung was a "subscriber" (and shrewd enough to later allow CNN a rare exclusive invitation to cover the visit to his country of former president Jimmy Carter, during North Korea's 1994 nuclear standoff with the West). To be in the know, one couldn't afford not to have CNN. Soldiers watched it. Spies watched it. And in newsrooms all around the globe, journalists clearly did as well: CNN had become a significant part of the news-gathering process.

In 1989, a tumultuous news year—from the San Francisco earthquake to the fall of the Berlin Wall—CNN scored consistently, arriving on the scene rapidly and following through with long-running, dramatic coverage. In China especially, at the Tiananmen Square uprising, CNN performed admirably, making the news-gathering itself part of the story (just as Schonfeld had always wanted it to be). The network stayed on the air after ABC and NBC were forced off, and then when Chinese government bureaucrats arrived to shut down the live feed, CNN broadcast these tense negotiations as they occurred, right up to the pulling of the plug. For the first time, government repression was seen live, in action, around the planet. "Karl Marx, meet Marshall McLuhan," announced *Newsweek*.

If 1989 was a landmark for the global village—the first world-changing year of the uplink era—it was especially momentous for CNN. Two events in December of that year would demonstrate CNN's emerging role in the new media world order. When George Bush and Mikhail Gorbachev met for a summit on vessels anchored in the waters off the island of Malta, the international news community, twenty-five hundred strong, was unable to get on board. Stuck in Valletta at the Knights of Malta infirmary, which had been converted into a news hall, reporters filed their stories as they watched the events on CNN monitors. As for Bush and Gorbachev, they both arranged to have a CNN feed sent out to the boats: they could watch themselves in action, see how their words were being played, and adjust accordingly. A new age of video self-consciousness, of instant feedback, was being born. The distinction between the news and the reporting of the news was dwindling—the two were now almost interchangeable.

As it turns out, only days later, when U.S. troops invaded Panama, the Soviet Foreign Ministry's first call was not to the U.S. Embassy but to the Moscow bureau of CNN. A CNN crew dashed over, and Soviet government spokesman Vadim Perfilyev unleashed a strongly worded statement deploring the invasion. "Who needs striped pants and diplomatic pouches?" asked the *Wall Street Journal*. "When world leaders want to talk to each other these days, they call CNN." Communiqués and red telephones were all very well, but when you wanted to make a statement to the world community, when you wanted to put a little *force de frappe* behind your words, you spoke to Ted Turner's channel.

Turner himself took the ascendancy of CNN with his characteristic modesty: "I wanted to use communications as a positive force in the world, to tie the world together," he said. "And you know something, it's working."

In three years, Turner's fortunes had been radically transformed. Once crippled with debt, he was now a billionaire, controlling a multibillion dollar communications empire, including arguably the most significant news vehicle on earth. And the future looked even better. Only one thing was missing. "If I had one wish for something that would give me personal joy and satisfaction, it would be to have my father come back and show him around," Ted said.

"I'd like to show him the whole shooting match. . . . I really would. I think he'd really enjoy it."

If Turner was thinking of his father and the milestones in his life, he could be excused. The late 1980s were not only a transitional time for his businesses but for Ted Turner personally. In November of 1988, he and a small group of his close friends had gathered at his log cabin outside Atlanta to celebrate his fiftieth birthday. Another, bigger party would be held at the CNN Center atrium, complete with music and a video wrap-up of Turner's five decades.

Like many who reach that landmark anniversary, he had mixed feelings. "Passing the fifty mark was a big trauma for Ted," Gerry Hogan remembers. "He was obsessed with turning fifty for months before it occurred. . . . He worried about how old fifty seemed—how little of life there was after that."

But for Turner, the birthday also came with extra baggage, because he was now fast approaching the age at which his father had shot himself—the age he never thought he would reach. For much of his life, he had talked about how he would die young, a suicide or the victim of an assassin's bullet. But now he was fifty and showed no signs of dying. "There was a time when he didn't expect to live that long, I suppose," says Hogan. Dee Woods, Turner's longtime assistant, agrees, adding, "Ted felt that his father had died tragically, and it was his duty to die tragically."

So here he was, lurching through his midlife crisis with a glass of champagne in hand, two marriages down the drain, and some minor health problems, like the nagging fatigue, that just wouldn't go away. While the specter of death clearly gave him pause, he had confronted his mortality before, confronted it on an almost weekly basis, talked about it so often that his friends and colleagues came to ignore the way he constantly worked it into conversations like background chatter. It wasn't death that loomed so large and unsettling for him; it was life. He had reached the half-century mark, and the territory up ahead was all terra incognita. He was traveling in a land where his father's road maps had run out.

A snapshot of Ted at fifty shows him standing erect, his jacket and tie still boyishly rumpled, his chiseled features and cleft chin still as handsome as ever. All around him are trophies. He is surrounded by trophies—Emmies and Aces and shelf after shelf of sailing cups—enough to fill a museum, and maybe just enough to convince him that he's as good as all that inscribed gold and sterling says he is. His hair has turned almost entirely to silver now, as well as his eyebrows and mustache. His face is creased with wrinkles, the tire tracks of time. But it is the directness of his glance that catches the moment most precisely. Narrowed, his eyes look out with a strange mix of confidence, youthful glee, and bone-wracked weariness.

Interior designer Kathy Leach had been working in planning and development at CNN since the early 1980s, helping to oversee the physical layout of the new offices when the news organization moved downtown into the CNN Center. A tall brunette with a striking figure and confident poise—it wasn't surprising that she caught Ted Turner's eye. If Leach was not exactly Turner's type (blond and energetic), still she was long-limbed, tailored, and stylish. As a divorcee in her thirties, she possessed a mature reserve that was unusual among Turner's women. The two dated occasionally through the mid-1980s, as Turner's relationships with Liz Wickersham, J.J. Ebaugh, and his wife Janie ran their course.

In 1987, when Turner decided he was tired of sleeping on the fold-out sofa in his office and sat down with Bunky Helfrich to design a penthouse on the roof of the CNN Center, he thought of Leach and asked her to come up with an interior plan for the space. She responded by giving the small upstairs area a masculine, corporate, nautical feel, with a huge freestanding fishtank dividing the living room and bedroom. Turner loved the look, and found himself more than a little taken with the designer. He would give her subsequent commissions—refurbishing the Hope Plantation and renovating the Avalon estate in Florida.

Turner and Leach quickly discovered that they had more in common than just a shared fondness for jewel tones, like emeralds and burgundies. Mixing work and pleasure, they had by mid-1988 started to date again more seriously, as his relationship with J.J. slowly petered out. While J.J. had claimed that Ted "gave up a whole harem of women" to be with her alone, his faithfulness, though much improved, was still patchy. By the time he had grudgingly agreed to join J.J. on the cover of *Atlanta* magazine in November 1988, and the magazine was labeling J.J. "The Woman Who Tamed Ted Turner," their relationship was already on the rocks.

Several women would come and go over the next two-and-a-half years, including Chinese TV star Yue-Sai Kan, the lovely young Gabrielle Manigault (the daughter of his South Carolina neighbor Peter Manigault) and feisty CNN reporter Shavaughn Darrow. All were attractive, though few were world-class beauties. The newly divorced, newly separated Turner simply hated to be alone, and he once again began notching up the female tally, as if in quest of some record number. His romantic style in pursuing all these different women simultaneously was, in the words of one of them, "go and go until you drop." But it was with Kathy Leach that Turner found himself spending more and more time going. They traveled together to London, Los Angeles, and New York, where he worked on the refinancing of his company and she attended meetings for her own business. Together they saw *The Phantom of the Opera,* visited the American Museum of Natural History. Their days were filled with a series of these lightning trips.

But Turner, worried about his health, intended to make several changes in his life. With J.J., he had spent too much time in California. The Big Sur property had made him feel trapped; there wasn't enough land there for the long quiet hikes he loved to take. And when he wasn't in California, he was in Georgia. As cities go, he had always liked Atlanta, but now he began to say that cities made him ill. Driving into downtown Atlanta, he would start to complain. He felt the pollution had a bad effect on him. He felt the metropolis itself, with its cars and its skyscrapers, was making him sick. He wanted to stay out in the country, where he could be relaxed, fishing and hunting and walking in the woods.

When Ted went to Wyoming to visit his old neighbor Peter Manigault, who was now spending his summers on a ranch on the Northern Great Plains, he was struck by the wide-open beauty of the landscape. The scale was so vast in these Western mountains, the horizon so distant, the terrain so rugged and pristine. "I like these ranches," he said to Manigault on his second day there. "How the hell do you go about getting one?"

"Well," said Manigault, "we've got a broker in Billings, Montana, who's really good."

The next day—one phone call and one Federal Express delivery later—Turner received a handful of brochures. He looked them over and announced, "I think I'll get this one." Calling the broker in Billings, he said, "I'll buy it." He would purchase the property sight unseen. The broker was flabbergasted. "C'mon, Mr. Turner," he said. "You gotta let me take you there, for godsakes! You might not even like it!"

So Turner and Manigualt jumped into Ted's private jet and flew over to the nearest airstrip. When they got to the ranch, the real-estate broker was waiting with a concerned expression on his face. "I gotta warn you," he said. "I've lost a couple of clients on this. . . . You're gonna see a hell of a lot of rattlesnakes on this place." He looked anxiously at Turner.

"Rattlesnakes?" Turner said. "Hell, I love rattlesnakes. That's a plus, as far as I'm concerned."

Without hesitation, Turner decided to buy the old Sixteen Mile Ranch, near Toston, Montana, halfway between Helena and Bozeman, adding a parcel next door before the year was out—in all, over 20,000 acres. He renamed his ranch the Bar None.

One day, several months later, after Ted and Kathy had been dating for a while, the pair was hurtling around the Montana property in his jeep. From time to time, Ted would stop and point out features of the landscape, and the two discussed plans for the Bar None house—rustic peeled logs and a stone fireplace. As they bumped along over the rocky terrain, Turner suddenly looked over at Leach and suggested a change in their relationship.

In no time, they had gone from being friends and colleagues to lovers. Ted seemed to like her quiet, low-key supportive style, but after Janie, after J.J., Leach hadn't been certain that Turner would want to settle down. Yet now, on the far side of fifty, he seemed to have a different, more serious attitude. In fact, he told Leach that if they were to live together, she'd have to give up her interior-design business and her outside clients altogether. According to Leach, "He wanted me to be with him all the time. He didn't want to be alone." Besides, he had plenty of work for her to do on all his properties. And then, like the high-powered businessman he is, Turner announced, "I'll give you one day to think about it."

Kathy Leach didn't need that long. Giving up her business, she stepped into the whirlwind. The two would do everything together—jetting around the world in a style more common to world leaders than ordinary folks. They traveled to England, to Cuba. They traveled to Newport for a pair of J-boat races. (Though Turner had, as seamen say, "swallowed the anchor," he came out of retirement to race and win these charity events.)

In May of 1990, a year after the Tiananmen Square uprising, they journeyed to China for the last great international conference of the Better World Society. They had come to investigate pollution and population control, and to sightsee a little. Together, Ted and Kathy climbed the Great Wall, strolled through the Forbidden City, went to the zoo to see the pandas. Above all, they attended meetings in Beijing with Jean-Michel Cousteau, Olusegun Obasanjo of Nigeria, and others, as the Better World Society researched China's zero-population model—the program that allotted money to those with one or two children, penalizing those with more.

A high point of their China visit came when Turner, continuing his effort to serve as a self-appointed ambassador with electronic and environmental portfolios, met with Li-Peng, the powerful premier. Although Turner had met with Gorbachev and Castro, he was nervous before this meeting, because—in the wake of Tiananmen Square—he was veering perilously close to real diplomacy here, and it was treacherous ground.

Li-Peng ushered them in, and offered Turner and Leach the traditional green tea. For the next half hour, they spoke of the environment, of population control, of the world situation. Leach watched Ted anxiously, not knowing what he might say next. But Turner was on his best behavior. As if by an unwritten mutual agreement, both men steered clear of any politically charged dialogue on CNN's coverage of the democracy movement or China's human-rights situation. Instead, they sipped their tea, talked hopefully of the world, and in the end, exchanging warm handshakes and scrutable smiles, parted.

A few months later, Turner and Leach flew to the Middle East. The trip had begun as a vacation, for Leach was interested in Egypt, its history, its pyramids. When Ted agreed to go, he managed to work in some business, and while visiting Tel Aviv for two weeks of sightseeing, he also met with government officials to discuss bringing CNN to the countries of the region. There for the first time he came face-to-face with some of the tensions underlying Arab-Israeli relations.

Traveling from Israel to Jordan, Turner and Leach were stunned to hear that they each needed two separate passports, since Jordan does not recognize documents with an Israeli stamp. In his inimitable Ted Turner style, Turner looked around this volatile region and declared that what the Middle East really needed was . . . peace. Pacing about in his hotel room or looking down at the desert from 20,000 feet, "he started thinking about it," Leach recalls. "The whole trip he was trying to figure out how to bring peace to the Middle East—how could he bring peace to the Middle East?" And as he met in turn with officials from Israel, Egypt, and Jordan, he brought them the good news: "There needs to be peace here," he'd say. "I wanna help bring peace. I wanna bring these countries together." Although not exactly an epiphany to the parties involved, his good intentions were doubtless appreciated.

Perhaps his most successful meeting came in Jordan, when King Hussein and his lovely blond American wife, Queen Nur, invited Turner and Leach to their palace in Amman for a long weekend. Together the foursome went out on the king's boat, cruising and snorkeling. Later, sightseeing took them to Petrus, Aqaba, and the Dead Sea, where Ted jumped in, as always, feet first and eyes open—only to realize how stingingly salty it was. He rocketed out, his eyes on fire.

While Turner had little patience for the mud wraps and facials that the queen suggested, he and King Hussein did have plenty of time for long, rambling political discussions that lasted well into the night. Turner talked of Israel and the West; the king explained the Jordanian position. Their conversations were friendly and wide-ranging, and for some of them they were joined by PLO leader Yassir Arafat. Once again, Turner was engaging in the same kind of exchange he had had with Castro and Gorbachev—a mix of politics and business. He was bringing CNN to the world; and they were bringing their world to CNN.

But for all the international travel, Turner was pulled back time and again to his new home in Montana. As his business was finally turning around, he had decided that he wanted—he *needed*—to spend more time serenely fly-fishing in the roar of a Western stream or wandering in the crisp cold air of the northern plains. One year after purchasing the Bar None, in the summer of 1989, he bought a second property an hour and a half away from the first, 20 miles southwest of Bozeman, in Gallatin Gateway, Montana.

Where Spanish Creek empties into the Gallatin River lie 130,000 acres of green rolling land and breathtaking ice-capped mountains, stretching all the way to the famous Spanish Peaks. Just over the border from Yellowstone National Park, the land shared much of the same rugged natural beauty. It was called the Flying D ranch, and it included one of Montana's largest cattle operations, the Shelton Ranch. It cost Turner $22 million, but it was worth it—one of the premier pieces of Montana real estate.

Ted and Kathy spent days together roaming the Montana properties with Ted's black labrador Sonny at their side. They'd climb the steep cliffs, discovering waterfalls and clambering up the rocks. Often Turner would point out birds on the wing. His love of wildlife was clear to her, his knowledge of it surprising. He could pick out almost any species at a hundred paces—mallard, wood duck, sandhill crane. As they hiked, he taught her, much as he had taught his children over the years.

But it was Leach who taught Turner how to ride a horse. While he loved animals, he had, in the words of his friend Peter Manigault, a grudging "armed truce" with horses. "He used to hate it when I'd go fast," Leach remembers, laughing. She'd gallop off, with Turner's horse dashing after her, trying to catch up. Hanging on for dear life, Turner would be shouting, "Stop!" at the top of his lungs.

Once when he and Kathy were elk hunting with friends, they had ridden out on horseback, and by the time he had actually bagged an elk, the sun was starting to go down. Fourteen miles away from the ranch house, Turner gave the elk and his horse to the guide, and said to Kathy, "Let's go back on foot." The hills were steep, and he thought they'd do better without the horses. In fact, he announced to the others by way of challenge that they'd get back first.

As night fell, the two of them hurried across the hills, helping each other over the stones, urging each other on—"You can do it," Turner exhorted. "We can do it." The moon and the stars were well up in the sky as they rushed over the hills. "It was exhilarating," recalls Leach. "We were running. We had to get back before anyone else did. We wanted to prove that we could do it on foot. It was unbelieveable. We didn't get back till one in the morning." But most important, they arrived first.

Out on the plains, Turner often pondered his businesses as well as his avocations. Occasionally, he'd have a brainstorm, and whenever one struck, Leach would dutifully jot it down. "I used to carry a pen and paper around," she says. "He would say something—

We need to do this, and this—and I would copy it down, and then when we got back to the house, we'd talk about it. He'd want to act immediately."

One day, when he and Leach were riding around in a jeep and looking over his Flying D ranch, Turner stopped the vehicle and stared out over the huge herd of grazing cattle, shaking his head angrily. "I want buffalo," he said. "As far as the eye can see, buffalo. I'm gonna tear all these fences down and buy a thousand buffalo." Ted's father had raised cattle at Cotton Hall, but Turner had never liked the creatures. The idea of bison had been percolating in his mind for some time as a replacement for these cows. Cattle involved more work and were more difficult to feed. "Buffalo just graze," says Kathy Leach. "They're a native animal. [Ted] thinks that buffalo are cleaner." As later events would show, Ted had no time for cattle.

Ted and Kathy spent many happy months together, roaming the world and his ranches in Montana. Theirs was an easy, relaxed relationship punctuated by wild moonlit races, but it was not destined to last.

For Ted Turner had met Jane Fonda.

IT WAS TO BE one of the great media romances of the nineties, a love story played out in the public eye between the prince of TV and the princess of film, Ted Turner and Jane Fonda. Both attractive, both immensely rich, both very public commodities, Ted and Jane would meet, logically enough, at a celebrity function in Los Angeles. At the time, Ted was on the rebound from his failed marriage to Janie and his breakup with J.J., and Jane was in the midst of a bitter split with her husband, Tom Hayden.

For Jane Fonda, it was a grim time. Although at the peak of her movie and aerobic fame, Fonda was miserable, her sixteen-year marriage to Hayden on the rocks. A politicized union born in the early 1970s between the SDS founder and the earnest young starlet of *Barbarella* turned would-be radical—the "Mork and Mindy of the New Left," in the words of the *Hollywood Reporter*—the marriage of Hayden and Fonda had always been about public gestures: clenched-fist salutes and anti–Vietnam War speeches, Black Power rallies and the trip to Hanoi. (Their son Troy was originally named Troi, after a Vietcong hero.) Over the years, Fonda had supported Hayden's various unsuccessful runs for office in California with the proceeds from her movies, such as *Fun with Dick and Jane* and *Coming Home*. But while he accepted her money, what had become clear was that this former Chicago Seven organizer had mixed feelings about his movie-star wife: "He was always ambivalent about her career and her wealth, for a couple of reasons," a longtime friend recalls. "One was that she was doing better than he was, and two, [her acting career] wasn't something he respected from a political or intellectual standpoint. . . . He often put her down and made her feel stupid. And she was for most of those years intimidated by him."

Hayden also wanted a much more "open" marriage than Fonda did. His philandering soon became the subject of common gossip. He seemed to picture himself as a Jack Kennedy or a Gary Hart, with women at each campaign stop. Fonda bore all of this with a sad, wounded dignity, trying to breathe life into their relationship even as Hayden snubbed her. As one friend says, "Tom couldn't acknowledge her [in public]. I remember Jane would go sit next to Tom sometimes and place his arm around her so it would look like they were together."

During the 1988 presidential campaign, when Hayden began a public romance with Vicki Rideout, a thirty-three-year-old Dukakis advisor and speechwriter—a romance that showed no sign of cooling off well past the heat of the election—Fonda finally had had enough. At Christmas in Aspen, she insisted that Hayden end the affair. He refused. And two months later, he loaded up the family's Volvo station wagon and drove away from their home in Santa Monica.

It was during this unhappy period that Ted Turner and several other would-be suitors rang up Jane Fonda, and while she was flattered, she was none too interested. She didn't really want to see anyone at the time, and Ted Turner perhaps least of all. One night at a screening at the home of her agent, CAA's Ron Meyer, she announced to the friends gathered there that Turner had been calling rather persistently, wanting to take her out. "Ted Turner!" She rolled her eyes and chuckled. "Can you imagine?" But Turner was nothing if not persistent. He had always talked about wanting to date an actress—like Charles Foster Kane, like most of America, he was fascinated by Hollywood royalty—and for Hollywood royalty like Jane Fonda, daughter of Henry, sister of Peter, and double Academy Award winner in her own right, he unpacked all his charm.

Finally, in the summer of 1989, they met at a Los Angeles fundraiser at which he was invited to speak. It was a typical Hollywood event, a political gathering in the main ballroom of the Century

City Hilton for the Show Coalition, hosted by Patricia and Mike Medavoy, who was then head of Tri-Star Pictures. In between Ted's speech and the last sip of coffee, they talked and had a chance to get to know one other a little.

Although Fonda seemed to find him pleasant enough, it was hardly love at first sight for her. But Turner, ever the romantic, was more than a little smitten by the woman seated next to him. Jane Fonda had both beauty *and* glamour, sex appeal *and* star power. Robert Wussler says, "Ted was extremely taken with Jane Fonda, right from the beginning."

He wanted to see Jane again, and while she put him off at first, when Ted discovered that her brother Peter had a ranch near Livingston, Montana, not too far from his Flying D ranch, he invited both Fondas over to visit. Ted and Jane and Peter spent an amusing, low-key afternoon together, fishing and horseback riding on the Montana range. The visit was more social than romantic, but as they rode across Turner's property, it was obvious that Ted had a closer relationship with Jane in mind.

Fonda really wasn't interested. She was in the process of remaking her life—a process that included psychotherapists and plastic surgeons, hairdressers, silicone implants, and more than one new boyfriend. So what if her husband had left her? She was going to be younger, sexier, more glamorous than ever. She had taken up with a handsome, well-built young Italian named Lorenzo Caccialanza. A soccer star turned movie actor (he was one of the good-looking terrorists Bruce Willis knocks off in *Die Hard*), Caccialanza had the extra attraction of being fifteen years Fonda's junior—the perfect revenge for Vicki Rideout. The fifty-year-old Turner had a lot going for him, but he was definitely *not* a muscular young Italian, and Jane Fonda was, by all accounts, very much in love with Lorenzo.

Fonda and her boyfriend traveled from St. Bart in the Caribbean across Europe to Milan, followed in hot pursuit by the tabloid paparazzi. But when Jane's fling eventually wound down, Turner did not hesitate to call her again repeatedly, wearing out the phone lines, even though he was still living with Leach. In March of 1990, only a few months after he escorted Leach to his *Gone with the Wind* ball, he showed up on the West Coast at that most public and star-studded of events, the Academy Awards, as Jane Fonda's escort.

Walking down the red-carpeted path through the gauntlet of fans and flashbulbs—he resplendent in tuxedo and she in a turquoise Versace gown with glittering décolletage—they seemed to be hinting that there might be more to this seemingly improbable relationship than anyone had at first suspected.

Through May and June of 1990, Ted and Jane were seen together across the country—sailing off Turner's St. Phillips Island, in South Carolina; attending Radio City's Night of 100 Stars in Manhattan; lunching with Mikhail Gorbachev at the Soviet Embassy in Washington to discuss American-Soviet film coproductions past and present. (Turner's TNT was filming a movie on the aftermath of Chernobyl, and Fonda had acted in the first coproduction with the Soviet Union—*The Blue Bird*—back in 1976.) And while Jane Fonda herself couldn't exactly believe it, Turner's persistence appeared to have paid off: a romance was clearly beginning.

One day, Ted and Jane showed up at CNN's New York offices, over Penn Station. Seeing his old friend, WTBS's former 3:00 A.M. news anchor Bill Tush, Turner called out, "Hey Tush! You know Jane Fonda? Give us a tour. I want to show her my stuff." So Tush guided Ted and Jane around the CNN offices as the couple held hands and nuzzled together. Spring was in the air and they showed every sign of being in love. "Ted's my buccaneer," Fonda told Tush.

Beaming with pride, Turner ambled down the corridors of this tiny part of his empire, showing off the equipment to Fonda: "Those are my cameras; that's all my stuff." But he was even more thrilled at showing *Fonda* off to his employees. "He had this big schoolboy grin," recalls Tush, "Like a kid who's got a hot date—the one who brings the best-looking girl to the dance."

As they strolled through the licensing and merchandising departments, Turner introduced his hot date to all the secretaries. "This is Jane," he said. "You know, Jane *Fonda!*" Before the stunned employees could answer, he burst into song: "Workin' nine to five, What a way to make a livin', Workin' nine to five . . ." The staffers, who had never seen Ted Turner in person before, much less Turner and Fonda together, stared openmouthed, and before they could get a word out, Ted and Jane had swept on past them down the hall.

———

During this time, Turner was still seeing Kathy Leach. His trips with her to China, Cuba, and the Middle East overlapped with his growing relationship with Fonda. Ted would return from one journey with Kathy and go off on the next with Jane. As always, he seemed to have the energy to juggle multiple careers and multiple women. In May of 1990, right after he and Kathy returned from the China trip, Ted left with Jane for the tiny town of Tougaloo, Mississippi, where he had been asked to give the commencement address at the predominantly black Tougaloo College by one of its trustees, his old sailing teammate from Brown, Charles Shumway.

On a bright, hot, sunny May 20, Ted took his place on the blue gazebo; seated on the lawn before him were Jane Fonda and the assembled graduating class. He stood ramrod straight in his black gown, his silvery white hair ruffled by the light breeze. How was Turner to know that this was, in a way, his own commencement, too—the end of one stage and the beginning of a new one? Ted listened as Tougaloo president Adib Shakir announced his honorary degree: "Robert Edward Turner has demonstrated through his life a belief in the cycle of birth and rebirth. He has exemplified mastery of these processes as he has navigated, sometimes literally, himself through periods of uncertainty and risk. He has demonstrated a belief in the power of renaissance. Today, in the spirit of rebirth, we affirm our belief in this spirit through recognizing a modern-day Renaissance man—Ted Turner."

After Turner was presented his honorary degree to the applause of the crowd, he strode up to the podium and launched headlong into his speech. In between his wide-ranging remarks on aiming high and treasuring the environment, Turner offered a surprisingly candid assessment of his own life, which seemed to be addressed to one member of the audience in particular: "Something I've learned—I'm sharing my deepest experiences with you—is when you get married, really get some books or get some counseling, because schools don't teach you about marriage. At least, my experience was I didn't get enough teaching. I had two failed marriages and it caused a lot of trouble for my children and everything else. Here I am at fifty, I'm now going and getting counseling and trying to learn. If I had done it earlier, I'm sure I would have had a lot happier life, and the women I lived with would have been a lot happier, too."

Reborn yet again, the Renaissance man looked out across the lawn at Jane Fonda and smiled.

To many it must have seemed an odd relationship—Hanoi Jane and the tele-Rhett Butler—a Southern good ol' boy and the Joan of Arc of Radical Chic. It was hard to imagine a spectrum they wouldn't be at opposite ends of: She wanted to liberate the world; he wanted to own it. She was politically correct before the term was coined; he was an equal-opportunity offender—blacks, Jews, Italians, the handicapped, everyone. She was Hollywood royalty; he was the evil captain of colorization. She was a passionate feminist; and he was, as he announced on one of their first dates, "a male chauvinist pig."

The only thing they appeared to have in common, besides their money, fame, and age, was a reputation for sexuality—Ted as the Don Juan of the South, Jane as the sex kitten of *Barbarella* and *Klute,* the goddess of sexual revolutions on-screen and off (according to the memoirs of her first husband, Roger Vadim, her all-consuming amatory voracity in their first encounter left him impotent for three weeks). Remade as the exercise queen of the 1980s, Jane became the Lycra-clad diva of aerobics, sensually toning and crunching and feeling the burn on TV sets all across America. With each passing year, he seemed more roguishly romantic, she more the emblem of body-sculpted perfection.

But what they actually shared was something else, buried deep in the past: "I'm sure those of us who've had powerful parental figures looming over our lives," said Jane Fonda, "whether famous or not, have had to find our way to a clearing of sorts, where we're not crowded by their shadow, and to work it through, to make our peace, and see them laughing in our dreams."

Like Ted, Jane had felt the chill of that long shadow. It is said that her mother, Frances, dearly wanted a son and was disappointed at her birth. Jane was raised primarily by a starchy nanny, who discouraged fondling and kissing the child; later she was shipped off to a series of boarding schools. Her father, Henry, a quiet, reserved man wrapped up in his brilliant film career, was no more attentive than her mother. "As a child, most of my dreams evolved from the basic need of being loved," Jane recalls. When she was twelve, her father left her mother for a younger woman. Frances Fonda, seeking relief from her depression, was treated at several private sanitariums, then went home, locked herself in the bathroom of their house in Beacon, New York, and slit her throat.

Perhaps it was the repercussions of that event, or a girlhood spent trying to please her cool, aloof, demanding father, that led to Jane's being bulimic from the ages of thirteen to thirty-five. According to Fonda, she sought refuge in food and ritually induced vomiting after each meal by sticking her finger down her throat. "I would literally empty a refrigerator," she recalls. "I spent most of every day either thinking about food, shopping for food, or bingeing and purging." But like Ted, she was somehow able to channel her childhood unhappiness into a drive for success. "Jane was a superachiever as well," Turner points out, "because of various things in her family background. She was insecure, too."

Odd as it may sound, Ted and Jane seemed to be able to build on this shared foundation of pain and insecurity. Charles Shumway, who lunched with the pair that day at Tougaloo, believed that he was finally seeing Turner in a deeply felt relationship—a relationship that freely acknowledged the tremors of the past. "I think that Ted's always been afraid that he might commit his father's crime, which is not unusual with people whose parents committed suicide," Shumway says. "One of the reasons he and Jane have done so well is that her mother committed suicide, too. That was a great bond for them. . . . I suspect that they really got in touch with each other's innermost thoughts by sharing this one tragedy that each had."

In June of 1990, Jane Fonda received her final divorce decree from Tom Hayden, and by July, at a TBS stockholders' meeting, one of the puckish shareholders was already proposing nominating Fonda to the board of Turner Broadcasting. "Whatever you want," Turner said, dismissing the idea with a wave of his hand. Then Ted turned to a colleague and muttered under his breath: "Miss Fonda has enough influence already." (The nomination never came to a vote.)

By the end of that summer, Ted had announced to old friends like Peter Dames that "he was courting—in the Southern style." He showered Jane with flowers, presents, and a genteel courtliness that was, according to her friends, a revelation to her. "By any definition," Dames says, "including the old-fashioned Southern one, he was trying to win her hand." On September 9, 1990, under cloudless skies, Turner and Fonda set off with Dames and his girlfriend to cruise the sun-flecked waters of the Aegean Sea. For a week they sailed through the Cyclade islands of Greece, from Mykonos to Seriphos, snorkeling and sunbathing during the day, anchoring off the islands and heading onshore for candlelit dinners at night. At each stop, as they strolled through the ruined temple of Poseidon on Cape Sounion or wandered through the street markets of Mykonos, Turner doted on Fonda, buying her keepsakes, mementos, whatever she wanted. Dames watched over the pair with a fond eye: "In previous relationships, Jane had obviously been the one in charge, picking up the tab, and so on. But here, for the first time, she was being treated in the old-fashioned style, like a lady. And she really seemed to like it."

Meanwhile, for Kathy Leach, whose relationship with Turner was coming to an end, it was a tough time. She had spent nearly two years, on and off, living on an international scale with Ted Turner, and now, suddenly, it was all over. "I learned a lot," she comments with a sigh. "I learned a lot about myself and relationships. It was very hard, too. . . . It was hard to step into the whirlwind and be in it for a couple of years, and then just step out of it." Leach was just one more woman that Ted was about to leave behind—for in a competition with Jane Fonda, how does one compete?

Ted Turner and Jane Fonda, two entrepreneurs in love, quickly realized that their romance had a certain extra allure, a value-added dimension. It wasn't just a romance; it was a publicity bonanza—not just a love affair but a merger of sorts. After all, Jane Fonda was, in the words of one wag, a trophy mate who came complete with her own trophies. And if those Oscars, those workout videos (a $300-million business), that international star power left Ted feeling a little chagrined at first, if he jokingly referred to himself as "Mr. Fonda," he soon came to be tickled by the fact that his wife was even more famous than he was. As a couple, they were a cross-promotional dream. With Jane Fonda at his side in Moscow for the Soviet premiere of *Gone with the Wind,* or in Washington, as he testified before a congressional panel on cable re-regulation, Ted was assured even more coverage, even more international attention. And when Jane Fonda wanted to promote her latest workout video, what better billboard than CNN?

On October 8, 1990, as Jane flogged her new "Lean Routine"

on *Larry King Live*, she was joined by Ted in their first joint national TV appearance. The ever jovial King instantly pronounced the boss and his new flame "America's most famous couple," but when he tried to get them to talk personally about one another or to announce nuptial plans, he came up empty. The relationship was still too formative for such public pronouncements. "The two of you look terrific," King gushed as he attempted to charm them into disclosure. "I mean, this is a happy . . . this is a getting-along thing, right?" he stammered. "I mean, the two of you are . . . in . . . ?"

"What does that mean?" Fonda asked, innocently batting her lashes.

"As far as I can see, it's going great," volunteered Turner. "Not a very long answer, but you know I'm not used to really talking too much about my personal life."

When it came to Jane Fonda's workouts, though, you couldn't shut him up. "I've lost eighteen pounds without really going on a diet as such," he enthused. "I'm working out a whole lot more than I did before, and I feel a lot better. I've got to say that. Regular exercise really is a good thing. . . . And everybody can work out with Jane with this tape. . . . It's a rare opportunity—it's an opportunity that can be shared by everybody, and it's really a heck of an experience." And a laughing Jane Fonda made sure to return the compliment, announcing that being fit and being a CNN couch potato were not necessarily mutually exclusive: "You can do both," she said, looking over at Ted. "Work out *and* watch television . . . *Yeah!*"

Ted grinned. "Watch CNN," he said.

Synergistically, they were the perfect couple, the very epitome of Turner's oft-proclaimed motto: "Early to bed, and early to rise, work like hell, and advertise."

Of course, there were certain disadvantages to a romance played out in the public eye. In early November, when Ted went to Tiffany's to pick out an engagement ring for Jane, the story leaked out. He had meant to surprise her with the opal-and-diamond ring—and a marriage proposal—on her birthday in December, but as it turned out, gossip columnist Liz Smith got the news before Fonda did. Nevertheless, as the pair embraced in public all the way from Moscow to Atlanta with the gusto of teenagers in love, their romance showed every sign of still being on track.

Thanksgiving found Jane and Ted on horseback at his Flying

D ranch, hunting duck and elk, and working on plans for a lavish log ranch house on the property. If they weren't precisely living together yet—Jane with houses in Brentwood and Santa Barbara, Ted with his residences all over the Southeast—in fact, both were spending an increasing amount of time at the Flying D, roaming down by the Gallatin River and up toward the glorious Spanish Peaks. Perhaps the clearest sign of intention came when Jane and Ted decided to merge their stables of horses, with Fonda moving her four prize Arabians to Montana. If their horses were living together, could marriage be far behind?

In the potential union of Turner and Fonda, each had found a soul mate—a powerful new ally in the causes nearest and dearest to their hearts. Fonda had come from the barricades of the radical left, Turner from the plantations of the conservative right, but for years before they met, each had been espousing many of the same do-gooder liberal-centrist positions on global brotherhood and clean air. Although painfully earnest and sincere, neither could be dismissed as a dilettante or a dabbler: both Fonda and Turner had dedicated significant time and money to the reforms they championed. This was the Jane Fonda of *The China Syndrome* and the Hollywood Women's Political Committee, the Jane Fonda who had spoken out since the early 1970s on Indian rights and women's rights and all sorts of environmental rights. And this was the Ted Turner who had founded the Better World Society. "We have a community of interests," said Turner. "She's certainly been working on these issues longer than I have, but I've been working on them very hard in the past decade." When Jane traveled to Long Island to address a celebrity crowd at Lorne Michaels' elegant beach house on behalf of California's environmental referendum, it was Turner (then recovering from hernia surgery) who shouted, "Listen up!" silencing Billy Joel, Christie Brinkley, Paul Simon, and Calvin Klein; and it was Turner who then put his money where his mouth was, proffering a check with six digits.

The Goodwill Games, fittingly enough, had brought Ted and Jane together shortly after her divorce. Ted's noble experiment in international brotherhood and sports programming had returned for a second edition in late July of 1990, even after losing $25 million the first time out. Ted had invited Jane to come to Seattle, the venue for that year's games, to bask in the spirit of global unity. It would be one of their first extended trips together.

The games were not universally loved. Even at the Turner organization, according to Robert Wussler, "there were a lot of people within TBS who hated the games—who wanted to see them killed." The board, too, was "somewhat skeptical" about this money-guzzling quasi charity, according to TBS senior vice president Paul Beckham, who had taken over the reins from Wussler. And when articles started appearing describing fistfights between TBS and Seattle officials, conflicts over $300,000 in fire protection, local fishermen's complaints about giving up days of prime salmon fishing in Shilshole Bay for yacht races, and above all, the worries of travel agents who were seriously underbooked, many of the organizers were concerned. They became even more concerned with published reports that the games might lose up to $20 million.

But not Turner. Although he hated to throw away an extra thousand dollars on a documentary budget, he didn't mind losing tens of millions of dollars if he believed in the cause. So while Paul Beckham was fretting and the board was grumbling, Ted was off with Jane cycling around Seattle. Rising early in the morning before the city was awake, they'd bike around Lake Washington and through the quiet streets, getting to know the city and putting in their requisite fitness time.

When the opening ceremonies finally got under way at Husky Stadium on the University of Washington campus, Turner was as thrilled as a youngster unwrapping the biggest present under the Christmas tree. As Native Americans danced their tribal dances and a Japanese contingent passed out thousands of origami folded-paper doves promoting peace, Turner beamed with pride.

"Good job, Beckham, good job!" he shouted, clapping Paul Beckham on the shoulder. And when the opening speeches by Armand Hammer and the mayor of Seattle and Turner himself were over, when the volley of guns went off to officially open the games, Ted was jumping up and down, his face cracked wide in a huge grin, his arms wrapped around Jane Fonda in a bear hug, his palpable joy and delight a glorious rebuke to cynics everywhere.

Unfortunately, the Seattle Goodwill Games ended up losing $44 million. The ratings were low (2.6), and once again make-goods were necessary to advertisers. But the exploits of track star

Carl Lewis and the U.S. basketball team established the games as a world-class sporting event. The board unanimously agreed to Turner's request for a 1994 Goodwill Games, but with a few provisos (no more surcharges for cable operators, and no more exclusivity—broadcast rights would also be sold to ABC, splitting the cost). Arm in arm with Jane Fonda, Ted left Seattle having achieved yet one more success. With any more victories like this one, however, he could well end up in the poorhouse.

The growing romance with Jane dovetailed neatly with Turner's increasing activism. Given their parallel ideologies and similar wide-eyed sincerity, Ted and Jane seemed poised to become the first couple of environmental salvation, proselytizing around the planet to rescue the trees, the wetlands, the indigenous peoples, anything that needed saving. Ted's credo was: Think globally, act globally . . . and especially, talk globally. Their mission: Save the Earth.

By the late 1980s and early 1990s, Turner had become increasingly sophisticated in his understanding of environmental issues. After years of talking to Jacques Cousteau and Lester Brown and reading about the ozone layer and sustainable development in articles and books (like Norwegian prime minister Gro Brundtland's report *Our Common Future*), Ted was not just a Green, he was a knowledgeable Green. And as he became more and more of a "Better World" activist over time, his understanding finally caught up with his rhetoric. He was giving dozens of speeches annually at colleges and professional forums on topics related in one way or another to the environment: on the necessity of stopping pollution to Save the Earth, or on stopping the arms race to Save the Earth, or even in favor of abortion to combat overpopulation to Save the Earth.

Saving the Earth was the right kind of mission for Ted Turner—large, bold, not all that well defined but full of good intentions and dramatic impact. He always seemed to do better with mega-issues and grand charges than with day-to-day minutiae.

"Projects that dealt on a most basic level with really human issues, Ted could never seem to embrace," remembers Ira Miskin, then executive vice president at Turner Entertainment Networks. "We did an incredible amount of environmental programming, which I loved. . . . But there were times when I felt that his concern about the ozone layer took precedence over things like whether or not we could feed people—the poor, the underclass. . . . He sees the bigger issues, the long-term ramifications of the global environment; he does not necessarily see its impact today on . . . a single human being. He sees the forest, but he doesn't see the trees."

It was in 1989 and 1990 that Ted began espousing his ten "voluntary initiatives"—or, as they were soon dubbed, "the Ted Commandments." From speech to speech, from meeting to meeting with business and political leaders as he and Jane circled the globe, he pursued his ecological agenda with a genuine passion, asking for pledges to cut down on everything from pesticides and nonrenewable resources to families with more than two children. (Of course, he himself already had five.) To some, Turner came across as a strange combination—part 1960s flower child, part 1980s tycoon.

For Turner, environmentalism was simply "a matter of survival, number one; and number two . . . a matter of good business"; it was idealism tempered by pragmatism. His global agenda just happened to mesh naturally with the business interests of a man who was trying to shape the first worldwide telecommunications empire. As a multibillionaire fighting for benevolent global interests like peace and clean water, he could be welcomed by heads of state all around the world. In the words of longtime colleague Mike Gearon, "The president of the United States doesn't get the treatment Ted gets in a lot of places. . . . There's no one else who is as much a citizen of the world as he is. . . . He's a one-world person, but his idealism has totally to do with his business interests. He's not sacrificing a single idealistic principle for his business. It's totally in sync."

Tom Belford, who headed the Better World Society for some five years, agrees: "Ted sees the angles on everything. . . . He's the consummate do-good, do-well guy. Ted cares deeply about issues, but he also understands the happy synergy: if you aspire to be a global communicator, it makes sense to be associated with world peace and the environment. He could raise the profile of his company by connecting an issue agenda to a business agenda. Most journalists bend over backwards to avoid that, but not Ted."

Ted Turner's agenda went far beyond mere speechifying. By 1990, his crusade for a better world carried over to all facets of his life. At CNN Center, he banned Styrofoam. Smoking, too, was *verboten:* Any new employee found lighting up would summarily get the boot. Ted didn't want any overweight employees either, and they were given the heave-ho as well. And he had yet one more pet peeve—the use of the word "foreign." To Turner, it smacked of something alien, a put-down, totally contrary to the better world spirit; "international" was the word of choice. Each time an employee slipped up and used "foreign," on or off the air, he or she was immediately slapped with a $100 fine. (Even Peter Arnett, fresh from the bombing of Baghdad, was not exempt. When he called himself a "foreign correspondent," *wham!*—$100.) At Christmastime of 1990, Turner made a personal gift to every member of Congress and every *Fortune 500* CEO in the entire country—a copy of World Watch's environmental report, *State of the World,* the perfect stocking stuffer.

Perhaps the most striking instance of his personal commitment to environmental issues was his use of conservation easements. Turner, who collects land the way other magnates collect Ferraris or Van Goghs, now had eleven homes and some 600,000 acres, the latest addition the 300,000-acre Ladder Ranch near Truth or Consequences, New Mexico, a sweeping expanse of juniper and pine and rolling hills. And on all his properties, he had placed legal restrictions, essentially converting them into a series of private wildlife reserves. The easements—established, ironically enough, with the Nature Conservancy, the same group he once outbid to purchase St. Phillip's Island, South Carolina—are simply legally binding pledges that the land will never be exploited. In other words, Turner is not giving away the property itself; he keeps the land but agrees to forsake forever certain property rights, like subdivision, development, and so forth, to keep the land forever wild. In a sense, he was on his way to creating the first private national-park system.

"Ted's one of the early users—and he is far and away the biggest user in the world—of conservation easements," says friend and neighbor Peter Manigault. "The biggest easement ever done anywhere was on the Flying D [before the New Mexico purchase], but every property he's ever gotten he's put an easement on. The minute he gets a property, he tells his in-house lawyer to crank out an easement."

On the work front, Ted was also cranking, turning out environmental and one-world programming in ever greater volume. On

WTBS alone, the 1990–91 schedule showed three hundred and sixty hours a year of ecologically oriented shows. Turner premiered *Network Earth*, *Earthbeat*, and (probably most important of all) the animated ecocartoon *Captain Planet*, a lively piece of environmentally superheroic propaganda dreamed up by Barbara Pyle and Turner (with the color scheme supplied by Kathy Leach) and aimed at winning over kids to the cause.

Stronger than Spiderman, cooler than Batman, Captain Planet proved that ecology could be fun. Other heroes might duke it out with run-of-the-mill thugs, but Captain Planet and his five multiracial Planeteers, under the guidance of the Earth Spirit Gaia, battled nefarious wrongdoers like overpopulation, global warming, and acid rain. And their uniforms were pretty neat, too. Speedy, strong, and pH-balanced, the captain was soon winning over the Sesame Street crowd to the joys of recycling.

Meanwhile TNT was offering environmentally oriented action-adventures for their parents, like *Nightbreaker*, a fictionalized account of evil atomic-weapons testing in the 1950s; *The Last Elephant*, one man's crusade to stop ivory poaching; and *Voice of the Planet*, a ten-hour docudramatic "autobiography of the planet," with William Shatner as a disillusioned ecologist who makes contact with "the mother goddess of the earth" through a computer in a Tibetan monastery (a show that sounds suspiciously as if it were conceived by Turner's ex-girlfriend J.J.).

And on CNN Turner introduced *World Report*, a weekly two-hour series that rebroadcast documentaries from around the world, emphasizing international reporters and international perspectives, and absolutely nothing "foreign." While the liberal *Nation* blasted Turner and the series as "a tool of right-wing fascists," and the conservative *National Review* termed him a "Soviet apologist," Turner himself emphasized the importance of a neutral, one-world forum: "As an American, I always thought of the world as 'President Reagan talked to Margaret Thatcher' or 'President Bush talks to Gorbachev,' but on *World Report* you'll see the president of an African country meeting with another African leader." Americans, he added, "never think that there's anything else going on in the world. The average American doesn't know from squat."

While few of these Turner programs could be counted as artistic successes, they each advanced his various causes. And whether one agreed or disagreed with Turner's philosophical positions, it was clear that his were the only networks on American television consistently espousing a philosophy, a sociopolitical agenda. Only the various televangelist channels came close.

In fact, Turner was and is prepared to stake out unpopular positions on television to further his beliefs. When Ford, Citicorp, and Exxon all threatened to pull their advertising from TBS over the Audubon Society's *Ancient Forest: Rage Over Trees*, a documentary critical of the logging industry, Turner chose to lose $250,000 in ads rather than back away from the program. He was equally unflinching in his decision to run *Abortion for Survival*, a pro-choice program. While every other network on the air nervously avoids such controversial fare for fear of alienating advertisers or viewers, Turner had no hesitation. There is something admirable, even heroic, in the Quixotic way he is prepared to charge at any and all windmills, to do something that bland American TV avoids at all costs—to take a position.

"You bet your bippy we're taking a position," he told a gathering of TV reporters. "We'll just take a lick for a night. So it'll cost us four or five hundred thousand dollars. . . . I don't want anybody else telling me what my daughter's got to have, or my wife, or my girlfriend. We live in a free country. There's absolutely no way that's anybody's business but the person that's involved. That's my opinion. We'll give the other bozos a chance to talk back. They look like idiots anyway." And as if that weren't enough, Turner couldn't resist one last parting shot: "That's fine if those people don't want to ever have sex. Swell. I happen to enjoy it. I don't get near as much as I want to."

Staking out these loud, confrontational positions—he once termed Christianity "a religion for losers"—Turner had to be prepared for both his share of controversy and the guffaws that followed when one of his cherished inspirations belly flopped. At the end of 1989, he established the Turner Tomorrow Award, offering a sizeable $500,000 prize for a book that contained "positive solutions to global problems." Ted announced, "The great minds of today need to focus on the problems of global significance if humanity is to see new tomorrows." But two years later, the prizewinning book, picked from twenty-three hundred submissions from fifty-eight countries, turned out to be *Ishmael,* a philosophical conversation between a man and an ape—a dialogue that many critics dismissed as simian. Despite the huge size of the award, the prize quickly fizzled away.

By July of 1991, the Better World Society had folded, too. Over six years, it had raised almost $12 million and produced some forty-eight documentaries, many of them well-meaning but turgid. In fairness, as both executive directors Tom Belford and Victoria Markell point out, the society had, in several ways, "done what we set out to do." Issues like the cold-war relations between the United States and the USSR that had led to its formation were now less pressing, and the environment had gone from an unacknowledged topic to the forefront of national consciousness—on *Time* covers and national news reports. But as Markell says, the fund-raising had been tough from day one; many of the corporations they had approached had asked, "Why should we give money to Ted Turner?" And even internal sources at Turner Broadcasting termed many of the produced programs "mediocre." It was on the flight back from the society's meeting in China that Turner talked privately of closing it down, and a year later he pulled the plug.

For these and other projects the brash Turner came under ridicule. The inflated quality of his rhetoric and his aspirations made him an easy target of satire, especially when things didn't work out. Sometimes lost in all his verbiage and posturing, however, was the extent to which he truly cared about these causes. "He poured his heart and soul into the Goodwill Games," says Robert Wussler. "He didn't understand that there were people out there who knocked the games as frivolous, unnecessary. . . . And he was truly hurt by the fact that people misunderstood his motives."

Ira Miskin says, "I think that Turner takes it very personally and feels very deeply criticism about things that he tries to do that people think of as 'image parading.' All of his involvement with the Soviet Union, for example. . . . The Better World Society, which many people felt was a joke, was a very, very serious endeavor for him. The Turner Tomorrow Awards. These are all outward manifestations that are very Turneresque, in which the execution of the ideas ran away with, or didn't live up to, the underlying vision of what he wanted to do. He feels, I think, very deeply about the world, and relationships among people. Yet he's not as articulate about saying things as one would like." Ironically, for a man who talks as much as he does, Turner's nonstop, high-decibel folksiness—his greatest asset—is also his greatest handicap in advancing his agenda. With his hyperactive mouth, he's impossible to miss but often hard to take seriously.

Campaigning for their vision of an improved planet became almost a full-time job for Ted and Jane. They appeared at more whistle-stop rallies and college talks than many political candidates. In New York, they launched the "Save the Earth" campaign; in Rio, they attended the Earth Summit and spoke out on global TV; and at Harvard's Kennedy School they appeared as a two-for-one "joint fellow," to talk about the fate of the Maya Indians and the future of environmentally positive careers.

Taking the Harvard podium first, Jane, in an austere but elegant brown pantsuit, looked surprisingly nervous for an international superstar, perhaps a little cowed by the sharp students in the audience. She spoke of how the mountains that ring Los Angeles had once been visible; she spoke of Judeo-Christian myths and mother goddesses. Sounding surprisingly like Ted, she declared: "We need the most major and profound social, economic, and spiritual transformation that has happened since the agricultural and industrial revolutions, and it has to happen much faster."

Ted, speaking after her, immediately had some pointed comments for the Kennedy School administrators in the audience: he complained about the lack of tenured women on the staff and demanded to know why CNN wasn't available in the dorm rooms. The students loved it. Then he got to his main topic. "I'm in so many businesses and I've made so much money, I'm thinking how to give it away," Turner said. But for young men and women coming out of school, he continued, there were all sorts of positive careers in almost everything from bicycles and photovoltaic cells to contraceptives. "Jane and I do need more user-friendly birth-control devices, and somebody's going to make several billion dollars off of it," he declared.

An audience member wanted to know if any of that money he was giving away might be for Harvard. "I tried to get in here," said Ted. "I wanted to go. I didn't get in. Let the goddamn people who went to Harvard take care of Harvard. . . . If they wanted my money, they should've seen I was going to be a success and admitted me."

When Ted asked for more questions, the audience responded with some barbed queries. One student wanted to know, "How do you square your interest in the environment with all the ads your stations run from polluters?"

Another jabbed at Jane: "Excuse me, but why should we care what any celebrity has to say?"

Jane flinched, but responded gamely that it was every person's duty to stand up for the things she believed in, celebrity or not. Ted leaned over to her and nudged reassuringly up against her shoulder. Almost imperceptibly, Jane leaned back into him for support. It wasn't easy saving the world. But at the end of the afternoon, they got a standing ovation.

With Ted pursuing Jane as well as his global agenda, the one thing he paid the least attention to was his bread-and-butter business. He began to spend less time in Atlanta, more time on the road and with Fonda in Montana or New Mexico. Staffers sensed a weakening of his interest in the day-to-day affairs of WTBS and TNT, especially when the only way they heard from Ted was via phone calls from the ranch. At a time when celebrations should have been the order of the day—after all, the company had turned the corner on its huge debt, CNN was marking its tenth anniversary, and the startup of TNT had been a financial success that surpassed all expectations—Turner Broadcasting seemed to be in the doldrums. The company's inspirational leader was clearly more wrapped up in his new romance than in new business. Both the *New York Times* and the *Wall Street Journal* noted a "middle-age malaise" at TBS. "We were like pirates before," Turner Entertainment president Scott Sassa told the *Times* reporter. The old derring-do had plainly faded, marking "the passage of TBS from swashbuckling entrepreneurship to stable *Fortune 500* elite . . . with the emphasis less on innovation, more on paring debt."

Within Turner Broadcasting, much of the focus seemed to be on spin-offs and repackaging of existing material, on creating a checkout channel or a fitness channel to air CNN footage in supermarkets or fast-food joints, post offices or health clubs—anywhere a person in line or in a seat might be turned into a captive audience. Turner Pictures was created to distribute TNT movies (such as *The Secret Life of Ian Fleming*) overseas as feature films. Turner Publishing was created to, among other things, turn out companion volumes (such as *Portrait of Great Britain*) to TBS programs. And while Turner was thrilled over the establishment of a book division ("He was like a kid in a candy store," recalls Ira Miskin of the delivery of their first volume. "He kept saying, 'God, this is *so* great! It's gonna be the greatest thing! We're gonna be publishers! It's just so great'"), even Miskin had to admit that Turner Publishing was "a two-bit operation" with an annual budget of $500,000. In short, the situation at Turner's empire was like the viewership of CNN: when there was a crisis or a news flash, like the Panama invasion, the audience soared, but in periods of calm it languished. So, too, had Ted's interest.

The boredom and chafing in the company was marked by the departure, after nearly a decade of management stability, of four top executives within a year, including the defection of a pair of successive number-two men at Turner Broadcasting. First Robert Wussler left (for Comsat), and then Gerry Hogan (for Whittle Communications). Both had complained that they spent their time consolidating past work, that they were "not learning a lot." Turner replaced them by promoting from within; his longtime aide and devoted friend Terry McGuirk became the new number two, but there were many both inside and outside the company who felt the new leadership team was composed primarily of "empty suits."

There was another, and much bigger, potential loss that had Turner worried as 1990 ended and 1991 began. He had centered himself on Jane Fonda, devoted himself to wooing her; he seemed determined to have her. Yet suddenly it looked as if the relationship was falling apart. Ted and Jane had decided at one point to marry in December 1990, but that date had come and gone. According to one friend, in mid-December, Jane had backed out because the two were unable to come to terms on a prenuptial agreement. In fact, there were a series of issues to be resolved.

The newly divorced Jane had certain trepidations about getting remarried, especially to a man who was a two-time loser, just as she was. She enjoyed Ted, loved him, but could she see spending the rest of her life with him? Then, too, how could they work out a lifestyle, with Ted in Atlanta and Montana and her own base in California? Much as he had with Kathy Leach, Ted was insisting that Jane give up most of her business activities to be with him. But she had worked hard to build a film career and an aerobics business from scratch, and wasn't sure she wanted to give them up.

And so in December 1990, the two had a falling-out and, according to friends, "Ted was heartbroken." For a brief period around Christmas, they stopped seeing each other, and "Ted became depressed." The night of the TBS Christmas party in Atlanta, he moped around and left early, by all reports looking like "a basket case."

Not long after, Turner was back on the phone to Fonda, asking for another chance to set things right, sending her cards and flowers. Once again, his persistence was persuasive. By March 1991, Jane was at his side as he testified in Washington. By June and July, they were living together full-time, and Jane had put her Santa Barbara ranch—complete with theater, dance pavilion, gym, pool, and riding trails—on the market. On October 22, 1991, when the Braves began to look like winners and the first World Series game *ever* finally came to Atlanta (albeit ten years later than their owner had promised), the fans spotted Ted and Jane smooching in the stands at Fulton County Stadium. Seated next to Jimmy and Rosalynn Carter, they cheered and chanted and did the tomahawk chop (later to be replaced by the politically correct palm-a-hawk peace gesture, after complaints from Native American groups) as the Braves went on to victory that day. By November, the pair had agreed to an unusual prenuptial agreement: $10 million in Turner Broadcasting stock would be bestowed on Fonda upon their marriage—almost a property settlement in advance. It now looked as if a wedding was in the offing. But no one was sure exactly where or when. The rumor was that it would take place on one of their properties, which could be anywhere in the United States. Apparently tipped off, British tabloid reporters set up camp outside the Montana ranch.

On the morning of December 21, 1991, Jane's fifty-fourth birthday, Peter Dames showed up at the white wooden fence along the majestic tree-lined drive at Avalon, Turner's 8,100-acre estate just east of Tallahassee. The day before, the workers had nailed up the "No Trespassing" and "Beware of Dogs" signs. Dames arrived with a copy of that morning's local newspaper, announcing the upcoming nuptials with the headline "TRESPASSERS AT TURNER WEDDING WILL BE PROSECUTED." Ted came out in his old bathrobe and grinned at the headline.

Meanwhile, Jane was upstairs getting ready, pinning up her golden brown hair and fitting herself into the full-length off-white linen and lace high-necked gown from her 1981 movie *Rollover*.

"Today's my birthday," she laughed with her friends. "And I'm getting Ted Turner."

Downstairs, surrounded by a small intimate group of about twenty-five friends and family (including all the children from all four previous marriages), Ted and Jane were wed by a local Baptist minister, A. I. Dixie. Since her father had died nine years earlier, Jane's eighteen-year-old son Troy gave her away. Ted, decked out in an all-white linen summer suit, had Jimmy Brown (who had been with the family for more than forty years now) as his best man.

Gathered in the large foyer of the old Southern plantation house at Avalon, the whole clan laughed and cried as Ted mugged through parts of the ceremony and embraced Jane at the end. Then everyone adjourned, heading out the French doors for photos (by Barbara Pyle) and dancing. Later, there was feasting on pheasant, sweet potatoes, and wild rice.

The marriage would mark another new beginning for Ted Turner. And if it was not exactly the lead story on CNN that night, where it was reported at 6:07 P.M., that was only because something else had come up—the official end of the Soviet Union and the beginning of the Commonwealth of Republics, a pact of almost equal importance. Actually, even as Turner marriage ceremonies go, this one wasn't the biggest of the year.

ON MAY 4, 1991, it had been raining all afternoon, and as the guests began to arrive at All Saints' Episcopal Church in Atlanta, it was pouring down harder than ever. They came dashing inside, couples in elegant tuxedos and ball gowns, sprinting in from the weather and shaking out their umbrellas, all come to celebrate one of the big events of the Atlanta social calendar—the wedding of John Rutherford Seydel II and Laura Lee Turner, Ted's eldest child. It was to be a marriage in the grand style, with twenty groomsmen, ten bridesmaids, eight hundred guests in all. The jeweled wedding gown came from France, the music was provided by the sixty-two-member Atlanta Boys Choir, and the reception to follow was slated at the posh old-society Piedmont Driving Club—the kind of snooty place that not so long ago wouldn't have admitted Ted Turner.

By 7:00 P.M., the guests had assembled, packed inside the great stone church, many standing along the walls when all the pews were filled. Out in the wood-paneled church library, the wedding party crowded together, Ted in tails, Laura resplendent in her champagne-and-gold wedding gown, the bridesmaids and grooms-men chatting excitedly. Turner's face suddenly opened into a huge grin as he looked over at his daughter and her attendants—the pride of the father of the bride. Perhaps at that moment he felt some other emotions as well, for almost all the significant people in his life had been invited into the church next door: not just business colleagues, board members, trusted staffers, and old friends, but also Ted's mother, Florence (now eighty-two and ailing, and in the last year of her life), Ted's first wife, Judy Nye Hallisy (surfacing after nearly two decades), and his own sweet Jane, plus all of his children and both of hers and the entire Seydel clan—a mixed group of siblings, half siblings, stepsiblings, and siblings-to-be. All in all, the turnout represented for Ted Turner a confluence of past, present, and future relations that must have seemed to him like *This Is Your Life*.

As the music began and the bridesmaids marched out, Ted and Laura Lee were left alone. Laura turned to her father. "It gives me the chills," she said. Turner stooped gently to help her with her train, and the two watched together with mounting emotions as the flower girl—young Pam Goddey, the daughter of the Hope Plantation manager—took her cue and made her way into the church. Then as the organ sounded its ceremonial flourish, the fifty-two-year-old, white-haired Turner took his daughter's arm in his and held it. Here was this lovely woman, his firstborn, setting off into married life.

"Dysfunctional?" he suddenly asked out loud, apropos of nothing and of a lifetime. "Dysfunctional?" He shook his head. "Hell no, man! *Functional!*" Patting Laura's hand, he gave her a quick smile, and together they marched off down the aisle.

For all the triumphs in Turner's life, his most remarkable success isn't to be read on any trophy or balance sheet. For this would-be missionary who has made a life's work out of lost causes—of rescuing broken birds and broken-down businesses—surely no rescue is as impressive as the way he seems to have salvaged and transformed himself within the past decade. Like many youthful hell-raisers, Turner has mellowed with age, but he has done something else as well: as a conscious act of will, Ted Turner has changed. Through years of therapy, reading, and intensive study, he has worked hard to remake himself from a self-centered, parochial Southern skirt-and-money chaser of the 1970s to a slightly less self-centered father, husband, and global citizen-magnate of the 1990s.

"There's something about Ted," says Mike Gearon. "He's so capable of transcending past habits and past patterns. He has an openness to change, and a willingness to evolve—whether it's his attitude toward women, or a whole lot of other things."

Mary Alice Williams says of her old boss, "I think Ted's grown up. He understood that now he represented something more than himself. It was more than the little boy who was trying to get everyone to like him and who had to always compete. Suddenly there was something more important in life, and he represented that—he took that on."

As Turner has matured and expanded in his interests, he has also, according to former aides like Gerry Hogan and Robert Wussler, "just gotten smarter and smarter." "Ted Turner learned more between the ages of forty and fifty than anybody else I have ever met in my life," Wussler says. "He asked a lot of questions, read a lot, looked at a lot of TV. Here was a guy who didn't know everything but was extremely eager to learn. He was a sponge, soaking everything up. I saw that man get more and more sophisticated. At a time when most people's learning curve goes down, his went up. He was learning a great deal about life, about style, about business. He became worldly, he traveled to China, Japan, the Soviet Union. . . . While his company was growing by leaps and bounds, he was, too."

Today, Turner has a newfound stability. He has finally settled comfortably into his own skin, making peace with himself, his kids, his demons. Perhaps lithium has made the difference. (Doctors, however, note that manic qualities never really go away, and one friend says that "he would become very manic without the lithium, kinda going crazy in a way, pacing back and forth, fidgeting, talking a lot, demanding a lot.") Perhaps having passed the age when his father had died, Turner could finally acknowledge his achievements to himself. "He doesn't have as much to prove," Gerry Hogan says. "If the age his father killed himself was an important milestone, he's past it now."

"I think Ted's finally convinced he's rich," Gearon says. "He used to say to me, 'What do you think about how I'm doing today? How would you measure it in, say, fifties economic terms?' I think that what he was really saying was, 'How do I match up to what my dad did?' But now, he's finally convinced—it's not all gonna go down the drain."

Most important, this man whom his father's accountant Irwin Mazo once described as "not a very warm person—he could never have a warm relationship," has finally begun to find it possible to do so. The process that began with J.J. has continued with Jane Fonda, and a textbook dysfunctional has become functional. "Jane was Dad's salvation," says his son Teddy. "They're cut from the same cloth. They're survivors. She really opened him up to all of us. You just can't believe how far Dad has come. He's worked hard to put back together the family he nearly tore apart." By way of highlighting the difference, Teddy told *Time* reporter Priscilla Painton, "You never thought of having fun with Dad before, but now you can. He does laugh a little more and play a little more."

There are some, however, who still find the old Turner in the new package. After not having seen him for years, Judy Nye says, "He's still not a very nice person. He's abrupt, rude." There are people in the business world who feel the same way. And not all of Turner's Montana neighbors are exactly thrilled with him either; at a banquet of Trout Unlimited, he announced that there'd be no trespassing on his property: "If you want to fish, go out and make a million and buy your own stream."

Turner's millions have bought him his own rustic San Simeon at the Flying D ranch. These days find Turner and Fonda home on the range more often than not. "I spend more time here than I should and less than I'd like to," he told a reporter for the *Bozeman Daily Chronicle.* The Montana house that Ted and Jane built in Gallatin Gateway is a comfortable but not palatial log ranch structure. Inside the front door with the embossed buffalo on it, this house is clearly a home for both of them: downstairs there's a workout room; upstairs, in the high-ceilinged living room, animal hides cover the floor and the heads of buffalo and elk adorn the walls. Although the big-screen TV is almost always on and tuned to CNN, Ted and Jane are more likely to be found taking in another view, the one out the huge floor-to-ceiling sliding glass doors.

Beyond the doors, beyond the deck that runs the length of the second story, is a breathtaking vista. A sparkling blue 14-acre reflecting pond mirrors the lush green rolling landscape and the jagged white Spanish Peaks that frame the horizon. It's a million-dollar view, a sweeping expanse of natural beauty that doesn't quite seem real—and, in a way, it isn't. In a fit of Hearstian extravagance, Turner created this vista by ordering the excavation of the pond and the shaving of a hilltop in the middle distance; tons of dirt and rock were lopped off and carted away so that the pond might reflect more of the mountains.

Scattered around the living room and the deck outside are a battery of binoculars and high-powered telescopes, which the couple use to watch the animals that come across the horizon. Turner likes to identify them, calling them off by name as they come into view: sandhill cranes, bald eagles, deer, elk, coyotes, and of course buffalo.

When Ted and Jane first began building their house outside Bozeman in 1991, they were not universally welcomed by their neighbors. In general, there has been resentment in the region about the influx of celebrities (among them Glenn Close, Meg Ryan, Michael Keaton, and Mel Gibson), who have driven Montana real-estate prices up, but the response to Fonda and Turner has had a more aggressive quality. It surfaced in the letters-to-the-editor column of local papers, as well as in the graffiti that appeared in the Western bar, in the tiny nearby town of Gallatin Gateway. The comments included "Take Hanoi Jane back to Atlanta" and "Screw You, Teddy. I'll shoot an elk on your land whenever I want and you too, if you get in my way." And then, of course, there were the various pornographic drawings that showed up on town walls—a veritable Kama Sutra of Ted and Jane and buffalo.

Part of the resentment stemmed from what was perceived as Turner and Fonda's elitist approach to environmentalism, in which everyone was to be saved, whether they wanted to be or not. Part of it came from Turner's sweeping and very public changes at the Flying D—not just his bad-mouthing of the cattle industry (a way of life for Montana ranchers for decades) but his selling off of most of his ranch equipment and cattle, tearing up miles of barbed-wire fence, taking down telephone poles, letting the pastures return to native grasses, and reintroducing the native buffalo.

But after a year or two, despite the flurry of publicity, Ted and Jane have settled into a simpler, quieter life in Montana. They can be found traversing their 130,000 acres in a jeep or on horseback or on foot, or strolling across the meadows on the lookout for Indian arrowheads, which Ted collects and proudly displays in a bowl in their living room. Turner has become quite serious about his trout fishing—labeling himself "trout devout"—and when the streams are running clear and the fish are rising to the lure, he's often out there "matching the hatch" and casting across his rivers, knee-deep in water, his line snaking across the sky. Sometimes his old friend Jimmy Carter helicopters in, and they spend a weekend fly-fishing together, reeling in the trout, then releasing them back into the currents.

These days Turner and Fonda share two parallel passions—buffalo and the American Indian. Typically, Ted approaches things pragmatically. While Fonda is interested in the lore and heritage of the Native Americans (North and South), Turner wants to make a business, a big business, of buffalo. He recently assembled the single greatest privately owned herd of bison in the world—approximately four thousand head, roaming free across his acres—and these days he's just as likely to champion the buffalo-meat business as CNN. "Everyone wants to come up with the new thing, the great new idea that lets 'em hit the big one," Turner says, and right now he believes that it's buffalo. Bison meat currently sells for about double the price of beef, and bison calves bring triple the price of beef calves at auction. He ticks off the reasons for that with a pitchman's zeal: Buffalo meat is lower in fat and cholesterol. Bison have a higher tolerance for cold weather; they don't need hay in winter; unlike cattle, they don't ruin riverbanks or trees; they're also cleaner than cattle; and not only that, they taste good. At a recent TV critics' gathering, Turner presented all the newspaper writers in attendance with buffalo steaks to show them just how good.

Montana's Native Americans from the Crow, Blackfoot, and Salish-Kootenai tribes have been impressed by what Turner and Fonda have done: Blackfoot tribal leader Curley Bear Wagner has dubbed Turner "Wowuka" after the nineteenth-century Paiute medicine man who shaped the Ghost Dance culture. Wowuka is said to have predicted that when the buffalo returned and the white men left, the Indians would flourish again. "Ted is the new Wowuka," says Wagner. "He's bringing back the buffalo."

Fonda and Turner have also gotten more involved in tribal history and economics: after visiting reservations and making a point of meeting Native Americans throughout the West, including several Crow friends of Peter Manigault, they decided to host Indian game wardens from various tribes. Fonda directs a program that has invited tribal members to the Flying D to learn how Turner manages his land and his vast herd. And the Flying D has also brought in archaeologists and Native Americans to inspect an ancient Indian village, some ten thousand years old, on the ranch, as well as prehistoric trading centers—all rendezvous sites along the early Great North Trail. Each of these important archaeological excavations is being funded by Turner, as is the recent purchase of the historic Ulm Pishkin buffalo jump site near Great Falls.

Together, Turner and Fonda—he in his blue jeans and jean jacket, she in her fringe coat and cowboy hat—have been working at their new, politically correct Montana life. He is on the board of the Greater Yellowstone Coalition; she is on the board of the Montana Nature Conservancy. And both are vocally committed to causes major and not so major that range from support for "sustainable development" to finding an environmentally proper way to wipe out the spotted ragweed and the leafy spurge.

But while Ted and Jane have been doing their share to preserve and change Montana, they seem to have been changing each other in the process. When Jane gave up alcohol on New Year's Day 1992, so too—after a lifetime of drinking—did Ted . . . at least for a while. When Jane suggested that Ted start working out, he did, and even began attending some of her classes. She has become something of a finishing school for her husband, teaching him, reading to him, giving him the aesthetic perspective that he never had: "One way Ted and Jane are synergistic," Peter Manigault says, "is that Jane still manages to do a lot of reading, in spite of their travels. Ted is so busy and into so many things that he almost has to get a lot of his reading predigested by somebody he trusts to give him the word. And Jane serves that role very well." Whether it's providing him with the Cliff Notes version of Romain Gary's *Racine du Ciel* or explaining the other side of the TV and movie business—the artistic side—Fonda is constantly striving to open her husband's horizons, to introduce him to the work of quality writers, directors, actors. "She's good eyes and ears for him," says one entertainment-industry CEO. As a result of his new life, some of Ted's avocations have changed. He has, for example, recently developed an interest in the visual arts. He's begun to do sketches— still lifes and landscapes, which friends say are surprisingly good. He's also started to collect Western art, buying up the work of frontier painters, and expedition artists who traveled the West with Emperor Maximilian in the 1860s. And these days, on occasion, he is even given to discussing qualities of style or brush stroke or chiaroscuro.

If Fonda has had an impact on her husband, Turner has clearly rearranged his wife's life as well. Feminist or not, she has consistently been attracted to one dominating man after another, first Roger Vadim, then Tom Hayden, and now Turner; according to her friends, each time she has mirrored the man she was with. Turner is perhaps the most forceful of the lot. And for him, the liberated Fonda has seemingly given up her career—given up a life's work as an actress—to be a wife. "People say to me, 'How could you have left moviemaking?'" Fonda says. "But . . . I can't imagine any movie that I ever made or could make in the future that would be worth giving up three months of being with Ted."

Perhaps, as cynical Hollywood types point out, there aren't a lot of big theatrical roles left for Jane, now that she's in her mid-fifties. Or perhaps the excitement of traveling as a world figure to meet heads of state around the globe has its compensations, especially for a woman who was always an activist and who now has the opportunity to be an activist on a grander scale, teamed up with a highly visible husband who shares her passions. But there is no denying that Fonda has become more of the dutiful spouse. Whereas she was once as unflinchingly outspoken as Turner himself, she now seems positively deferential, sounding more like Mrs. Ted Turner than a millionaire entertainment mogul. "Ted said to me in the beginning, he said, 'Cut what you're doing in half,' and I did," says Fonda. "And then he said about six months later, 'Try cutting it in half again,' and I did."

Fonda seems dedicated to making their relationship succeed. If Ted wants to go hunting, she goes. Knowing his history, perhaps she feels that to leave him alone for any length of time is to invite female trouble; at any rate, she has made a point of never being away from him for more than two days at a time since they've been married. "She's going at it as a career, a commitment of a lifetime for her," a family friend says. "She's making an incredible effort to make all this work. She doesn't want it to fail."

In addition to being in love, the two of them are also in business together. Ted asked Jane to work on a docu-biography of her father—*Fonda on Fonda*—for TNT. She has narrated National Audubon Society documentaries on TBS. In 1994, she produced *Lakota Woman* and is currently working on *Pigs in Heaven,* both for Turner Pictures. And no doubt she encouraged Ted to bring a series like TNT's *A Century of Women* to the small screen.

But what Turner and Fonda really offer each other is a kind of safe haven. Since both are famous, rich, and successful, they have reached a balance, an equilibrium. Unlike their previous mates, neither one of them is overshadowed. Each understands the bumpy road the other has traveled, a road potholed with family tragedy and failed relationships. Perhaps this background is what brought them together. According to some—like Dr. Frank Pittman—it may ultimately tear them apart. Pittman has been heard to say that "those two have so much baggage, they're not going to last." (Pittman may be drawing from his text on incurable philanderers here, or it may just be sour grapes: Fonda reportedly convinced Turner to take their psychiatric business to the West Coast.)

Together for the moment, and perhaps a lot longer, Ted and Jane have seemingly found a refuge in each other, and in their home on the range.

Economically, too, Turner and his business have come of age, both settling comfortably into their middle years. In 1992, with Turner Broadcasting worth over $6 billion (and with revenues of $1.8 billion), the company did something it hadn't done since 1975, before the superstation, before CNN, before MGM and TNT—it paid dividends to its stockholders, a sure sign of prosperity. Turner himself was reported variously as being worth $1.8 or $1.9 billion. Of course, it was only on paper, with most of the money tied up in company stock, but the day in 1992 when the stock went up and Turner topped the $2-billion mark, there was a very real explosion of joy from his office.

Perhaps the most startling indicator of Turner's extraordinary and unexpected success was the lawsuit filed against him by MGM investors. Once inclined to chortle about how their leader Kirk Kerkorian had fleeced Ted Turner, they now angrily declared that, given how well Turner had done with the film library, MGM had actually been sold too cheaply. It was they, not Turner, who had been cheated in the deal, they insisted.

In fact, Ted Turner had merely squeezed value out of his company and continues to do so. He has packaged and repackaged the same material over and over again, showing the same John Wayne movies, for example, on TBS and TNT, then selling them on Turner Home Entertainment (home video) and again to local stations via Turner Entertainment. Over and over he has replayed and resold the same James Bond films, proving that—environmentalist that he is—no one can recycle like Ted Turner. A classic example was a three-part documentary about Metro-Goldwyn-Mayer called *MGM: When the Lion Roared*—three nights of more or less free programming, derived from cutting and pasting the footage he already owned. The documentary also served as a huge six-hour, in-house promotion—a massive advertisement for the glamour and appeal of MGM movies, of which Turner is the sole supplier. And as a final grace note, his publishing division was even able to turn out an MGM coffee-table book with the same title, which sold for $50.

Turner is not embarrassed to compare his approach to that of chicken farmers: "They grind up the feet to make fertilizer, they grind up the intestines to make dog food. The feathers go into pillows. Even the chicken manure they make into fertilizer. They use every bit of the chicken. Well, that's what we try to do over here with the television products—use everything to its fullest extent." Part of that strategy included spinning off the same programming overseas—starting in 1991, and continuing, no doubt, into the next century—by establishing TNT Latin America and TNT Europe, with TNT Asia on its way.

An equally impressive source of international expansion is CNN. Turner's goal has long been to become *the* source of TV news for the entire world, and while he obviously has quite a way to go, two decisions in late 1990 and early 1991 set him on the right path. The first was to hire a new president from outside instead of advancing one of CNN's vice presidents. Defying expectations at the last moment and shaking up the company, Turner appointed the respected Times-Mirror vice chairman W. Thomas Johnson Jr. While some at CNN took the appointment as "an unbelievable offense," it has strengthened the company.

The second decision came only a few weeks later, following the Iraqi invasion of Kuwait, when international troops began to mobilize in the Middle East. At Johnson's first major meeting, the new CNN head had tried to get a sense of how much he should be spending for the coverage. Did his new boss want the economy-size $5-million approach or the lavish $32-million one? "You spend whatever you think it takes, pal," Turner said.

Not long afterward, Johnson started to get calls from the White House. He had mentioned to President Bush that CNN had a team, including Bernard Shaw, in Iraq, and Bush had told him, "I certainly hope you get Bernie in and out of there soon." But as the January 16, 1991, launch date approached for the U.S.-led multinational attack on Iraq, Johnson began to get urgent signals from many sources, including Vice President Dan Quayle and Press Secretary Marlin Fitzwater. "Your people are in serious danger," they warned. Johnson gathered his top aides together for a conference call and phoned Turner in Montana. They explained that Peter Arnett and producer Robert Werner wanted to stay in Baghdad, but that the White House itself had been pressing them to move everyone out. Unspoken was the fact that Johnson had lost two *Los Angeles Times* bureau chiefs, one killed in Tehran and the other in Nicaragua. Everyone on the conference call felt the weight of the decision and expressed grave doubts about the journalists' remaining in Baghdad.

Then Turner's voice came crackling down the line: "We have a global job to do and we should do it. Those who want to come out can come out," he said. "But anyone who wants to stay can stay." According to Johnson, "Ted was emphatic. I think he was taking the responsibility off our shoulders. 'They can stay,' he said. 'You will not overturn me on this.'"

As the bombs rained down on Baghdad, and the Scud missiles fell on Saudi Arabia and Jerusalem, Turner's courage and his international perspective bore fruit. CNN scooped all three "major" networks, broadcasting the first accounts of the war when no one else could even get a picture out of Iraq; NBC was reduced to interviewing CNN correspondents. In a breathtakingly modern update of the famous World War II eyewitness reporting of the London blitz, CNN stamped the Gulf War story as its own. As the allied planes swept in, Peter Arnett, Bernard Shaw, and John Holliman huddled under a table in their hotel room in downtown Baghdad, describing the scene "here at ground zero, feeling the ground rumble, hearing the thunder." For the first time, live pictures of a war in progress were being broadcast from behind enemy lines. Turner's 1990 trip to the Middle East with Kathy Leach had yielded new friends for CNN who now proved invaluable—the network had access to satellite hookups in Iraq and Jordan that no one else had.

Some critics scoffed that CNN was "Saddam's link to the outside world," but Turner remained firm. He said that the Iraqis "were the enemy of the United States, but CNN has had, as an international global network, to step a little beyond that. . . . We try to present facts not from a U.S. perspective but a human perspective." Then Turner, once an avowed flag-waving, America-First sympathizer, added quite coolly, "I'm an internationalist first and a nationalist second."

The conflict in the Persian Gulf was a watershed for the Cable News Network. As the dust settled and the hundred-hour war drew to a close, the real victor was Turner's CNN. While everyone else had been stuck reporting from in front of the odd blue domes of the Dhahran International Hotel, CNN had established itself, once and for all, as the network of record—*the* place to turn for breaking news. Its global audience swelled to seventy-five million, and a select group of world leaders including President George Bush would later acknowledge that they learned more from CNN than from their top intelligence sources.

In the afterglow of great ratings and reviews, Turner's news network began a major expansion drive, with more news, more sports, more updates, more bureaus and reporters around the world. By 1994, the network's international bureaus had expanded from twelve to twenty. And since 1991, CNN has not only been more profitable than any of the network news divisions, it has been more profitable than any of the other networks. In a stunning reversal of fortune, CNN alone is now bringing in more profits than all of CBS.

But it should be noted that if Turner's networks are today more lucrative than ever, if they are received in more homes than ever before—as of the latest (1993) figures, TNT, CNN, and TBS are all received in about two-thirds of America's households with TVs, and CNN international is currently viewed by over forty-five million homes around the world—if they are, in short, economic success stories, the quality of their programming still lags behind. CNN, which has come a very long way, continues to be weaker on analysis and context than on facts and pictures. It can show the news on the air, but it has a hard time telling viewers what the news means. CNN also lacks differentiated programming—shows that stand out from the great gray river of news items. The recent acquisition of Judy Woodruff (from PBS's *MacNeil/Lehrer NewsHour*) is, however, plainly a step in the right direction.

The programming on TNT and WTBS is more flawed. Perhaps under the influence of Jane Fonda, Turner has recently attempted to do more serious, high-minded films like *Heart of Darkness,* hiring skilled actors like John Lithgow, Barbara Hershey, and John Malkovich, among others. And announcements of new projects by Paul Verhoeven, Ron Shelton, Philip Kaufman, and Francis Ford Coppola lend hope for the future. But so far, it's clear that despite noble aspirations, few people at Turner Broadcasting have any sense of how to put together quality television. Series like *A Century of Women* or *The Native Americans* are typically Turneresque—politically correct in inspiration, large and grandiose in ambition, and short on artistic focus.

Mostly, TBS and TNT continue to fill their airwaves (and cash registers) with *Beverly Hillbillies* marathons, "Three Stooges" marathons, and "Men of Iron" weeks. In one wonderful back-to-back announcement, the superstation declared that it would air the Nobel Prize Awards Dinner and a show titled *Harley Davidson: The American Motorcycle.* But surely the most startling publicity release of the 1990s was the proud declaration that Turner Broadcasting had bought up all the episodes of *Dallas*—the series that Turner himself had once denounced as the symbol of all that was rotten on network TV.

On August 30, 1992, at the Pasadena Civic Auditorium, television's finest gathered to celebrate their annual TV ritual, the Emmy Awards. It was to be a night full of bad jokes and windy discourses on Dan Quayle, Murphy Brown, and potatoes. But then a serious Candace Bergen came to the microphone to introduce Ted Turner, who was being honored with the lifetime-achievement Governor's Award. "Without using a single vowel," she said, alluding to CNN, "one man has changed not only the face of television . . . he has created more jobs for more newsmen than anybody since Charles Foster Kane. . . . Ladies and Gentlemen, it gives me great pleasure . . ."

Ted Turner, in a tux and a scraggly white beard, vaulted up the steps and grasped the award awkwardly. "Please terminate the applause, the show is running long," he sputtered, and no one was sure if they should laugh or not. Fumbling with the statuette, he dedicated the award to his employees, declared that "it's been a blast," and then abruptly wrapped up: "My last line is, Give my regards to Broadway." The camera cut away to Jane who laughed and smiled knowingly, but there was an odd pause as the rest of the audience, baffled and trying to figure out what the hell he was talking about, collectively shrugged and applauded.

As a sign that Ted Turner has finally made it, the now white-haired CEO has, in the 1990s, entered his honorary awards-gathering phase. Racking up the statuettes and parchments as if he were still in need of outside validation, Turner over a three-year span notched honorary doctorates from Lehigh, Tufts, the University of North Carolina, and—thirty-three years late—from Brown. He also collected, among others, the Jackie Robinson Foundation's "Robby" award, the Radio-Television News Directors Association's "Paul White" award, and UCLA's "Neil Jacoby" award. Handed the latter, he declared, "I can guarantee you that this is the most important award that I'll get this month."

On January 6, 1992, *Time* named Turner "Man of the Year." He followed a long line of newsworthy winners, dating back to 1927, that included Lindbergh, Roosevelt, Churchill, Hitler, Stalin, Gandhi, and Madame Chiang Kai-shek. *Time* wrote: "Visionaries are possessed creatures, men and women in the thrall of a belief so powerful that they ignore all else—even reason—to ensure that reality catches up with their dreams. . . . For influencing the dynamic of events and turning viewers in 150 countries into instant witnesses of history, Robert Edward Turner is *Time*'s Man of the Year for 1991." Not quite so impressed by his visionary credentials, the *Village Voice* pointed out somewhat cynically that in the week after Turner appeared on *Time*'s cover, TBS stock hit a fifty-two-week high, and "Time Warner, which has a 21.9 percent stake in TBS, has earned about $50 million since it named Turner its Man of the Year."

In this heady period of accolades, there was even talk once again of Turner running for president of the United States. "In 1990 and 1991," Robert Wussler recalls, "Ted held a number of meetings with various Democratic leaders. Through Jimmy [Carter], certain doors were opened, certain conversations held." A handful of Southern power brokers—including Carter aide Hamilton Jordan and John Jay Hooker, the Kentucky Fried Chicken magnate—unhappy with the slate of likely Democratic nominees, began to put together a list of potential third-party candidates. Eventually, they narrowed their selection down to two candidates, two no-nonsense, pragmatic billionaire businessmen with reputations for getting things done—Ted Turner and H. Ross Perot.

It was at about this time that Turner's lithium use was disclosed, and he announced that he was just as happy to be out of politics, declaring himself "a citizen of the world, not a politician." Besides, his wife had vetoed the idea of politics: "Jane was married to a politician, and she's had enough of that." Some former aides chuckled that while Perot was going after the Oval Office, the way Turner saw it, the job of America's president was too small for him—he was after a global position. And, in a comment that sounded surprisingly like Perot, Turner admitted that he was too impatient to be a politician: "I'm used to moving quickly and getting things done, and governments move so slowly. If we wait for the governments to do it, it ain't gonna get done."

In an effort to get things done privately, Ted in 1991 formed the Turner Family Foundation, with a board composed of his wife and five children and headed by Peter Bahouth. Twice a year, the foundation allocates money to charitable concerns, focusing primarily on Turner's current hobbyhorse, the notoriously unsexy cause of sustainable growth and overpopulation. "I want to get my children in the habit of giving money away," he has told friends, "not spending it." He himself has recently given a total of $75 million to Brown, The Citadel, and the McCallie School, all the institutions that have played a major role in his own and his family's life.

On January 14, 1993, a fifty-four-year-old Ted Turner returned to McCallie to address the student body inside the new gymnasium his money helped to build. As he looked out over the young faces, he was thinking back to when he had first come there more than forty years ago: "I love this school a lot," he told them. "It did a lot of good for me. . . . A lot of times you don't appreciate things as much when you are there as when you have the opportunity to look back on them from a number of years."

In rapid summary, Turner retraced his own experiences, from high-school debating and sailing, through his early years in his father's business: "I didn't really set out to make a lot of money, I really didn't, and that has not been my major motivation. . . . I just wanted to see if these things could be done. I feel that had I been born four or five hundred years ago, I would have wanted to be an explorer, like Columbus or Magellan. In fact, I probably would have preferred to do that. . . . But I did get to be a kind of pioneer or explorer in the television business."

After putting in his customary pitches for turning off lights, strengthening the United Nations, and making the world a better place in which to live, he rambled into some final words of advice: "With technological change coming as fast as it is in virtually every industry, it just comes so fast that it's really hard to keep up with it. It's like a run-and-gun basketball game. You've got to keep your eyes open, because the ball might be coming at you from any direction and you want to get those uncontested layups. That's basically what I did. I just broke away from the field in television and was out on the corner and they got the ball to me and I was able to take three steps and dunk the ball. It wasn't hard at all." In the warm glow of nostalgia, Turner's hard-fought career had now become a slam dunk. Of course, everything looks easier when you're sitting on the winner's bench after the buzzer has sounded.

At this point, it may have seemed that Ted Turner, in his mid-fifties and leafing through his memory album, had successfully wrapped up his life, that he was now going to unofficially retire to his ranch with his movie-star wife and spend the rest of his days collecting arrowheads and awards and issuing valedictories. But Turner had no intention of riding into the sunset. Lest anyone think so, he declared acidly at a recent company Christmas party that he had three wishes for the new year: "Peace on earth, goodwill to men . . . and to get my company back."

And if Time Warner and TCI had no intention of giving up their shares of Turner Broadcasting, Ted had other irons in the fire. He had been holding discussions for many months with Hanna-Barbera Productions, Inc.—home of the Flintstones, Yogi Bear, and Huckleberry Hound—and in December of 1991 he completed the purchase of the company for $320 million. The animation studio was interesting to Turner, but what he really wanted was the library—three thousand half-hours of cartoons that he could slice, dice, and resell, just as he had with MGM's movies. All these episodes would in fact become the basis of a new Cartoon Network, a twenty-four-hour-a-day animated channel—round-the-clock slam-bang Yabba Dabba Doo!, including everything from Hanna-Barbera's *The Jetsons* to MGM's *Tom and Jerry*. The service debuted on October 1, 1992, and Turner declared, "We're opportunistic. . . . While the timing might not be good from a channel capacity or legislative point of view, we don't play for the short term. There's no question this is a long-term play." In fact, the Cartoon Network began making money almost immediately.

Looking ahead to the multimedia of the future, Turner in February of 1992 also created an interactive unit that included everything from CNN news compilations on CD-ROM to interactive games based on Hanna-Barbera cartoons—in other words, still more ways to process his chicken parts. Meanwhile, he was also aggressively pursuing the same strategy overseas, establishing the TNT and Cartoon Network Europe, Cartoon Network Latin America, and TNT and Cartoon Network Asia, as well as purchasing stakes in overseas stations like Russia's brand-new Channel 6 in Moscow and Germany's NTV news network in Berlin—ball carrying the same core of Turner programming around the world.

But most important, as 1992 ended and 1993 began, Turner was once again feeling the need to expand either his distribution (via a broadcast network) or his program and film capacity (via a studio), just as he had back in 1985. So—once again—he was meeting with CBS, ABC, MGM, Orion, and Paramount Studios, and talking about mergers, buyouts, and partnerships.

It was John Malone who suggested that Turner meet with Paramount CEO Martin Davis, and they did in fact talk together on at least three occasions to see if they could combine their companies (all this well before the 1994 Viacom deal was under way). Davis didn't know what to make of Turner at first: "As I got to know him," he told *Vanity Fair*'s Bryan Burrough, "I learned to make sure I got him in the right mood swing, or he'd be crazy. . . . When he got to talking, he'd get so excited I thought he'd have an unregulated orgasm."

For Davis, the conversations seemed fairly pointless: "Ted has no authority," he said. "Any expenditure over $2 million he has to go to the board for. He's a minor player. He's got no control over his destiny, and he doesn't even know it. [Malone] sends him up in a blimp and tells him he's got an air force. Malone controls everything."

According to Davis, Turner had nothing but the highest praise for the shrewd Malone. He talked about him "in glowing terms," almost as a mentor. But meanwhile Malone was himself negotiating with the Paramount head. Discussing all the ways that TBS could be cut into pieces and recombined with parts of TCI, Paramount, and the TCI subsidiary Liberty Media. Malone was cavalierly deciding Ted's fate. "I could see that Turner had respect for this guy that John Malone was already abusing," Davis said. "He'd say, 'We'll get Turner contained and then do anything we want.' . . . Here I am seeing him sell Ted down the road. And I know full well I would be next."

But Turner, in fact, was not oblivious to his predicament. He was chafing at the way he was continually being blocked. TCI and Time Warner had goals different from his, and he knew it. They didn't want him to team up with a broadcast network like ABC. "John Malone and Time looked at ABC as the enemy," Turner says. "They would have to get out if ABC came in." They didn't want him to link up with a studio either: that would be competition for Time Warner films. And any possible combination for Turner, with partners like TCI and Time Warner in place, ran the risk of antitrust problems for all concerned. It wasn't that his board actually vetoed many serious proposals, but that everyone in the entertainment business seemed to realize Turner wasn't in a position to put any major deals together. "When you have two ten-ton gorillas controlling him, he can have all the conversations he wants, but nothing ever goes anywhere," said one top executive who had negotiated with Turner.

It was in this context that Ted Turner sat down privately with John Malone and Gerald Levin at their March 5, 1993, board meeting and dropped something of a bombshell: If he couldn't control his destiny, he wanted out. He was frustrated. He couldn't take it anymore. Turner let word leak to the press about how he had "chains" and felt "hemmed in." "TURNER'S FRUSTRATIONS RAISE ODDS OF A SPLIT WITH CABLE PARTNERS," the *Wall Street Journal* revealed. "TED TURNER'S SEASON OF DISCONTENT," was the headline in the *New York Times*.

Time Warner advisor Oded Aboodi was given the assignment of crafting a breakup proposal, and all sorts of various arrangements were suggested: Perhaps Turner would sell out and get stock in TCI and Time Warner and sit on their boards; perhaps he would sell off certain pieces of his companies (just CNN or TNT, for example) and retain the others. Throughout the cable industry, there were rumblings of concern. Almost everyone in the industry hated this kind of destabilizing development. They wanted Ted Turner to remain in place and his company to stay as it was: "Virtually nobody thought it was a good idea," Turner acknowledges. "We're kind of neutral ground for the cable industry. TCI and Time are both two-billion-dollar cash-flow companies. And there's increasing friction between them. There have been times over the past six years that they've come really close to going at each other. We're kind of the watering hole between them."

But those around Turner soon realized that his breakup proposal was less a serious intention than, as he would later term it, "a trial balloon." "Ted's not a seller; Ted's a buyer," said Paul Beckham, then president of Turner Cable Network Sales. "I didn't think that he was going to sell. I thought, Ted's probably positioning for something."

As it turned out, Turner was indeed positioning for something. And on August 17, 1993, it became clear what that something was. While Time Warner had opposed almost any merger with a film-making studio, Turner's threat had cleared the way for him to win one of his most coveted goals—one that had eluded him since he had sold off most of MGM. He was about to enter the motion-picture business. For approximately $600 million in cash and stock (plus about $230 million in debt), Turner announced that he would buy New Line Cinema and Castle Rock Entertainment, a movie studio and a film-production company.

While both were modest-size businesses, between them the pair had turned out hits like *Teenage Mutant Ninja Turtles* as well as critically well-received films like *A Few Good Men* and *In the Line of Fire*. Turner seemed not to care that Castle Rock lacked a library. And though he continued to decry violence in film (refusing to show even *The Godfather* on his networks), he seemed not to care that New Line was the company that had produced slasher films like *Nightmare on Elm Street* and *Friday the 13th*. He was able to embrace all those contradictions, and embrace them happily, because at last, after seven years, he was going to Hollywood. Finally, he was going to be able to sit behind a tinsel-town desk and have the power to green-light movies of his own—the power to make his own *Gone with the Wind,* his own *Citizen Kane*.

So where is Ted Turner now? While he may occasionally slow down enough these days to find a certain peace along his Montana streams, he is still a man in motion looking for opportunities. The Turner qualities that transcend all others are active ones: his enthusiasm, his excitement, his nonstop drive. Every day for Ted Turner, the world begins again.

It is these qualities that set him apart from the Murdochs and the Berlusconis, for while they may seem on occasion to tally greater returns—with the expansion of the Fox Network, or victory in the Italian election—Turner is finally a leader in a class by himself, impatient to be getting about his business and going on to the next thing, charging ahead pell-mell and running roughshod over people's lives as he plunges into the next battle, seeking the next enemy who will help to define him, the next victory to be won. "It's what binds you to him and that company," says Ira Miskin. "You finally believe that what he's telling you to do comes from every molecule of his fabric. And that he will stick to it until his last drop of blood."

Turner has certainly not given up his quest to expand his property, or his desire to acquire a major Hollywood studio or broadcast network like CBS or NBC, or his hope that one of his teams will one day present him with a championship trophy in basketball or baseball (both the Hawks and the Braves have come tantalizingly close). With his latest acquisitions of land in New Mexico and Montana, he currently owns an estimated 768,000 acres in the Rocky Mountain region (as a point of comparison, the Grand Teton National Park encompasses 310,000 acres). His most recent purchase, in February of 1994, was the Pedro Armendaris Ranch, 300,000 hardscrabble acres of grassland and cactus and creosote bush on which his buffalo can roam, across from his other property along New Mexico's Jornada del Muerto trail—the dry, seared region known as the Journey of Death. Since the Armendaris includes the entire Fra Cristobal mountain chain, this acquisition makes Turner one of the few private citizens anywhere in the world who has his very own mountain range.

Turner also now owns between a quarter and a third of a company currently valued at $8 billion, and which continues to grow some 15 percent a year. That makes him worth approximately $2 billion, on his way to $3 billion. But accumulating money was never Ted Turner's primary goal.

Accepting the fact that time is running out for him, Turner, in his mid-fifties, appears eager to stuff the hourglass with quality sand: his marriage to Jane, his improving relationship with his children, his aspirations for his media empire.

As for his personality, the contradictions are still apparent. Consider, for example, the white hair and youthful energy, charm and solipsism, sentimentality and ruthlessness, seriousness and hot air. And what you see is most definitely what you get. In that sense as in so many others, Ted is very much his father's son.

On a warm, breezy spring afternoon, the Turner clan assembled once more at All Saints' Episcopal Church in Atlanta. In the wooden pews up front sat Ted and Jane, Rhett, Beau, Jenny, and Teddy, plus dozens of friends and family, from the manager of the Hope Plantation to Judy Nye Hallisy—all gathered to celebrate the christening of Ted Turner's first grandchild, John Rutherford Seydel III. The late afternoon sun came streaming through the stained-glass windows of the old stone church as the priest went through the ritual. Then dipping his fingers in the holy water, he traced the sign of the cross on the baby's forehead. As the brief ceremony drew to a close and the family milled about, Jane held the baby and Ted looked on over her shoulder, the proud grandfather admiring the future. The baby gave a little cry—the next generation of the Turner clan staking out his turf. And for Ted Turner, once more the world began again.

EPILOGUE
(SEPTEMBER 18, 1993)

OUTSIDE ON PARK AVENUE and 61st, it's a drizzly Saturday morning. At 9:00 A.M., the still half-empty lobby of the Regency Hotel soothes the early rising guest with the muted elegance of its huge vases of fresh-cut flowers, its gilt-edged mirrors, its tapestry-covered walls. The desk clerk, his head lowered, looks as if he may be dozing. Then the high-pitched raucous voice of Ted Turner rolls into the room like a luggage cart with squeaky wheels.

"I didn't sleep at all last night," he complains, striding along with a large phalanx of people marching behind him. He was up late watching his Braves lose to the Mets 3-2 in extra innings. And the Giants are gaining on them in the tightening pennant race. Atlanta's lead is now cut to three games. What the hell, they're still ahead!

But Turner has other things on his mind as well. He recently acquired New Line Cinema and Castle Rock Entertainment. And only the previous day he met with his board of directors to discuss buying Paramount or one of the other seven remaining major Hollywood studios. He's currently negotiating with Doordarshan, the Indian TV network, to provide progamming aimed at the subcontinent. And he's in the planning stages for two new networks, Turner Classic Movies and an American edition of CNN International. In addition to TBS, CNN, Headline News, TNT, and the Cartoon Network, they will be his sixth and seventh American cable channels. And now, here he is in New York attending a press conference to celebrate the launching of Turner Pictures' first feature film, *Gettysburg*.

Turner calls it "a noble project." It's hard to imagine another moviemaker who would claim nobility for his or her work. But words like "noble" roll off Turner's tongue like birdsong. At a running time of four hours and eight minutes, it's either one of the longest movies ever made or, according to some of the reviewers gathered at the Regency this morning, just seems to be.

Despite his lack of sleep, Turner is positively crackling with energy as he enters the ballroom followed by the movie's director—Robert Maxwell—its stars, and a host of PR people. There's Sam Elliott, who plays Buford; Jeff Daniels, who plays Chamberlain; and Martin Sheen, who plays General Robert E. Lee. Turner is predicting Academy Awards for everybody. "It's going to be one of the great classic movies of all times," he states flatly, not wanting to leave any doubt about what he thinks of their achievement.

At the half-dozen or so tables spread around the room are the representatives of the media, still fiddling with their tape recorders and their notepads. The fifty-four-year-old Turner plunges right in, moving from table to table shaking hands. He is wearing a beige plaid sport coat and a white shirt open at the neck—expensive clothes that look rumpled on his lanky frame. The longish white hair, thick eyebrows, mustache, and heavily lined forehead give him the look of an elderly riverboat gambler who has sat in on more than his share of big games. But there is no hint of age in the way he moves around the room. "Ted Turner," he says to an attractive young woman, shooting out his hand. "Who are you?" Coincidentally, she happens to be named Turner, too. "One good Turner deserves another," he says with a wink and, pleased with himself, moves on.

"It cost about twenty million," he announces to those at the next table. "If a major studio had made it, it would have cost sixty million." At another table he says, "As far as I'm concerned, it's a lifetime achievement. I did it against strong advice from a lot of quarters. And I *never* wavered." As always, it's Turner versus the multitudes. "I hope to be a major force in the movie business," he adds frankly. Then out of sheer irrepressible ebullience, he suddenly cries, "I *love* movies!" Not satisfied with that, he goes on: "I love life, I love the planet, I love my wife, my kids. Animals." There's no stopping him: "I love albatrosses, eagles—chipmunks! I love trees. The redwoods in California!" Everybody within earshot seems momentarily stunned at his outburst. Then, almost touchingly, he lowers his voice and adds, "I want to be remembered as somebody who made a difference."

Sitting down at the rear table in the corner, Turner faces the handful of journalists waiting for him there. He leans forward earnestly in his chair. "There wasn't a family in the United States that was untouched by Gettysburg," he reminds everyone at the table. "Both sides, the North and the South." He lets that sink in, then rubs his hands together with satisfaction. "Everybody else turned this film down. *Everybody* . . . for fifteen years. But I liked this film. I didn't like it, I *loved* it. I know the Gettysburg Address almost by heart. It's the most significant three days in the history of our country . . . probably. If it had gone the other way, we'd probably be two countries today. Like Bosnia-Herzegovina, we'd still be dukin' it out."

From around the table, the questions start coming. The reporters want to know why he's taken a TV film and placed it in the movie theaters. "No, no," says Turner, shaking his head. "This is our first movie. I had MGM in '85 but only for two months. I never got to green-light a movie or anything. I had to sell it. I knew I had to sell it because I was gonna go broke if I didn't. . . . I didn't wanna go out. I wanted to make movies. . . . And now we're into theatrical movies. I wanna make as many great films as possible."

"Does that mean you're going to buy Paramount?" the person sitting next to him asks. Failing to hear, Turner bends forward and cups his left ear. The question is repeated. From under bushy white eyebrows, Turner looks up at him slyly. "You read the article in this morning's paper? I can't say much more. I mean, as much as I'd *love* to say a lot more, I can't. You understand?"

Asked why he needs Paramount when he already has New Line, Castle Rock, and Turner Pictures, he says, "Well, it never hurts to have a major studio. There's only seven of 'em, and . . . and . . . I can't talk too much more about it."

"So why did you pick *this* movie to make?" he's asked. "Why did you green-light this one?"

"It was a noble project," says Turner. "All of us have seen men die."

"There are other noble subjects besides war," a reporter observes.

"Sure," Turner snaps back. "There are a lot of noble ones. And I can tell you what they are. *Gone with the Wind* is noble. *All Quiet on the Western Front* is noble. *Ben-Hur* is noble. And *Dances with Wolves* is noble. *Lawrence of Arabia* is noble. *Dr. Zhivago* is noble. *Wuthering Heights*—" Abruptly, he stops and smiles. "I try to play Heathcliff when I'm with my wife. That haunting look. You know up there on the cliff. I'm *soooo* romantic," he croons.

Turner not only gave the financial go-ahead to this movie, he also acted in it—donning the Confederate gray to die as an extra at Pickett's charge. Someone asks why he chose to make his acting debut in this film. Ted explains how he moved to the South when he was only nine, and went to a school where they were still fighting the Civil War "big time." He says, "I became a Confederate to survive. And for a Confederate—after seeing *Gone with the Wind;* that made a big impact on me—the only place you wanted to die was Pickett's charge at Gettysburg, where they thought they were gonna win the day. . . . So I said, well I've gotta have a cameo when we're gonna do Pickett's charge. . . ."

"I've seen the movie twenty-six times," Turner announces. He looks around the table, fixing each of the reporters with an emphatic glare. "It'll be a travesty if this movie isn't a smash hit," he says. "It can go in a thousand theaters. The theater owners will want to show it if it does well. A lot of it depends on you-all. I know that. We took a chance. But I'm proud of it. And remember this. *The Last Emperor* didn't do squat. . . . *Citizen Kane,* which we own, was a complete flop when it came out. So what!"

Turner insists that movies are art. And he's learning the art. He says he's got a book called *How to Make Movies,* and he's read it. "You've got to start with a story. Great story, great book, great script. It starts with the writing." Turning to the reviewer next to him, he says, "Right?" Then laughing, he shouts, "Give me five!" and slaps palms with the dumbfounded fellow.

"I wish I had been at the screening with you-all last night," he says, "but my wife said we've seen it so many times. I've seen it twenty-seven times. I know every line by heart just about. She said we're goin' tonight. And then we've got to go to Washington next week. We're gonna go to Washington, Los Angeles, and the Atlanta premiere. So we'll already have seen it five times theatrically at least. But I wanted to be there last night. I heard you-all applauded after it was over. That's what happened in Charleston. Now when critics are applauding, buddy, you've got something. I'll be disappointed if this movie is not nominated for an Academy Award."

Leaning across the table, Turner confides, "My own people said we can win the Emmy if we go on television. I said we can win the Academy Award if we go on movies. They said yeah, but that's a big risk. I said, listen, pal, this movie is *big*. And this is the kind of movie you gotta see on the big screen, with all those cannons and armies. I mean, I love television. It's my basic business, but to premiere this baby on television with Pampers commercials every ten minutes—I mean, there's nothing wrong with Pampers commercials, but this thing needs to be in the theater."

Asked whether Jane has a role in his moviemaking plans, Turner does a double take. "Absolutely!" He shakes his head in disbelief. "I'm telling you! Does Jane have a role? She's my Scarlett O'Hara. She's my sweetheart. And she knows a lot about movies. Forty-nine of 'em she acted in. She produced ten or eleven. One of which was *On Golden Pond*. One of which was *China Syndrome*. They were movies that made a difference. I'm inspired by her and her dad, and Humphrey Bogart and Sam Elliott, and everybody in this room. Martin Sheen. We watched *Apocalpyse Now* last night. We wanna make good movies of all kinds. I like comedies. I like adventures. I like epics. I like Westerns. I like all kinds of movies. I like science fiction. I loved *The Day the Earth Stood Still*. I love *The Thing*."

The reviewer next to him asks if he likes *2001,* one of his MGM movies. *"2001!"* Turner cries jubilantly. "Give me five!" he orders, and slaps his palm again. "I like it all, I like it all!"

"Can you tell what's a good script and what's not?" he's asked. Turner admits, "I don't read the scripts very much. I read them occasionally. I let Jane read them, and she can tell me 'Good,' 'OK,' 'Needs more work.' She's producing a movie right now called *Lakota Woman,* which is about the more recent battle at Wounded Knee. We're doing this series of television movies on the American Indian. Some of them may turn out to be theatricals. She has worked on that script with the writer and director for six months. *Every* day."

Asked how many strong women's parts there are in *Gettysburg,* Turner says, "None." Then he hurriedly adds, "But women like this movie because they like the men in it."

Turner leans back in his chair and recounts how he and Jane went together to Gettysburg and watched the battle scene from a cherry picker. Dozens of feet up in the air, they looked down on the scene and cried. It was a terrifically moving experience for both of them. "I'll tell you how I felt," he says. "It's like the way you felt on your wedding night or when your father died." When it comes to strong emotions, those experiences are the touchstones for Turner.

Another recent major touchstone for him is becoming a grandparent. "Would you like to see my grandchild?" he asks eagerly, and without waiting for an answer pulls the baby's picture from his wallet. In the candid photo, the little naked blond boy on the bed looks up at the camera with a dazed expression. Elated, Turner hands the picture around the table like a choice hors d'oeuvre.

"A new grandchild, a new movie company," someone says, noting his latest acquisitions. "Does this mean you're entering a new phase in your life?"

"What?" says Turner, looking off distractedly to the next table.

"Are you entering a new phase in your life?"

"I hope so," he says.

"And what is that phase?"

Leaning forward to hear the question, Turner cups his hand around his left ear.

"What is that phase?"

Turner's face lights up. "Show biz," he replies jauntily, and all at once, in a loud tuneless voice, he sings, "There's no business like show business. No business I know. Everything about it is appealing. Da-da, da-da, da-da, da-da, daaaaa."

Gettysburg opened to generally mixed reviews. It was shown in a relatively small number of theaters across the country. By its sixth week of release, it had grossed about $6 million, more than Turner said he anticipated but modest by any standard for a movie that cost $20 million to make. It received no Academy Award nominations. *Gettysburg,* the coffee-table book, was brought out concurrently by Turner Publishing, and sold moderately. The movie premiered on television on June 26, 1994, as a two-part TNT mini-series, and aired repeatedly throughout the summer.

Though Turner met with QVC's Barry Diller to discuss joining him in the bidding war for Paramount, he was in the end not one of the principals. After a protracted battle between QVC and Via-com, the latter finally won control of the studio in 1994.

Castle Rock, New Line, and Turner Pictures are set to make thirty-nine pictures in the coming year. "We're gonna do Joan of Arc next," Turner says. "It's been done before, but we're gonna use a fifteen-year-old girl, which she really was at the time." And unlike *Gettysburg, Joan of Arc* is definitely going to be a winner when it comes to Academy Awards. Turner is bubbling with enthusiasm about his upcoming project. All they need is a script. "Great story, great book, great script!"

For Ted Turner, defeat is merely a glitch—a temporary setback, never an end.

ACKNOWLEDGMENTS

To those listed below who gave of their time, memories, documents, and private papers, and in other ways supported our work, we are truly grateful:

Gerry Hogan, Irwin Mazo, Dr. Irving Victor, Dr. Jules Paderewski, Dr. Fenwick T. Nichols Jr., Charles H. Wessels, Frederick Wessels III, James Roddey, Bill Tush, Mrs. Carl Helfrich, Julie Mazo, Don Farmer, Chris Curle, Elliott Tourett Schmidt, Craig Houston Patterson Jr., Shirley Patterson, Spencer J. McCallie III, John Sharp Strang, William R. Steverson, Robert Wussler, Priscilla Painton, Charles Shumway, Martha Mitchell, Bill Zimmerman, David Schneider, Zeke Siegel, Ira Miskin, Michael J. Gearon, Mal Whittemore, Carl A. Wattenberg Jr., Billy Roe, Nancy Roe, Reese Schonfeld, Lewis Holland, Dr. Lawrence Brenner, Keith Humphrey, Carolyn Steadman Whittemore, Edward Perlberg, Andrew Inglis, Herbert Schlosser, Carl Cangelosi, Andrew Goldman, James Trahey, Farrell Reynolds, William S. Sanders, Dr. Robert Browning, James Hardee, Bob Bavier, Bob Crane, John McIntosh, Tom Allen, Douglas Woodring, Judy Nye Hallisy, Jane Dillard Greene, Lee McClurkin, Sidney Topol, Daniel Schorr, Peter Manigault, Donald King Anderson, Peter M. Homberg, Mrs. Ernest Risley, Edward Taylor,

James Traub, Matthew Sappern, Bob Hope, Paul Beckham, W. Thomas Johnson Jr., Lisa Fischberg, Dr. Bernard Cassidy, John Gisondi, Peter Dames, J.J. Ebaugh, Mary Anne Loughlin, Kathy Leach, Leiston Shuman Jr., Barbara McIntosh, Vera G. Rambo, John H. Baker, Mary Alice Williams, Don Lachowski, Jean Walker, Robert R. Pauley, Carmen Baumgarden, Thomas Belford, Dr. Adib Shakir, Keith J. Shuler, Marcia Terrones, Cassandra Butler, Doug Lester, Lewis Manderson, Robert Shuman, Julie Roberts, Frances Morgan Wood, Marty Koughan, David Baker, Tim Smith, Ray Sokolov, Valerie Lyons, Dan Burkhart, Nick Taylor, Tom Houck, Jesse Kornbluth, Vince Coppola, Robin McDonald, Laura Landro, Dick Williams, Leslie Linthicum, Kevin Goldman, Emma Edmunds, Kevin Gleason, Jim Kennedy, Martin Lafferty, Raymond Butti, Anita Sharpe, and Vicky Markell.

And to the many who out of modesty or natural hesitancy preferred to remain anonymous, our understanding and gratitude.

To those who helped with the preparation and publication of this manuscript, our sincere thanks for their expertise. We wish to especially acknowledge: Anne Freedgood, Cork Smith, Walter Bode, Sara Lippincott, Kathleen A. Bursley, Celia Wren, and Adam Gussow.

To Georges Borchardt, our agent nonpareil, warmest appreciation.

A special thank you to Henriette Goldberg, who at ninety-one is still eager to see what her sons and grandson and great-grandson will do next.

And finally, for those hardy few who for years lived intimately with this book on a daily basis—suffering the highs and lows of the biographical mode—we reserve our love and dedication.

NOTES ON SOURCES

Abbreviations

CW   Christian Williams, *Lead, Follow or Get out of the Way: The Story of Ted Turner* (New York: Times Books, 1981).
DF   Ted Turner interviewed by David Frost, October 25, 1991.
HW   Hank Whittemore, *CNN: The Inside Story* (Boston: Little, Brown, 1990).
PB   Porter Bibb, *It Ain't As Easy As It Looks: Ted Turner's Amazing Story* (New York: Crown, 1993).
PRR   Ted Turner interviewed by Peter Ross Range, "Playboy Interview," *Playboy,* August 1978.
PRR-2   Second Turner interview by Peter Ross Range, "Playboy Interview," *Playboy,* August 1983.
RAF   Robert Ashley Fields, *Take Me out to the Crowd: Ted Turner and the Atlanta Braves* (Huntsville, AL: Strode, 1977).
RVGG   Roger Vaughan, *The Grand Gesture* (Boston: Little, Brown, 1975).
RVTT   Roger Vaughan, *Ted Turner: The Man Behind the Mouth* (Boston: Sail Books, 1978).

Prologue

Most of the material in this book comes from interviews by the authors. In some instances, sources were interviewed more than once. The date cited is that of the first interview. Sources are identified in the footnotes in their first appearance; subsequent references will be made only when necessary for the sake of clarity.

Page 2   "*Fortune*'s most recent estimate": "The Billionaires," *Fortune,* September 7, 1992, p. 114.
2   "one of the three": Herbert Schlosser interview, July 1, 1992.
3   "Richard Cheney said": *Wall Street Journal,* January 21, 1991.
3   "'the best source of information'": Ibid.

3   "'Our foreign policy'": Bush administration figures Brent Scowcroft and Marlin Fitzwater told the diplomatic editor of Britain's *Channel Four News* that "CNN was decisive" in the sending of U.S. troops to Somalia in 1992. *New York Observer*, December 8, 1994.

4   "'Ted would crap out'": Irwin Mazo interview, August 11, 1992.

6   "'my buccaneer'": Bill Tush interview, October 6, 1992.

7   "'to own everything'": Jerry Adler, "Jane and Ted's Excellent Adventure," *Esquire,* February 1991, p. 72.

7   "like the time some twenty years earlier": Jim Trahey interview, August 13, 1992.

Chapter 1

Page 9   "unusually warm": *Savannah News,* March 5 and 6, 1963.

9   "Over six feet tall": John McIntosh interview, March 27, 1993.

9   "a fifth of liquor a day": DF.

9   "scotch with a buttermilk chaser": RVTT, p. 159.

9   "a new man": Jane Dillard Greene interview, April 3, 1993.

9   "his breakfast": Ibid.

9   "waited on by Jimmy Brown": Irwin Mazo interview.

9   "nobody like Mr. T": RVTT, pp. 160–161.

10   "He loved the old hunting lodge": Jane Dillard Greene interview; Dr. Irving Victor interview, August 11, 1992.

10   "had remade Binden": Jane Dillard Greene interview.

10   "a dirt farmer in Sumner, Mississippi": RVTT, p. 151.

10   "at 9:40 A.M.": *Savannah News,* March 6, 1963.

11   "'has damaged himself'": Jane Dillard Greene interview.

11   "Sixty miles away": This account of events on the day of Ed Turner's suicide is gleaned from our interview with Dr. Irving Victor.

11   "'such a vibrant person'": Judy Nye Hallisy interview, April 7, 1993.

11   "a $4-million purchase": Irwin Mazo interview.

12   "'Having a million dollars'": Jane Dillard Greene interview.

12   "'the best deal'": Irwin Mazo interview.

12   "the $800,000 down payment": Ibid.

12   "a second call from Atlanta": Ibid.

12   "Mazo drove up": Ibid.

13   "'Looking back'": DF.

13   "Their arguments": CW, p. 21.

13   "what did the boy really know": Irwin Mazo interview; Jane Dillard Greene interview.

13   "a drinking problem": DF.

13   "In July 1961": Silver Hill staff member interview, September 24, 1992.

13   "his will": Dated July 5, 1961 (Savannah, Georgia) and witnessed by Thomas H. Adams.

13   "three months later": Silver Hill staff member interview.

13   "he was astonished": Irwin Mazo interview.

14   "'It's a long way'": RVTT, p. 173.

14   "'not only will we give him his money back'": Irwin Mazo interview.

14   "Turner left Silver Hill in January of 1963": Silver Hill staff member interview.

14   "all kinds of prescribed medicines": DF.

14   "with a check for $50,000 as a down payment": RVTT, p. 173.

14   "so shocked she started to cry": Jane Dillard Greene interview.

14   "at the DeSoto Hotel": Ibid.

15   "'I don't want to hurt anybody'": Ibid.

15   "Forty-eight hours later": The night before he killed himself, Ed called several of his top aides to reassure them that, despite recent events, their jobs were secure. Vera Rambo interview, February 25, 1995.

15   "Sipple's Mortuary": *Savannah News,* March 6, 1963.

15   "knew the Turner family well": Mrs. Ernest Risley interview, September 5, 1992.

15   "suggested to the widow": Dr. Irving Victor interview.

15   "'I never will understand'": Mrs. Carl Helfrich interview, November 16, 1992.

16   "Not far away": Charles Wessels interview, August 11, 1992.

16   "lost confidence in himself": Mrs. Irwin Mazo interview, August 11, 1992.

16   "'Nothing has been worthwhile'": RVTT, p. 174.

16   "seriously depressed": Dr. Irving Victor interview.

16   "he phoned her": RVTT, p. 173.

16   "'a nervous breakdown'": DF.

16   "'I begged him'": Gary Smith, "What Makes Ted Run?" *Sports Illustrated,* June 23, 1986, p. 88.

16   "tranquilizers and quaaludes": DF.

16   "'Nobody had any control over him'": Ibid.

16   "he loved his father": CW, p. 21.

17   "'same gun'": Curry Kirkpatrick, "Going Real Strawwng," *Sports Illustrated,* August 21, 1978, p. 78.

17   "'She was sweet'": William A. Henry III, "Shaking up the Networks," *Time,* August 9, 1982, p. 55.

17   "died in 1960, at the age of nineteen": Death certificate of Mary Jean Turner (Date of death: December 15, 1960), Cincinnati, Ohio.

17   "'The only time'": Billy Roe interview, August 12, 1992.

17   "'At the end'": Kirkpatrick, "Going Real Strawwng," p. 78.

17   "marked 'insufficient funds'": Lee McClurkin interview, May 24, 1993.

17   "'It was devastating'": DF.

Chapter 2

Page 18   "In 1938, the Carew Tower was the tallest building": *Cincinnati, a Guide to the Queen City and Its Neighbors* (Cincinnati: Wiesen-Hart, 1943) p. 177.

18   "Cincinnati the sixth largest city": Ibid., p. 69.

18   "Sumner, Mississippi": Jane Dillard Greene interview.

18   "easy-credit hardware store": Ibid.
18   "one bird for every shell": Ibid.
18   "Southwestern at Memphis": *Savannah News,* March 6, 1963.
19   "traffic counts": RVTT, p. 151.
19   "Queen City Chevrolet Company": Ibid., p. 150.
19   "By 1937": *Cincinnati, a Guide,* p. 133.
19   " 'good-time Charlie' ": Mrs. Carl Helfrich interview.
19   "convent school": RVTT, p. 150.
19   "Catholic population": Wesley A. Hotchkiss, *Areal Patterns of Religious Institutions in Cincinnati* (Chicago: 1950), p. 95.
19   "Henry Sicking": RVTT, p. 150.
19   " 'He really was witty' ": Ibid., p. 152.
20   " 'I told him' ": Ibid., p. 151.
20   "August 17, 1937": Marriage date cited on divorce decree No. 167915, Washoe County, Reno, Nevada (1957).
20   "they became Episcopalians": RVTT, p. 152.
20   "918 Dana Street": *Greater Cincinnati Telephone Directory, 1938–1939.*
20   "at 8:50 A.M.": Robert Edward Turner III birth certificate.
21   "By the end of 1939": *Greater Cincinnati Telephone Directory, 1940– 1941.*
21   "There is a photograph": *Time,* January 6, 1992, p. 35.
21   "The Clopay Corporation": *Cincinnati, a Guide,* pp. 140–141.
21   "born on September 18, 1941": Death certificate of Mary Jean Turner.
21   "he spread mud": RVTT, p. 152.
22   " 'to be so bad' ": Ibid.
22   "In 1944": Ibid.
22   "a commission in the Naval Reserves": *Savannah News,* March 6, 1963.
22   "booted out of the class": Judy Nye Hallisy interview.
22   "especially unhappy": RVTT, p. 153.
22   " 'I got beaten up' ": CW, p. 21.
22   "Central Outdoor Advertising": John McIntosh interview.
23   "and in 1947": *Savannah News,* March 6, 1963.
23   "He intended": RVTT, p. 158.

Chapter 3

Page 24   "topping all previous records": "Outdoor Advertising National Sales Volume up," *OAAA News,* January 1947, p. 1.
24   " 'on the threshold' ": Frank Dunigan, "Opening the Conference," *OAAA News,* October 1949, p. 1.
24   "fifty million cars": Howard Menhinick, "Outdoor Advertising and the Significance of Urban Developments," *OAAA News,* January 1960, p. 12.
25   "a man named Shuman": RVTT, p. 158.
25   " 'Outstanding Young Man of the Year' ": Shuman obituary, *OAAA News,* November 1962, p. 33.
25   " 'absolutely bitter rivals' ": John H. Baker interview, February 12, 1993.

25   "or the IRS": RVTT, pp. 158–159.

25   "The one story": Leiston Shuman Jr., interview, June 10, 1993.

26   "a reputation": Lewis Holland interview, August 13, 1992.

26   "a fifth-grade student at GMA": letter to authors from William R. Steverson, director of public affairs at McCallie, February 1, 1993.

26   " 'pretty alone' ": Ted Turner interviewed by Tom Snyder, *Tomorrow Show*, October 24, 1977.

26   " 'It was pretty rough.' ": CW, p. 21.

26   "Rumor quickly had it": Ibid., p. 22.

26   " 'I hid in my locker' ": Ted Turner interviewed by authors, September 18, 1993.

26   " 'I became a Confederate' ": Ibid.

27   " 'My father' ": DF.

27   " 'that old adage' ": Ibid.

27   "wire coat hangers to razor strops": RVTT, p. 152; DF.

27   " 'a lot of hostility' ": RVTT, p. 154.

27   " 'I'm beating you' ": DF.

27   " 'ninety percent of the arguments' ": RVTT, p. 152.

27   "In the summer": Dr. Jules Paderewski interview, August 11, 1992.

27   " 'I dreamed of dying' ": Ted Turner interview.

28   " 'Ay, tear her tattered ensign down!' ": Oliver Wendell Holmes was an early favorite of Turner; he often recited lines from "Old Ironsides" and "The Chambered Nautilus." Shirley Patterson (wife of McCallie instructor Houston Patterson) interview, November 11, 1992.

28   "named it after his daughter": James Hardee interview, February 18, 1993.

28   "the yacht club's board of stewards": John H. Baker interview.

28   *"The Black Cat"*: John McIntosh interview.

29   *"Pariah"*: Ibid.

29   "paid $300 for his son's craft": RVGG, p. 99.

29   "the majority of the time": Both Baker and Hardee emphasize that the course they devised stressed practical experience in sailboat racing; interviews.

29   " 'My first eight years of sailing' ": Smith, "What Makes Ted Run?" p. 78.

29   "not the easiest boat to handle": John H. Baker interview.

29   "regularly fish Ted": John McIntosh interview.

30   " 'a little boy' ": RVGG, p. 287.

30   " 'Turnover Teddy' and 'The Capsize Kid' ": CW, p. 24.

30   " 'I didn't have the ability' ": "Rebel with a Cause," *Broadcasting,* May 19, 1980, p. 40.

30   "something he knew he could do": John H. Baker interview.

30   " 'I just kept working and working and working' ": Smith, "What Makes Ted Run?" p. 78.

30   "One club member": Mrs. Carl Helfrich interview.

30   "some 'secret' way to beat the competition": Ibid.

31   "Though seventeen-year-old Jimmy Brown": RVTT, p. 160.

31   "He was dedicated to Ed": James Hardee interview.

32   "A hard, demanding father": Ed could be rough even with his friends' children. Irving Victor recalls: "Ed taught my son how to tie his shoelaces by shaking his fist at him."

32   "He personally handled": James Hardee interview.

32   "began placing bets": Charles Wessels interview.

33   " 'an eye for beautiful gals' ": Mrs. Carl Helfrich interview.

33   "enrolled in Charles Ellis": William R. Steverson letter.

33   "McCallie had high academic standards": Lewis Holland interview.

33   "Ed made an appointment": Spencer J. McCallie III interview, November 12, 1992.

34   "McCallie's only seventh-grade boarding student": *The Pennant* (McCallie Yearbook), 1951.

34   " 'There was nothing I could do' ": RVTT, p. 153.

Chapter 4

Page 35   "his face was as white as bleached bones": John Strang interview, November 12, 1992.

35   "forty new seventh-graders that year": *The Pennant,* 1951.

35   " 'I was certainly' ": Turner address at McCallie, January 14, 1993.

35   "He was not insensitive to the fact": Elliott Schmidt interview, November 11, 1992.

36   "getting into trouble": Ibid.

36   " 'Ted Turner's name' ": Houston Patterson interview, November 11, 1992.

36   " 'Terrible Ted' ": RVTT, p. 157.

36   " 'to have grass *outside*' ": Elliott Schmidt interview.

36   " 'Ted tried to take in dogs' ": RVTT, p. 155.

37   "12-to-14-pound Army rifles": John Strang interview.

37   "his little wooden rifle": Ibid.

37   " 'I had to buy him new shoes' ": RVTT, p. 153.

37   "Jimmy Brown would be sent": Elliott Schmidt interview.

37   " 'Ted, you *are* going' ": Charles Wessels interview.

37   " 'as honest as he could be' ": John Strang interview.

38   " 'I thought the education' ": Turner address at McCallie, January 14, 1993.

38   "he has expressed reservations": DF.

38   "two such different graduates": A list of "Notable McCallie Alumni." In addition to these two, the list includes Howard H. Baker Jr. (former U.S. Senate majority leader and White House chief of staff under President Reagan), Carroll Campbell Jr. (governor of South Carolina), and Ralph McGill (late editor and Pulitzer Prize columnist for the *Atlanta Constitution*).

38   "Founded in 1905": historical marker, The McCallie School; and "School Description and Profile," The McCallie School, 1993.

38   "Of the four courses": John Strang interview.

38   "John Strang likes boys": Ibid.

38   " 'He was frail-looking' ": Ibid.

39   "by the boys as 'Yo' ": Jon Meacham, "We Call Him 'Yo,' " *McCallie: In Our Words,* The McCallie School, n.d.

39   " 'I don't think' ": John Strang interview.

39   " 'a very religious person' ": DF.

40   "evangelical preachers who arrived at chapel": Ibid.

40   "Ted stepped forward": John Strang interview.

40   " 'a pretty clear-cut choice' ": DF.

40   " 'Just average' ": Houston Patterson interview.

40   " 'I was content' ": CW, p. 26.

40   "interested but mediocre": Turner address at McCallie, January 14, 1993.

40   " 'termite' football and boxing": *The Pennant,* 1956.

40   " 'I tried and I tried' ": CW, p. 23.

40   "Ed Turner arrived": John Strang interview.

41   " 'I never considered' ": RVTT, p. 153.

41   "went to work": CW, p. 28.

41   "$25 a week": Ibid.

41   "If he could find a better deal": Ibid.

41   "Ted loved him": DF.

41   " 'sympathy for me' ": Frederick Wessels interview, August 11, 1992.

42   " 'holy hell-raiser' ": Charles Wessels interview.

42   "He'd sit in the back row": Ibid.

42   "he climbed out the window": John Strang interview.

42   " 'Breakfast was' ": Ibid.

42   "which was rather unusual": Houston Patterson interview.

43   " 'like my second home' ": Turner address at McCallie, January 14, 1993.

43   "a terrible cadet": CW, p. 25.

43   " 'I turned it around' ": RVTT, p. 156.

43   "in the student yearbook": *The Pennant,* 1954. (The issue is dedicated to Craig Houston Patterson).

43   " 'You're the skipper' ": Houston Patterson interview.

44   "grown to nearly fifty": *Tornado,* September 30, 1955.

44   "After extensive testing": Dr. Fenwick Nichols interview, August 11, 1992.

45   "encephalitis": RVTT, p. 157.

45   "Mary Jean's screams": John McIntosh interview.

45   " 'God, let me die' ": Henry, "Shaking up the Networks," p. 55.

45   " 'her brain destroyed' ": Kirkpatrick, "Going Real Strawwng," p. 78. [It's possible that Turner is here referring to the movie *Dark Victory,* in which Bette Davis is slowly dying of a brain tumor.]

45   "Ed couldn't bear it": Irwin Mazo interview.

46   "what he had to do": Charles Wessels interview.

46   "Schmidt made Ted study": Ibid.

46   "reading about the sea": CW, p. 26.

46   "rank of corporal": *The Pennant,* 1956.

46   "a political liberal": Elliott Schmidt interview.

46   "newest additions in 1955": *The Pennant,* 1956.

46   "tournaments in Nashville": Ibid., 1955.

46   "promoted to sergeant": Ibid., 1956.

46   "the Neatest Cadet": Ibid.

47   "a young Alexander": CW, pp. 26–27.

47   " 'That stuff' ": Ibid., p. 27.
47   "contemplated suicide": RVTT, p. 34.
48   "*Endymion*": Shirley Patterson interview; Mrs. Carl Helfrich interview.
48   " 'If that's the type of God' ": Smith, "What Makes Ted Run?" p. 86.
48   " 'I especially urge' ": *Tornado,* September 30, 1955.
49   "to become a missionary": DF.
49   "He was named Captain": *The Pennant,* 1956.
49   "The debating team had begun the year": The prospects for the next year are adumbrated in *The Pennant,* 1955.
49   "The debate question for that year": Ibid., 1956.
49   " 'I want a case' ": Elliott Schmidt interview.
49   " 'I beat a girl in the finals' ": CW, p. 25.
49   "Cook and Turner": *The Pennant,* 1956.
50   "Miss Nancy Drake": *Tornado,* October 21 and November 18, 1955; March 23, 1956; May 4, 1956; June 4, 1956.
50   "the United States Naval Academy": CW, p. 29.
50   " 'as long as he was interested' ": James Hardee interview.
50   "finally rejected the idea": CW, p. 29.
50   "His IQ was 128": *Providence Journal-Bulletin,* May 13, 1986; John Kennedy's IQ is to be found in *JFK: Reckless Youth,* by Nigel Hamilton (New York: Random House, 1992), p. 88.
51   "Harvard turned Ted down": CW, p. 29.
51   "the Holton Harris Oratorical Medal": *The Pennant,* 1956.
51   "a new Lightning-class sailboat": CW, p. 27.
51   "half the price": Ibid.
51   "his old Lightning": Houston Patterson interview.
51   " 'When I left there' ": RVTT, p. 156.

Chapter 5

Page 52   " 'shoot those guys down' ": DF.
52   " 'knight in shining armor' ": Ibid.
52   "*Greased Lightning*": John McIntosh interview.
53   "Frederick Wessels and Bunky Helfrich": Frederick Wessels interview.
53   "a governor on the engine": Barbara McIntosh interview, April 13, 1993.
53   "how to disable the governor": John McIntosh interview.
53   "the hills of West Virginia": Frederick Wessels interview.
53   "shared a room for $15": Ibid.
53   " 'Ted was very cheap' ": Ibid.
54   " 'Oh, my daddy's gonna get mad at me' ": Ibid.
54   "other forty-four boats": There were forty-five in all—International Lightning Class Association (Worthington, Ohio).
54   "jumped the gun": John McIntosh interview.
54   " 'Frederick,' he called elated": Frederick Wessels interview.
54   "twenty-third place": International Lightning Class Association.
55   " 'quite a country boy!' ": Robert Crane interview, March 29, 1993.

55   "he had almost drowned": "Rebel with a Cause," *Broadcasting,*
      p. 42.

55   " 'a lot of bull at Brown' ": CW, p. 33.

55   "one ugly-looking building": Ted's old room on the third floor of
      Maxcy is now the office of a sociology professor. "I understand
      the students used to hate this building," she said when inter-
      viewed. She knew nothing about Ted Turner once having
      roomed there.

55   "310 Maxcy": Ted's room as listed in the 1956 *Student Directory*.

56   " 'I'm Ted Turner' ": Carl Wattenberg interview, October 7, 1992.

56   "highway road signs": Douglas Woodring and Carl Wattenberg in-
      terviews.

56   "to have a good time": Douglas Woodring interview, April 6, 1993.

56   "he'd give him $5,000": CW, p. 29; RVTT, p. 166.

57   "for heavy drinking": Carl Wattenberg interview.

57   "five-card stud with him for nickels and dimes": Donald Anderson
      interview, April 8, 1993.

57   "To another freshman": Edward Perlberg interview, October 9,
      1992.

57   "the squeaky-clean way he groomed himself": Douglas Woodring
      interview.

58   "likely to hear about it": Carl Wattenberg interview.

58   "he could be very funny": Ibid.

58   "liked popular history": Donald Anderson interview.

58   "conservative opinions": Ibid.

58   " 'defending the South' ": Ibid.

58   "attacking H. L. Mencken": Ibid.

58   " 'a raw plutocracy' ": H. L. Mencken, "The Calamity of Appo-
      mattox," in *Mencken,* ed. Alistair Cooke (New York: Vintage,
      1955), p. 199.

58   " 'not five percent' ": Ibid.

58   "the efficiency of war": Priscilla Painton, "The Taming of Ted
      Turner," *Time,* January 6, 1992, p. 39.

59   " 'Hitler,' he told a classmate": RVTT, p. 147.

59   "walked out on him": Douglas Woodring interview.

59   "firing it out his dorm window": Donald Anderson interview.

59   " 'a complete wild man' ": Edward Perlberg interview.

60   "an impressive structure": Charles Shumway interview, October
      16, 1992.

60   "the breeze was shifting": Edward Perlberg interview.

60   "famous for its parties": *Liber Brunensis* (Brown University Year-
      book), 1960.

61   "number-one freshman sailor in New England": *Brown Daily Herald,*
      October 17, 1957.

61   "*very* conscious of the upheaval": Houston Patterson interview.

61   "one of his new friends": Malcolm Whittemore interview, October
      13, 1992.

61   " 'I was happier at military school' ": CW, p. 33.

62   "the 22nd day of August, 1957": Divorce decree.

62   "Ed called Florence": RVTT, p. 166.

62   "$15,000 per year": Property Settlement between Robert Edward
        Turner and Florence Marie Rooney Turner, August 16, 1957,
        p. 7.

62   "custody of the children": Ibid., pp. 5–6.

63   " 'In the happy event' ": "Trust Agreement," Ibid., Exhibit A,
        p. 4.

63   "the word 'thirty' ": Ibid., p. 5.

63   "a growing drinking problem": Of his college days, Turner says,
        "I learned mainly about drinking and sex . . ." RVGG, p. 29.

64   "and throw one of his informal shindigs": John McIntosh interview.

64   "to be redecorated": *Liber Brunensis,* 1958.

64   "number 138": 1958 *Student Directory.*

64   "to drink a few beers and shoot pool": Dr. Lawrence Brenner in-
        terview, October 2, 1992.

64   "visited a tattoo parlor": Carolyn Steadman Whittemore interview,
        October 13, 1992.

64   "as a gambler": Malcolm Whittemore interview.

64   "chugalug an entire fifth": Ibid.

65   " 'some kind of trick' ": Ibid.

65   " 'never ever met a Jew before' ": Dr. Lawrence Brenner interview.

65   "Some put him down as a cracker": Roger Vaughan, "Ted Turner:
        'I Didn't Fail College; College Failed Me,' " *Brown Alumni
        Monthly,* September 1975, p. 11.

65   " 'Ted was also a bigot' ": Kirkpatrick, "Going Real Strawwng,"
        p. 79.

65   "the fraternity had to have a permit": Malcolm Whittemore in-
        terview.

66   "a girlfriend from Pembroke": Dr. Lawrence Brenner interview.

66   " 'She was a nice girl' ": Ibid.

66   "it was suspected": Ibid.

66   "a pledge brother named Mal": Malcolm Whittemore interview.

66   " 'The suit would be rumpled' ": RVGG, p. 22.

67   "Brown placed seventh": *Brown Daily Herald,* October 17, 1957.

67   "extraordinarily cold": Malcolm Whittemore interview.

67   "such sailing conditions are not unknown in Chicago": Secretary
        interview, Chicago Yacht Club, February 12, 1993.

67   "as low as twelfth": *Brown Daily Herald,* December 3, 1957.

67   "going full tilt": Malcolm Whittemore interview. Though Shumway
        doesn't recall that much ice, Whittemore insists on his version
        of events, and he was in the boat.

67   "toes were freezing": Ibid. Carolyn Steadman Whittemore also men-
        tions this. On the subject of Ted's daring, she says, "And Mal
        said another time Ted sank a boat to win."

68   " 'Well, that did it' ": Charles Shumway interview.

68   "a junior sailing instructor": Bob Bavier interview, March 29, 1993.

68   " 'A nice fellow' ": Edward Perlberg interview.

69   " 'to get drunk and chase women' ": CW, p. 32.

69   " 'no lights-out at ten o'clock' ": Ibid., p. 33.

69   "knocked out his front teeth": Donald Anderson interview.

69   "signed in before 11:00 P.M.": Carolyn Steadman Whittemore interview.

69   "Windows were smashed": Donald Anderson interview.

69   "drove out of Providence together": Ibid.

Chapter 6

Page 70   "Ed turned up Bill's daughter, Jane": Jane Dillard Greene interview.

70   "how he had learned to fly": Ibid.

71   "Ed had once recommended": See Ed's letter to his son on p. 75.

71   "not generally called up by draft boards": *Brown Daily Herald,* October 4, 1957.

71   "a small ceremony": Jane Dillard Greene interview.

71   "to get back together": Ibid.

71   "now had a social status": Irwin Mazo interview.

72   "On February 12, 1958": Ted Turner's Coast Guard Military Record.

72   "scored respectably": Ibid.

72   "at Cape May, New Jersey": Ibid.

72   "the rank of fireman's apprentice": Ibid.

72   "a member of the class of '61": 1958 *Student Directory.*

72   "a new room on the second floor": Turner is listed as occupying room 238, Kappa Sigma, in the 1958 *Student Directory.*

72   "go regularly to Lincoln Downs": Dr. Lawrence Brenner interview.

72   "without paying the bill": Ibid.

73   "One woman friend": Carolyn Steadman Whittemore interview.

73   "'Ed bragging one night'": RVTT, pp. 167–168.

73   "pass most of them": Douglas Woodring interview.

73   "Greek tragedy": Charles Shumway interview.

73   "'His courses'": Vaughan, "Ted Turner: 'I Didn't Fail,'" p. 15.

73   "interest in athletics": Douglas Woodring interview.

73   "'one of the easier departments'": Ibid.

74   "'colorful' young man": RVTT, p. 163.

74   "'great, fiery convictions'": Ibid.

74   "'to catch hell'": Carolyn Steadman Whittemore interview.

74   "'My dear son'": *Brown Daily Herald,* April 15, 1959.

76   "'Ed was simply furious'": RVTT, p. 166.

77   "'One homecoming weekend'": Vaughan, "Ted Turner: 'I Didn't Fail,'" p. 16.

77   "instant dismissal": *Brown Daily Herald,* November 17, 1959.

77   "switched to economics": Vaughan, "Ted Turner: 'I Didn't Fail,'" p. 15.

77   "'commie professors'": Ibid.

77   "the Schell trophy": *Brown Daily Herald,* December 16, 1958.

77   "watch *Columbia* defend the America's Cup": Carolyn Steadman Whittemore interview.

77   "The Cup races had been discontinued": Librarian interview, New York Yacht Club, March 21, 1994.

78   "Boston on May 31": Ted Turner's Coast Guard Military Record.

78   "to Charleston": Jane Dillard Greene interview.
78   "she liked keeping house": Judy Nye Hallisy interview.
78   "He loved the outdoors": Jane Dillard Greene interview.
78   "hired a manager": Ibid.
79   "outfitted the back of it": Ibid.
79   "go for long walks together": Ibid.
79   "the same term that John Workman": RVTT, p. 163.
80   "'Commodore and Team Captain-elect'": *Liber Brunensis,* 1959.
80   "third floor of Goddard House": Turner is listed as occupying room 332, Goddard House, in the 1960 *Student Directory.*
80   "still working hard at": Peter Dames says, "So Turner and I had to look around for something to excel at, and we settled on drinking and lechering. We really worked hard at it, too, and we excelled." CW, p. 33.
80   "a few women skippers": Judy Nye Hallisy interview.
80   "Judy found him funny, attractive, and cute": Ibid.
80   "Harry Nye, the famous sail maker": John McIntosh interview.
80   "at loose ends": Judy Nye Hallisy interview.
80   "he could change": Ibid.
81   "going into his father's business": Ibid.
81   "walking a tightrope": Ibid.
81   "another woman in his room": RVTT, p. 167.
81   "'a rebel ahead of my time'": *East Bay Window* (Phoenix, AZ), September 14, 1977.
81   "News of what happened": Malcolm Whittemore interview.
81   "a car for $125": RVTT, p. 168.
81   "He had a plan": Ibid., p. 167; and Peter Dames interview, May 12, 1994.

Chapter 7

Page 83   "that smashed into": Peter Dames interview.
84   "cut off his son's allowance": Judy Nye Hallisy interview.
84   "above a bar": RVTT, p. 168.
84   "sheets of the Miami phone book": Ibid.; CW, p. 34.
84   "'We were miserable'": CW, p. 34.
84   "On February 12": Ted Turner's Coast Guard Military Record.
85   "commissioned in 1927": Dr. Robert Browning (U.S. Coast Guard historian) interview, February 8, 1993.
85   "'dream cruise'": CW, pp. 34–36.
85   "never left its home berth": USCGC *Travis* log.
85   "His marriage proposal": Judy Nye Hallisy interview.
86   "Judy had her engagement ring": Ibid.
86   "five hundred people": Ibid.
86   "thinking how alike they were": Jane Dillard Greene interview.
87   "the Reverend Robert May": Turner-Nye marriage certificate.
87   "the traditional kiss": Judy Nye Hallisy interview.
87   "a small one-bedroom apartment": Ibid.
87   "going to have to earn it": Jane Dillard Greene interview.
87   "she ironed on a card table": Judy Nye Hallisy interview.

87   "basement apartment": Ibid.
88   " 'the best dog ever' ": Ibid.
88   "named Homer": Ibid.
88   " 'fifteen hours a day' ": RVTT, p. 169.
88   "the Red Cross": Ibid., p. 168.
89   "in some ways spoiled": Judy Nye Hallisy interview.
89   "Ted was cheating": Ibid.
90   "On December 15, 1960": Mary Jean Turner's death certificate.
90   "July of 1961": Judy Nye Hallisy interview.
90   "Ted was too busy": Ibid.
91   "screaming bouts": Ibid.
91   "dump ice water": Mrs. Irwin Mazo interview.
92   "laid him out": Mike Gearon interview, August 14, 1992.
92   "hated every minute": Judy Nye Hallisy interview.
92   "she was pregnant again": Ibid.
92   "from mid-1961 on": Jane Dillard Greene interview.
92   "one of the most charming men": Judy Nye Hallisy interview.
93   "his long-running feud": Irwin Mazo interview.
93   "one of the biggest": "Personnel and Plants," *OAAA News,* November 1962, pp. 30–31.
93   " 'He had me out' ": CW, p. 38.
93   "a first-class apartment": RVTT, p. 172.
93   "He took a small apartment": Judy Nye Hallisy interview.
93   "a stylish modern house": Ibid.
94   "he bought finches": Ibid.
95   "He called him a quitter": Jane Dillard Greene interview.

Chapter 8

Page 97   " 'I'm not going through' ": Irwin Mazo interview.
97   " 'it wouldn't have been difficult' ": Ibid.
97   " 'My father didn't leave me all this money' ": Ibid.
98   "Though informally handwritten": Ibid.
99   "Turner and McIntosh met": John McIntosh interview.
99   "paced about the hotel room": Ibid.
99   " 'Dad just got anxious' ": Ibid.
99   " 'take these people to court' ": Ibid.
100   " 'not in his right mind' ": Ibid.
100   "As Turner tells the story": CW, p. 42.
101   "would later suggest": Ibid.
101   "flew out to Palm Springs": Irwin Mazo interview.
101   " 'the biggest facelift' ": "The Biggest Facelift," *OAAA News,* July 1963, p. 6.
101   "recently paid $10 million": Ibid.
101   " 'You know, Ted' ": Irwin Mazo interview.
101   "approximately $200,000": CW, p. 43.
101   "but also self-interested": Irwin Mazo interview.
102   " 'I wish you'd asked me' ": CW, p. 43.
102   " 'to personally loan Ted thirty thousand dollars' ": Ibid.
102   "He sold the plantations": *Current Biography,* 1979, p. 410.

102   "'a hard-nosed businessman'": RVTT, p. 175.
102   "'stall, stall, stall'": Irwin Mazo interview.
102   "A gold Cadillac": Ibid.
102   "He loved this Caddy": Ibid.
103   "'a different ball game'": Ibid.
103   "According to psychoanalysts": Robert Goldberg, "Under the In-
           fluence," *Savvy,* July 1986, p. 51.
104   "no time for grief": Judy Nye Hallisy interview.
104   "'I want to thank you'": "General News," *OAAA News,* May 1963,
           p. 5.
105   "'The Turner booth'": "Sales Promotion," *OAAA News,* January
           1964, p. 31.
105   "'I'd like to point out'": R. E. Turner III, "Leasing: A Key Factor
           in Operations," *OAAA News,* January 1964, p. 25.
106   "There is a photograph": Ibid., p. 24.
106   "'In this soaring Space Age'": Harry O'Mealia Jr., "Inside Out-
           door," *OAAA News,* September 1962, p. 3.
106   "'Think BIG'": "The Story of Supersaturation," *OAAA News,*
           April 1959, p. 8.
106   "a giant billboard sign": "Local Campaigns," *OAAA News,* De-
           cember 1964, p. 12.
106   "featured on the cover": "Turner Sales Contest Produces
           $147,814.30 in New Local Business," *OAAA News,* July 1964,
           p. 4.
107   "'Ted Turner of Savannah Georgia'": *Savannah Morning News,* Au-
           gust 19, 1963.
108   "Judy will never forget": Judy Nye Hallisy interview.
108   "'I wish my daddy'": Ibid.
109   "strike a pose": Ibid.
109   "He was wonderful": RVTT, p. 160.
110   "'we weren't really compatible'": CW, p. 49.
110   "'Atlanta was a young town'": Nancy Roe interview, August 12,
           1992.
110   "the one-million mark": Menhinick, "Outdoor Advertising," p. 12.
111   "Many who would later run for office": Billy Roe interview.
111   "'fundamentally conservative'": Mike Gearon interview.
112   "'I'm lookin' for a cute gal'": Nancy Roe interview.
112   "Jane was pregnant": Irwin Mazo interview; Turner family friend
           interview, August 14, 1992.
112   "'the same day'": *Atlanta Constitution,* January 13, 1976.
112   "married Judy on June 22": Cook County marriage certificate.
113   "'I didn't really want to get married'": Irwin Mazo interview.
113   "a comfortable home": Nancy Roe interview.
113   "'Ted named them'": *Atlanta Constitution,* January 13, 1976.
113   "physically abusive": Judy Nye Hallisy interview.
113   "'This looks *real* bad'": Ibid.
114   "Judy went into the bathroom": Ibid.
114   "she impulsively decided to fly to Atlanta": Ibid.
114   "spending $100,000 a year": RVTT, p. 184.
115   "'I'm gonna have my own boat'": John H. Baker interview.

115   "it was 'hairy' ": RVGG, p. 104.
115   " 'Give me a million bucks' ": *Providence Journal-Bulletin,* May 13, 1986.
115   "by the biggest margin ever": RVGG, p. 105.
115   " 'You can't win' ": Ibid., p. 106.
115   " 'Some races are so beautiful' ": Kirkpatrick, "Going Real Strawwng," p. 80.
116   " 'Stick with me' ": Billy Roe interview.
116   "On July 1, 1964": "Personnel and Plants," *OAAA News,* August 1964, p. 20.
116   "His company billings": Ibid.
117   " 'When the original idea' ": "Turner's Two Million Dollar Sales Contest," *OAAA News,* September 1965, p. 15.
117   " 'you borrow all you can borrow' ": John H. Baker interview.
117   " 'the secret' ": Ibid.
117   " 'like a Ponzi scheme' ": Billy Roe interview.
118   " 'a dark side' ": Nancy Roe interview.
118   "his fascination with Howard Hughes": Ibid.
119   " 'Keep pressing me' ": Ibid.

Chapter 9

Page 120   " 'You would have thought' ": Jim Roddey interview, May 24, 1993.
121   "Ted had grown bored": Ibid.
121   "Ted's father used to tell him": Gerry Hogan interview.
122   "not a very good businessman": Jim Roddey interview.
122   "losing $100,000 a year": CW, p. 64.
122   "an entire roomful of computers": Jim Roddey interview.
122   "He and Roddey and Dick McGinnis": CW, p. 54.
122   " 'We wound up' ": Ibid.
122   " 'for Go-Go radio' ": Jim Roddey interview.
122   " 'Guess what' ": Charles Shumway interview.
122   "At one of their meetings with TIAA": Irwin Mazo interview.
123   " 'When Ted came to see me' ": Lee McClurkin interview, May 24, 1993.
123   " 'I had never watched the station' ": CW, p. 65.
124   " 'I told him' ": Don Lachowski interview, July 30, 1993.
124   "had lost $900,000": Irwin Mazo interview.
124   "about $2.5 million in stock": Henry, "Shaking up the Networks," p. 51.
124   " 'Watch This Channel Go' ": "Rebel with a Cause," *Broadcasting,* p. 37; PB, p. 76.
125   " 'When I bought Channel 17' ": HW, p. 13.
125   "a huge, 1093-foot": RVTT, p. 225.
125   "tallest freestanding tower in the country": Ibid.
125   "One night on a dare": Jim Roddey interview. R. T. Williams says that he himself climbed the tower and it took him nine hours to climb down: CW, p. 89.
125   "the station had no de-icer": Jim Trahey interview.
125   " 'We'd put on the helmets' ": Ibid.

125   "'trade deals'": Gerry Hogan interview, July 22, 1992.

126   "an underground station": Tom Bradshaw, "How an Indie 'U' Made
          It Big with 'Good Old Days' Programming," *Television/Radio
          Age,* June 24, 1974, p. 26.

126   "'Mainly hippies'": Ibid.

126   "'We had absolutely no equipment'": HW, p. 15.

126   "'playing banjos and smoking pot'": Ibid.

126   "'I just love it'": Ibid., p. 14.

127   "'the whole damn thing was a challenge'": Bradshaw, "How an
          Indie 'U'," p. 26.

127   "a company called U.S. Communications": Ibid.

127   "'One of the first things I learned'": Ibid.

127   "'Never look back'": Gerry Hogan interview.

127   "his 'five Ps'": Bradshaw, "How an Indie 'U'," p. 26.

127   "'a lot of time with the film people'": Ibid.

127   "a number of disaffected salesmen": Jim Trahey interview.

127   "'You gotta come'": Ibid.

128   "Sanders turned him down": William Sanders interview, December
          15, 1992.

128   "a good handle on things": Ibid.

128   "about to go out of business": Lee McClurkin interview.

129   "'Irwin and the others'": CW, p. 67.

129   "'This company ain't going to make it'": Irwin Mazo interview.

129   "had a habit of mixing": Jim Roddey interview.

129   "'sunk our ass'": Roger Vaughan, "Ted Turner's True Talent,"
          Esquire, October 10, 1978, p. 35.

129   "moving much too fast": Lee McClurkin interview; Jim Roddey
          interview.

129   "a normal part of his conversation": Jim Roddey interview.

130   "heated arguments": Ibid.

130   "'We're both competitive guys'": CW, p. 53.

130   "'I didn't just eliminate'": *Providence Journal-Bulletin,* May 13, 1986.

130   "Roddey remembers": Jim Roddey interview.

130   "he didn't think it was very funny": Ibid.

130   "'now worth five hundred million dollars'": Ibid.

131   "Its highest rating": HW, p. 16.

131   "even he had his doubts": Bradshaw, "How an Indie 'U'," p. 26.

131   "'And they would say'": Vaughan, "True Talent," p. 36; HW,
          p. 15.

132   "a 'Begathon'": William Sanders interview.

132   "In March of 1971": Bradshaw, "How an Indie 'U'," p. 50.

132   "'Turner was making no headway'": Vaughan, "True Talent,"
          p. 35.

132   "'It was really embarrassing'": CW, p. 69.

132   "'Well, Gerry'": Gerry Hogan interview.

133   "'I felt the people of Atlanta'": Bradshaw, "How an Indie 'U',"
          p. 26.

133   "'we run the FCC minimum'": Ibid., p. 51.

133   "'an "escapist" station'": Ibid.

134   "rights to professional wrestling": Vaughan, "True Talent," p. 36.

134	"'It's a big hit'": RVGG, p. 119.
134	"more programming": Bradshaw, "How an Indie 'U'," p. 26.
134	"'THE NBC NETWORK'": Ibid., p. 51.
134	"'a whole bunch of phone calls'": Ibid.
134	"$1.3 million": Ibid.
135	"the three-year package": Ibid.
135	"he got all his salespeople together": Jim Trahey interview.
135	"In Alabama in 1972": Andy Goldman interview, September 22, 1992.
135	"'we'd like to get your signal'": Ibid.
135	"a man named Frank Spain": Ibid.
135	"'a product'": Ibid.
136	"'There's a helluva business here'": Ibid.
136	"thirty-three markets": Jim Trahey interview.
137	"'We were in it'": Vaughan, "True Talent," p. 36.
137	"the number-one UHF": Bradshaw, "How an Indie 'U'," p. 24.
137	"first comes sailing": RVTT, p. 128.
137	"eighteen- and nineteen-hour days": Andy Goldman interview.
137	"'I programmed the whole station'": CW, p. 88.
137	"'He's almost never here'": *Atlanta Constitution*, January 13, 1976.
137	"'I cried a lot'": RVTT, p. 186.
137	"'It's great'": RVGG, p. 120.
138	"Her social life was limited": *Atlanta Constitution*, January 13, 1976.
138	"'gettin' to be a good cook'": Jim Roddey interview.
138	"everything be shipshape": RVTT, p. 191.
138	"Janie's marital fantasy": Turner family friend interview, March 22, 1994.
139	"Janie was drinking": Billy Roe interview.
139	"'I was miserable'": CW, p. 50.
139	"'kind of a joke'": Ibid., p. 87.
139	"rolled on the floor": Billy Roe interview.
139	"'I enjoyed it'": RVTT, p. 56.
140	"'We did everything bad'": CW, p. 61.
140	"considered a shoo-in": Charles Shumway interview.
140	"'he presented a commanding image'": RVGG, p. 31.
140	"having qualified": Charles Shumway interview.
141	"'Ted said he had the money'": RVGG, p. 20.
141	"When Ted finally called": Ibid.
141	"thought it was nuts": CW, p. 63.
141	"an offer of $70,00: RVGG, p. 108.
141	"on December 15, 1968": Ibid.
141	"the *Eagle*'s mast": Ibid.
141	"Within three days": RVTT, p. 185.
141	"World Ocean Racing Championship": RVGG, p. 109.
142	"'Sailing is like screwing'": PRR, p. 90.
142	"'I was still very insecure'": DF.
142	"The big thrill for Turner": PRR, p. 83.
142	"began on a January afternoon": Red Marston, "*Eagle* Wins S.O.R.C. Opener," *Yachting*, March 1970, p. 243.
142	"'Usually just in the back'": PRR, p. 78.

142   " 'sailed erect as a New England church spire' ": Marston, *Eagle Wins,*" p. 243.

142   "flew sixteen hours": RVTT, p. 185.

143   "spending about $20,000 a year on plane tickets": *Providence Journal-Bulletin,* May 13, 1986.

143   " 'the brightest star' ": Bob Bavier, "From the Cockpit," *Yachting,* May 1970, p. 52.

143   " 'giving aid and comfort to the enemy' ": *Providence Journal-Bulletin,* May 13, 1986.

143   "about three thousand members": Peter M. Homberg (general manager of the New York Yacht Club) interview, September 10, 1993.

144   "a member of the Capital City Club": PRR, p. 87.

144   "in order to become a member of the New York Yacht Club": Peter M. Homberg interview.

144   "he was blackballed": Bob Bavier interview.

144   " 'he's a jerk' ": *Daily News,* August 19, 1980.

144   "sailed regularly with him": RVGG, p. 39.

145   "Hinman checked": Ibid., p. 21.

145   " 'he said he would' ": Ibid.

145   " 'you want to know the faults' ": Bob Bavier interview.

145   " 'I remember' ": Kirkpatrick, "Going Real Strawwng," p. 80.

Chapter 10

Page 146   " 'the sailing capital of the world' ": *New England,* Michelin Guide Series, 4th edition, p. 181.

146   "nineteenth-century mansions": Ibid., pp. 183–186.

147   "a falling-out": RVGG, p. 20.

147   " 'a supergroup' ": Ibid., pp. 3ff.

147   "a complex formula": "Defending the America's Cup," *Time,* September 19, 1977, p. 84.

147   " 'He loves to battle' ": "Staging a Battle Royal on the Briny," *Sports Illustrated,* July 4, 1977, p. 20.

148   "losing most of the starts": Bob Bavier, *The America's Cup* (New York: Dodd, Mead, 1986), p. 34.

148   " 'An equivalent score' ": RVGG, p. 189.

148   " 'to rearrange his face for him' ": Ibid., p. 200.

148   "differences in temperament and style": Ibid., p. 12.

148   "Chance didn't think much of his record": Ibid., p. 21.

148   "little or no Cup experience": Rich du Moulin and Conn Finley were the only two crew members who had previous Cup experience.

148   "Chance wanted him replaced": RVGG, p. 165.

148   " 'Turner isn't steering the boat right' ": Ibid., p. 190.

149   "no square-tailed fish": *Providence Journal-Bulletin,* May 13, 1986.

149   " 'Damn it, Brit' ": Ibid.

149   "a nervous wreck": RVGG, p. 256.

149   "he fouled his rival": Ibid., p. 263.

149   " 'This has been worse' ": Ibid.

149   " 'Sure it hurt' ": CW, p. 149.

150 " 'Tomorrow after we beat *Mariner*' ": RVGG, p. 270.

150 " 'To Dennis' ": Ibid., p. 274.

150 "Commodore Morgan, a short man": Ibid., pp. 282ff.

150 " 'We're gonna drink until we pass out' ": Ibid., p. 286.

150 " 'You're not sailing' ": Ted Turner and Gary Jobson, *The Racing Edge* (New York: Simon & Schuster, 1979), p. 28.

151 " 'the single greatest design disaster' ": *Providence Journal-Bulletin,* May 13, 1986.

151 " 'chains of gold' ": RVTT, p. 8.

151 "a Confederate challenger": Henry, "Shaking up the Networks," p. 57.

151 "she had been the cook": RVGG, p. 76.

151 " 'But my ass is better' ": RVTT, p. 123.

151 " 'She ain't just pretty' ": Taylor Branch and Eugene M. Propper, *Labyrinth* (New York: Viking, 1982), p. 390.

151 "but he did poorly": RVGG, p. 289.

152 " 'broken up and a little crocked' ": Ibid., p. 298.

152 " 'I'm like the grass' ": Ibid., p. 154.

152 " 'I'm a man without a country' ": Ibid., p. 298.

152 "In the background": Ibid., p. 296.

152 "living in Conley Hall": Ibid., p. 173.

153 " 'I'm so busy' ": Ibid., p. 33.

153 "to show his kids": Ibid., p. 200.

153 "Father of the Year": Ironically, Ted was once named Father of the Year, according to Christian Williams, p. 8.

153 "a necessary evil": Painton, "The Taming," p. 37.

153 "harder for them": Smith, "What Makes Ted Run?" p. 84.

153 "watching a television program": Mike Gearon interview.

153 " 'If he caught you crying' ": Smith, "What Makes Ted Run?" p. 84.

153 " 'yelled and screamed' ": Painton, "The Taming," p. 37.

154 " 'I had a little problem with him' ": Lewis Holland interview.

154 " 'coaching a Little League team' ": *Savannah Morning News,* August 5, 1984.

154 "feeling badly neglected": RVTT, p. 83.

154 "having problems": Kirkpatrick, "Going Real Strawwng," p. 75.

154 " 'He never had to worry' ": *Atlanta Constitution,* January 13, 1976.

155 " 'Sometimes it's boring' ": Ibid.

155 " 'I've seen them' ": Harry F. Waters et al., "Ted Turner Tackles TV News," *Newsweek,* June 16, 1980, p. 64.

155 " 'Janie can't play too well' ": *Atlanta Constitution,* January 13, 1976.

155 "liked to play topless": RVTT, p. 123.

156 " 'This is the year' ": RVGG, p. 33.

156 "Channel 17 profits had dipped": Vaughan, "True Talent," p. 36.

156 "One of the subjects": Vaughan, "Ted Turner: 'I Didn't Fail,' " p. 13.

157 " 'the Bobby Orr of his profession' ": Program, Fourth Annual Athletic Hall of Fame Induction Dinner, November 1, 1974.

157   "'You can always change skippers'": "The Comeback of the
        'Courageous,'" *Brown Alumni Monthly,* November 1974, p. 34.
157   "already approached McCullough": RVTT, p. 12.

Chapter 11

Page 158   "having his boss away from Atlanta": William Sanders interview.
158   "'but we all loved it'": PB, p. 82.
159   "He was expecting Sid Topol": Sidney Topol interview, December
        5, 1993.
159   "had first become interested": Reese Schonfeld interview, July 6,
        1992.
159   "Schonfeld was faced with a problem": Ibid.
159   "didn't make much of an impression": Ibid.
159   "'We'd never do news'": Ibid.
159   "'No News is good news'": CW, p. 87.
160   "'chintzy plywood desk'": Bill Tush interview.
160   "his 'low-budget Walter Cronkite'": PRR, p. 86.
161   "After talking to Levin": Andy Goldman interview.
161   "he'd think about it": Years later, Turner would attribute his early
        interest in satellites to another source: While flying to Atlanta
        with a high-ranking cable executive, Turner was asked where
        he first got the idea to go up on satellite and said, "I picked up
        Broadcasting magazine, and I saw that HBO was going on sat-
        ellite": Cable executive interview, April 1, 1993.
161   "Turner's knowledge about satellites": Ed Taylor interview, De-
        cember 3, 1993.
161   "geosynchronous (stationary) orbit": Vaughan, "True Talent,"
        p. 38.
162   "went to talk it over with Andy Goldman": Andy Goldman in-
        terview.
162   "that cost between $80,0 and $100,000 each": Ibid.; Vaughan, "True
        Talent," p. 46.
163   "'I wanna buy an uplink'": Sidney Topol interview.
163   "With the help of these banks": William Sanders interview.
163   "who turned him down": Andy Goldman interview.
163   "to two or three others": Ed Taylor interview.
164   "'always talking business'": Vaughan, "True Talent," p. 46.
164   "his $65,000-a-year job": Ibid.
164   "more than a million dollars": RVTT, p. 64.
164   "December 13, 1975": *Jane's Encyclopedia.*
164   "HBO was up on the bird": Matthew Sappern (HBO executive)
        interview, March 11, 1994.
164   "'Ted was horrible to Will'": Mike Gearon interview.
164   "'almost a year'": Ed Taylor interview.
165   "'Of course I did'": CW, p. 100.
165   "a 30-foot-high transmitter-receiver" Vaughan, "True Talent,"
        p. 46.
166   "'even Ted wondered'": CW, pp. 99–100.

166   " 'I am unique' ": *CIS Index* (Bethesda, MD: Congressional Information Service, 1976).
167   "Turner acknowledged his indebtedness": Ibid.
169   "On December 17, 1976": CW, p. 98.
170   "again by half": CW, pp. 98–99; PB, pp. 102–103. Today, a commercial dish is available for as little as $1,000.
170   "Two years later": Vaughan, "True Talent," p. 46.
170   "seen in twenty-seven states": *CIS Index,* 1976.
171   " 'It's gonna be tough' ": Jim Trahey interview.
172   " 'sic Ted on him' ": Ed Taylor interview.
173   " 'Look what's happened' ": Vaughan, "True Talent," p. 48.

Chapter 12

Page 174   "cable owners were eager": Ed Taylor interview.
174   " 'We televise all the Braves' away games' ": RVGG, p. 96.
174   " 'really stank' ": CW, p. 94.
175   " 'Hey, Dan' ": Ibid., pp. 94–95.
176   "tax loopholes": "Rebel with a Cause," *Broadcasting,* p. 42.
176   " 'something we *have* to do' ": William Sanders interview.
177   " 'using its own money' ": CW, p. 95.
177   "is probably worth": A spokesperson in the National League Office pointed out that the Baltimore Orioles recently sold for $175 million, and they don't have superstation exposure.
178   " 'We're gonna operate' ": RAF, pp. 29–35.
178   " 'doing it primarily for the city' ": Ibid.
178   " 'I got a deal' ": PRR, p. 74.
178   " 'any more headlines' ": RAF, p. 34.
179   " 'what my financial backing was' ": PRR, p. 70.
179   " 'need to know about baseball' ": Smith, "What Makes Ted Run?" p. 82.
179   " 'He'd never done anything' ": HW, p. 25.
179   "running wind sprints": RAF, p. 45.
179   " 'the negative stuff' ": Ibid., p. 42.
179   "Janie thought he was kidding": *Atlanta Constitution,* January 13, 1976.
179   " 'burns a different fuel' ": RAF, p. 170.
179   "a while to learn": CW, p. 138.
180   " 'I was down at spring training' ": PRR, p. 76.
180   "the tobacco stains": RAF, p. 184.
180   "a singing commercial": Ron Fimrite, "Bigwig Flips His Wig in Wigwam," *Sports Illustrated,* July 19, 1976, p. 25.
180   "ticket sales soared": RAF, p. 74.
180   "more than thirty-seven thousand": Ibid.
181   " 'I never could understand' ": RVTT, p. 47.
181   "led the crowd in singing": RAF, p. 74.
181   " 'Here comes Ted' ": Ibid., p. 173.
181   "a photograph": Ibid., p. 79.
181   " 'Come on, Pablo' ": *Atlanta Journal,* April 14, 1976.
182   " 'I think it's still there' ": RAF, p. 176.

182   " 'incredibly intense' ": Waters et al., "Ted Turner Tackles," p. 65.

182   "leaning on a cane": RAF, p. 79.

182   " 'That first year' ": CW, p. 141.

182   "the amount of baseball knowledge": RAF, p. 220.

182   "lineup was completely altered": Fimrite, "Bigwig Flips," p. 28.

182   " 'sugarcoat the pill' ": RAF, p. 216.

183   " 'It's over, Davidson' ": Ibid., p. 87.

183   "Frank Hyland": Ibid., pp. 87–88.

183   " 'The sportswriters' ": CW, p. 142.

183   " 'I rarely dined' ": *Atlanta Journal,* May 19, 1976.

183   " 'Give me a break' ": CW, p. 142.

183   " 'screw around with his money' ": Mike Gearon interview.

184   " 'Hey, throw that ball back' ": Fimrite, "Bigwig Flips," p. 29.

184   " 'The Braves had a reputation' ": PRR, p. 70.

184   "really in the entertainment business": RAF, p. 44.

184   " 'The first thing I did' ": PRR, p. 76.

185   "a tiny jockey's cap": RAF, p. 163.

185   "would later complain": Kirkpatrick, "Going Real Strawwng,"
        p. 75.

185   " 'why do I do this' ": RAF, p. 153.

185   " 'I beat Tug' ": PRR, p. 74.

185   " 'When you're little' ": RVTT, p. 227.

185   " 'just want to thank you' ": Fimrite, "Bigwig Flips," p. 30.

186   " 'A COMMON LOVE' ": Sportswriter Jesse Outlar wrote, "Ted
        Turner is the most popular citizen in the history of Atlanta pro
        sports." *Atlanta Constitution,* January 16, 1977.

186   " 'fairly anonymous' ": *Atlanta Constitution,* January 13, 1976.

186   "become a public figure": Fimrite, "Bigwig Flips," p. 25.

186   " 'hardly anyone knew who I was' ": Ted Turner interviewed by
        Tom Snyder, *Tomorrow Show,* October 24, 1977.

186   "has seen *Citizen Kane*": Painton, "The Taming," p. 38.

186   " 'I'm the little guy's hero' ": *Atlanta Constitution,* April 5, 1976.

186   " 'leave here a loser' ": Ibid., June 1, 1976.

187   "*Time* printed a picture": April 26, 1976, p. 76.

187   "*Sports Illustrated* emphasized": Fimrite, "Bigwig Flips," p. 26.

187   "*Playboy* quoted him": PRR, p. 77.

187   "enjoy having people read it aloud": Reese Schonfeld interview.

187   " 'Owners don't play poker' ": PRR, p. 70.

188   "the nickname 'Channel' "; RAF, p. 96.

188   " 'the things I was doing' ": PRR, p. 70.

188   " 'see the change' ": RAF, p. 187.

188   " 'Call me Ted' ": PRR, p. 77.

188   "there had never been such harmony": RAF, p. 171.

188   " 'We were in New York' ": Ibid., pp. 192–193.

189   "to hurl a no-hitter": Ibid., p. 136.

189   " 'the finest bunch' ": Ibid., p. 144.

189   "costing Turner his shirt": Roy Blount Jr., "Losersville, U.S.A.,"
        Sports Illustrated, March 21, 1977, p. 77.

189   " 'losing as much money' ": PRR, p. 71.

190   "Turner was disappointed": *Atlanta Constitution,* October 22, 1976.

190   "already been fined $10,000": Ibid., April 23, 1977.
190   "on his fifth or sixth vodka": Kirkpatrick, "Going Real Strawwng,"
       p. 77.
190   " 'Everybody is going to be there' ": *Atlanta Constitution*, October
       22, 1976.
191   " 'Go shake hands' ": Ibid., October 25, 1976.
191   "as well as to Ford's": RVTT, p. 127.
191   "asked him to come to Atlanta": White House internal memo, March
       22, 1977.
191   " 'all psyched' ": Bob Hope interview, February 8, 1994.
192   "a pornographic movie": Kirkpatrick, "Going Real Strawwng,"
       p. 75.
192   " 'a couple of things he could do' ": Bob Hope interview.
193   " 'hold up on Gary Matthews' contract' ": *Atlanta Constitution*, De-
       cember 8, 1976.
193   " 'The commissioner of baseball is going to kill me!' ": PB, p. 112.
193   "he was going to get a gun": *Atlanta Constitution*, December 10,
       1976.
194   " 'for once in his life' ": Ibid., January 9, 1977.
194   " 'spreading myself a little bit thin' ": *New York Times*, January 4,
       1977.
195   "fired in memorable fashion": Mike Gearon interview.
195   " 'He could be very cruel' ": Ira Miskin interview, August 15, 1992.
196   " 'Ted came down hard' ": Farrell Reynolds interview, October 5,
       1992.
196   " 'didn't order me shot' ": *Atlanta Constitution*, January 9, 1977.
196   " 'Great White Father' ": CW, p. 145.
196   " 'The thrust of our argument' ": *New York Times*, January 19, 1977.
197   "He had earlier made preparations": Bob Hope interview.
197   " 'The world has gotten along' ": *New York Times*, January 19, 1977.
197   "secretly hired": Bob Hope interview.
197   "on March 8, 1977": Blount, "Losersville, U.S.A.," p. 77.
197   " 'I think any member' ": *Atlanta Constitution*, April 23, 1977.
197   "Steinbrenner of the Yankees": *Atlanta Journal*, May 4, 1977.
197   "court costs": Ibid.
198   " 'if you wanna get to the top' ": Bob Hope interview.
198   " 'scared to death,' ": *Atlanta Journal*, April 23, 1977.
198   " 'After this is over' ": Kirkpatrick, "Going Real Strawwng," p. 78.
198   " 'Ted Turner's statement' ": *New York Times*, April 30, 1977.
198   " 'baseball's bad boy' ": *Atlanta Constitution*, April 30, 1977.
198   " 'I just don't understand' ": Bob Hope interview.
199   "to stay out of the dugout": CW, p. 144.
199   " 'like everybody else' ": Kent Hannon, "Benched from the Bench,"
       Sports Illustrated, May 23, 1977, p. 68.
199   " 'In their encounter' ": *Atlanta Constitution*, May 27, 1977.

Chapter 13

Page 200   " 'I had as much chance' ": *East Bay Window*, September 14, 1977.
     200   "turned him down in favor of Ted Hood": RVTT, p. 11.

201   "Loomis felt": Ibid., p. 13.

201   "He proposed": Ibid.

201   " 'count me out' ": Ibid., p. 14.

201   " 'nice to have one amateur' ": Ibid.

201   "Hood had provided some sails": Hood Sailmakers News Release, 1977.

201   " 'I got a lot of advice' ": RVTT, p. 13.

202   "It was his hope": Ibid., p. 12.

202   "the most computer-analyzed 12-meter": "The Best Defense," 1977, Part IX of ESPN series *The America's Cup;* PRR, p. 78.

202   " 'We didn't use tank tests' ": PRR, p. 78.

202   "seven of the eleven": RVGG, p. 24.

202   " 'no way we could lose' ": CW, p. 150.

203   " 'Isn't it the *greatest* thing' ": "The Best Defense," 1977, ESPN.

203   " 'a no-good liar' ": PB, p. 121.

203   " 'I am tolerant' ": RVTT, p. 99.

203   "he hoped 'the Australians' ": *East Bay Window,* September 14, 1977.

203   "He threw up": RVTT, p. 3.

203   " 'Trim, damnit!' ": "The Best Defense," 1977, ESPN.

203   " 'The most fun' ": PRR, p. 77.

204   " 'I can make eleven guys work harder' ": Ibid., p. 71.

204   " 'He can really psych us up' ": *East Bay Window,* September 14, 1977.

204   "By the end of June": *Providence Journal-Bulletin,* May 13, 1986.

204   " 'They came here' ": "The Best Defense," 1977, ESPN.

204   " 'wearing a muzzle' ": *Providence Journal-Bulletin,* August 20, 1977.

205   " 'I told him' ": PRR, p. 82.

205   " 'I could watch the Braves' ": Ibid., p. 86.

205   " 'We didn't panic' ": *New York Times,* September 4, 1977.

205   "and saw *Rocky*": PRR, p. 78.

206   " 'not a single really wealthy kid' ": Ibid.

206   " 'a poor boy being plotted against' ": RVTT, p. 19.

206   " 'Boy, did we blow that' ": "The Best Defense," 1977, ESPN.

206   " 'It was to the death' ": PRR, p. 90.

207   " 'We got 'em' ": "The Best Defense," 1977, ESPN.

207   " 'Don't just stand there' ": *Providence Journal-Bulletin,* September 19, 1977.

207   "almost 'monklike": *New York Times,* September 19, 1977.

207   " 'he flirted with every girl' ": "Defending the America's Cup," *Time,* September 19, 1977, p. 84.

207   " 'Show me your tits!' ": CW, pp. 153–154.

207   " 'low tolerance for alcohol' ": PRR, p. 79.

208   "barred from the place": Kirkpatrick, "Going Real Strawwng," p. 78.

208   " 'boozy bellowing' ": Jonathan Black, "Blackbeard Among the Bluebloods," *New York Times,* September 16, 1977, p. 56.

208   "As recounted by Jonathan Black": "Blackbeard," p. 57.

208   " 'a downright lie' ": Ted Turner interviewed by Tom Snyder, *Tomorrow Show,* October 24, 1977.

208   "Turner told *Playboy*": PRR, p. 79.

208   " 'When we got to their house' ": RVTT, p. 22.

209   " 'I'm horny as hell' ": PRR, p. 79.

209   "Turner's hostess": Ibid., p. 80.

209   "letter of apology": RVTT, p. 26.

209   " 'I'm the one' ": Branch and Propper, *Labyrinth,* p. 395.

210   " 'I'm gonna do something' ": "The Best Defense," 1977, ESPN.

210   "tears of empathy": *Providence Journal-Bulletin,* May 13, 1986.

210   " 'She showed no weakness' ": "Defending the America's Cup," *Time,* p. 85.

210   " 'You have been selected' ": PB, p. 123.

210   "Turner praised . . . Robbie Doyle": PRR, p. 78.

211   " 'There will never be a time' ": *Providence Journal-Bulletin,* May 13, 1986.

211   "he himself had funded": RVTT, p. 109.

211   "to invite loan officers": William Sanders interview.

211   "that 'stubborn bastard' Alfred Lee Loomis": RVTT, p. 129.

211   "a secret trip": *Providence Journal-Bulletin,* May 13, 1986.

211   " 'I've been fair to you' ": RVTT, p. 128.

211   "a personal note": Public Papers of the President, Carter Library.

212   "At the local airport": RVTT, pp. 127–128.

212   " 'There's another man' ": Public Papers of the President, Carter Library.

213   " 'the most dangerous' ": "The Best Defense," 1977, ESPN.

213   " 'I sure as hell' ": *East Bay Window,* September 14, 1977.

213   " 'A precious possession' ": "Defending the America's Cup," *Time,* p. 85.

213   "Together with Gary Jobson": *America's Cup Race,* May 9, 1992, ESPN.

213   "On September 13": "The Best Defense," 1977, ESPN.

214   " 'Sometimes I think my father is somewhere' ": Kirkpatrick, "Going Real Strawwng," p. 80.

214   " 'It's like saying' ": *Atlanta Constitution,* September 18, 1977.

214   " 'She ain't much' ": RVTT, p. 201.

215   "Turner would sail conservatively": Bavier, *America's Cup,* p. 48.

215   " 'broke out the six-packs' ": *Atlanta Constitution,* September 19, 1977.

215   "Turner recalls": Ted Turner interviewed by Tom Snyder, *Tomorrow Show,* October 24, 1977.

215   "more than three thousand well-wishers": *Atlanta Constitution,* September 19, 1977.

216   " 'a bottle of aquavit' ": Ted Turner interviewed by Tom Snyder, October 24, 1977.

216   "One reporter present": RVTT, p. 205.

216   " 'up against the best defender' ": "The Best Defense," 1977, ESPN.

217   " 'Wouldn't the old man' ": Mrs. Carl Helfrich interview.

217   "Following a farewell luncheon": *Providence Journal-Bulletin,* September 20, 1977.

217   " 'I had just gone on the air' ": CW, pp. 150–151.

217   "arrived at Atlanta's Hartsfield Airport": *Atlanta Constitution,* September 20, 1977.

218   "The next morning": Ibid.
218   " 'Of all the pictures' ": RVTT, p. 190.
218   " 'one of those times' ": Ibid., p. 214.
218   " 'the most obnoxious' ": Mike Gearon interview.
218   "an unprecedented": PRR, p. 68.
219   "Martine Darragon had just arrived": Branch and Propper, *Laby-rinth*, pp. 390–397.

Chapter 14

Page 222   " 'What would you guys think' ": HW, p. 30.
222   " 'a twenty-four-hour news network' ": Ibid.
222   "four discrete half-hour segments": Ibid., pp. 30–31.
223   " 'TURNER BE SURE' ": *Atlanta Constitution,* February 24, 1980.
223   "a reality check": Gerry Hogan interview.
224   "claims to have been": Robert Pauley, letter to authors, January 1992.
224   " 'the last person I'd expect' ": Stephen Banker, "The Cable News Network Sets Sail," *Panorama,* April 1980, p. 44.
224   " 'an unflagging contempt for news' ": Ibid.
224   " 'I grew up hating newspapers' ": Ibid., p. 46.
225   " 'tongue in cheek' ": *New York Times,* May 25, 1980.
225   "Turner fired him": Mike Gearon interview.
225   " 'Anything you don't want' ": Reese Schonfeld interview.
225   "about twenty-five independent TV stations": CW, p. 246.
225   "the answer was always the same": Reese Schonfeld interview.
225   "making about $200,000": "Rebel with a Cause," *Broadcasting,* p. 40.
225   "he was hoping": Unpublished letter of Ted Turner to Andy Gold-man, dated September 27, 1978.
226   " 'Land is a good investment' ": *Atlanta Journal,* April 12, 1978.
226   "their father could be broke tomorrow": *Atlanta Constitution,* January 13, 1977.
226   "a symbol of attainment": CW, p. 16.
226   " 'Ted has no sense' ": William Sanders interview.
226   "He had heard rumors": HW, pp. 6–7.
226   "a contract with ITNA": Mike Gearon interview.
227   "On the back of an envelope": Reese Schonfeld interview.
227   "inexpensive nonunion labor": In 1989, Natalie Hunter, former CFO of NBC News, told the authors, "The main difference between the three major networks and CNN is our wage rate. We're much higher because we're unionized."
227   " 'There are only four things' ": Reese Schonfeld interview.
227   " 'now ESPN's got that' ": ESPN would go on the air in September 1979.
228   " 'I'm gonna call it Cable News Network!' ": HW, p. 34.
228   " 'bring them all into the tent' ": Reese Schonfeld interview.
228   "wanted to be able to go live": Ibid.
229   "he called Russell Karp": HW, p. 34.

229   "According to one cable-industry executive": Cable executive in-
            terview, September 29, 1992.
229   "Turner spoke of stars": HW, pp. 35–36.
230   " 'I'm Scorpio, too.' ": Reese Schonfeld interview.
230   " 'To make a go of it' ": HW, p. 32.
231   "Turner presented his idea": HW, pp. 36–39.
233   "Ted had rigged up a stereo system": Gerry Hogan interview.
233   " 'We have a retail approach' ": CW, p. 104.
234   "The provision was backed": CW, p. 169.
234   "Paramount claimed": "Rebel with a Cause," *Broadcasting,* p. 38.
235   "Schonfeld got the call": CW, p. 246.
235   " 'you're going to run it' ": Reese Schonfeld interview.
235   " 'the most powerful man in America!' ": HW, p. 43.
235   " 'what big name can we get' ": Ibid.
236   "Schorr had reservations": Daniel Schorr interview, June 5, 1993.
236   "a furious discussion on the phone": Reese Schonfeld interview.
239   " 'let's sign something' ": HW, pp. 48–49.
239   "Schorr drafted a letter": Daniel Schorr interview.
240   "among the well-known personalities": HW, p. 50.
240   "the price for CNN": Ibid., pp. 50–51.
240   " 'how many hours of TV *news*' ": Ibid., p. 54.
241   " 'competition is what journalism is about' ": Howell Raines on *Fresh
            Air,* National Public Radio, September 16, 1993.
241   " 'Experts,' he once said": CW, p. 245.
241   "restoring old buildings": HW, p. 58.
242   " 'Bunky turned out to be terrific' ": Reese Schonfeld interview.
242   " 'It's just Rhett and me' ": CW, p. 158.
242   " 'Just surviving' ": *Charleston News & Courier,* July 17, 1979.
243   "Turner purchased": *Savannah Morning News,* August 2, 1979.
243   "There were 303 boats": CW, pp. 208–209.
243   "a sudden furious storm": CW, p. 233.
243   " 'a completely new concept' ": HW, p. 60.
244   " 'Just turn around' ": Reese Schonfeld interview.
244   " 'we used to call him Captain Panic' ": Ibid.
245   " 'Sailing in rough weather' ": John Skow, "Vicarious Is Not the
            Word," *Time,* August 9, 1982, p. 57.
245   " 'the big regatta in the sky' ": CW, p. 16.
245   " 'The king is dead' ": Smith, "What Makes Ted Run?" p. 84.
245   "They were expecting sympathy": *Atlanta Constitution,* September
            27, 1979.
245   "the featured speaker": CW, p. 247.
245   "one of the ten sexiest men": Banker, "The Cable News Network,"
            p. 44.
245   "He talked about 'bad news' ": Ibid., p. 50; "Rebel with a Cause,"
            Broadcasting, p. 44.
246   " 'I'll let Reese' ": HW, p. 56.
246   " 'But until the fire is over' ": Reese Schonfeld interview.
246   " 'very important that we got George Watson' ": Ibid.
246   "temporary quarters": HW, p. 72.

247   "no more than $18,000": Banker, "The Cable News Network,"
            p. 50.
247   "'a pushcart operation'": Bill Zimmerman interview, July 8, 1992.
247   "'a chance of a lifetime!'": Mary Alice Williams interview, Septem-
            ber 24, 1992.
247   "contract be with TBS": Don Farmer interview, August 20, 1992.
247   "had had to be fired": Reese Schonfeld interview.
248   "'I'll get it done'": Ibid.
248   "'the greatest achievement'": CW, p. 252.
249   "'don't even know the problems'": Ibid., p. 254.
249   "drop in the company's earnings": "Rebel with a Cause," *Broad-
            casting,* p. 40.
249   "'our entire startup funds'": CW, pp. 254–255.
249   "'increase our service to the cable-TV industry'": "The Saga of
            Satcom III," unpublished memoir by A. F. Inglis.
250   "'I didn't know that satellites failed'": HW, p. 82.
250   "Turner got the news": Ibid.
251   "from HBO or Showtime": Ibid.
251   "'charm the snakes'": A. F. Inglis interview, July 15, 1992.
251   "Turner told them": RCA Americom executive.
251   "Inglis genuinely liked": A. F. Inglis interview.
252   "seriously worried": Carl Cangelosi interview, July 15, 1992.
252   "regarded it as 'a disaster'!": CW, p. 255.
252   "'just got off the boat'": Ibid., p. 256.
253   "to 'go nuts on them'": HW, p. 90.
253   "by the shirt": A. F. Inglis interview.
253   "the only RCA lawyer there": Carl Cangelosi interview.
253   "'got no legal ground'": HW, p. 91.
253   "'You're gonna kill me'": Reese Schonfeld interview.
253   "'I'm a small company'": HW, p. 91.
254   "'You don't know'": Reese Schonfeld interview.
255   "'sweating blood'": HW, p. 92.
255   "'a friendly lawsuit'": Carl Cangelosi interview.
255   "'Ted, shut up!'": A. F. Inglis interview.
255   "'a secret deal!'": HW, p. 91.
255   "Coxe filed a suit": Ibid., p. 96.
255   "'In no event'": *Atlanta Constitution,* March 5, 1980.
256   "At lunch with the lawyers": CW, p. 257.
256   "'I don't blame you guys'": Mike Gearon interview.
257   "'they agreed to stop opposing'": CW, p. 258.
257   "'we're gonna take the news'": HW, p. 124.
258   "Burns said": Banker, "The Cable News Network," p. 50.
258   "would eat crow": HW, p. 105.

Chapter 15

Page 259   "'I'm going to do news'": Waters et al., "Ted Turner Tackles,"
                p. 59.
      260   "a little frayed around the edges": *New York Daily News,* June 8,
                1980; CW, pp. 4–6; HW, pp. 142–143.

261   "'an opening statement!' ": HW, pp. 143–144.

261   "critics tended to scoff": Mary Alice Williams interview.

261   "glitches were bound to happen": HW, pp. 145–150.

262   "a dozen blooper reels": Ibid., pp. 156–157.

262   "short of material": Ibid., pp. 160–161.

262   "losing a million dollars": Ted Turner on *Donahue,* April 1, 1981.

262   "closer to $2.2 million": "Rebel with a Cause," *Broadcasting,* p. 42.

262   "penny-pinching operation": Waters et al., "Ted Turner Tackles," p. 59.

263   "the official figure": Ibid.

263   "filled with anxiety": Zeke Siegel interview, August 15, 1992.

263   "counting foul balls": Waters et al., "Ted Turner Tackles," p. 62.

263   "'The boss'": Banker, "The Cable News Network," p. 50.

263   "Fred Friendly noted": Waters et al., "Ted Turner Tackles," p. 66.

264   "'Ted Turner has been a major supporter'": Public Papers of the President, Carter Library.

265   "less than a quarter": Waters et al., "Ted Turner Tackles," p. 59.

265   "'I've got everything'": Turner speech to advertisers shown on *Donahue,* April 1, 1981.

266   "'the worst enemies'": PRR-2, p. 68.

266   "'a cheap whorehouse'": as cited in PRR-2, p. 68.

266   "'bunch of pinkos'": Waters et al., "Ted Turner Tackles," p. 59.

266   "as he did on *Donahue*": *Donahue,* April 1, 1981.

266   "'making Hitler Youth'": Nick Taylor, "The American Hero as Media Mogul," *Atlanta,* September 1982, p. 64.

267   "'exposés on the networks'": Waters et al., "Ted Turner Tackles," p. 59.

267   "'I'm up for it'": *Atlanta Constitution,* May 31, 1980.

268   "'No excuse to lose'": Dennis Conner, *No Excuse to Lose* (New York: W. W. Norton, 1978).

268   "'He brags'": *Atlanta Constitution,* May 31, 1980.

268   "'We own Newport'": *Newport News,* June 10, 1980.

269   "on Bannister's Wharf": PB, p. 189.

269   "'akin to having a Russian admiral'": *Daily News,* August 19, 1980.

269   "'like a schoolboy'": Ibid.

270   "'Turner's last hope'": Ibid.

270   "'a middle-aged man'": Diane K. Shaw, "No Excuse to Lose," *Newsweek,* August 25, 1980, p. 54.

270   "'We do appreciate'": *Atlanta Constitution,* August 28, 1980.

270   "'Three times is enough'": Bavier, *America's Cup,* p. 48.

270   "'It hurts worse'": *Atlanta Constitution,* August 28, 1980.

270   "'I'm through with sailing'": *Washington Post,* May 12, 1981.

271   "'You know he's serious'": Ibid.

271   "a $24-million working deficit": John N. Ingham and Lynne B. Feldman, "Robert Edward 'Ted' Turner," *Contemporary American Business Leaders: A Biographical Dictionary* (New York: Greenwood Press, 1990), p. 713.

271   "Staffers remember": Robert Wussler interview, August 6, 1992; Mary Anne Loughlin (CNN anchor) interview, January 7, 1994.

271   "a little creative": Mike Gearon interview.

272   "only one problem": Reese Schonfeld interview.
272   "'in desperate trouble'": *Donahue,* April 1, 1981.
272   "'Ted's killed it'": Mike Gearon interview.
273   "'the gorilla keepers'": Senior TBS executive interview, August 12, 1992.
273   "'nuts' is a word": Bill Zimmerman interview; Don Farmer interview.
274   "'you lost half your brain'": Mike Gearon interview.
274   "how boring it was": Billy and Nancy Roe interviews.
275   "'like a hyperactive kid'": Gerry Hogan interview.
275   "tales circulated": Ibid.; Robert Wussler interview; Mike Gearon interview; et al.
276   "'My biggest job'": Taylor, "The American Hero," p. 62.
276   "Turner donated": PB, p. 201.
276   "suit against the White House": Reese Schonfeld interview; *Atlanta Constitution,* May 11, 1981.
277   "'to deny CNN'": HW, p. 197.
277   "'a detrimental effect'": *Atlanta Constitution,* May 11, 1981.
277   "'to be hunted down'": HW, p. 197.
277   "'And we're winning'": *Atlanta Journal,* May 11, 1981.
277   "'Westinghouse and ABC'": HW, p. 199.
277   "a real challenge": Ibid., p. 205.

Chapter 16

Page 279   "'a preemptive first strike'": HW, pp. 202–203.
280   "'only money to lose'": Ibid., p. 203.
280   "'They're fifty times bigger'": Ibid., p. 206.
280   "more than a little nervous": Farrell Reynolds interview.
280   "'When I cast my lot'": HW, pp. 207–210.
281   "'basically a defense": Ibid., p. 204.
282   "'Is this enough?'": PB, pp. vii–viii.
283   "deliberately courting disaster": Mike Gearon interview.
283   "arrived in Havana": HW, p. 233; *Atlanta Constitution,* February 20, 1982.
283   "the network's live reporting": HW, p. 198.
283   "'I just wanted to let you know'": as quoted in Peter Manigault interview, April 5, 1993.
283   "'I'm a very curious person'": *Atlanta Constitution,* February 20, 1982.
284   "'as Citizen Turner'": "The Man from Atlanta," May 1982, BBC-TV.
284   "the only two people on the island": Peter Manigault interview.
284   "the best marksman": *Atlanta Constitution,* March 5, 1982; PB, p. 224; Taylor, "The American Hero," p. 61.
284   "'I could've shot him'": Smith, "What Makes Ted Run?" p. 76.
285   "'When there's trouble'": HW, p. 233.
285   "'After three drinks'": Ted Turner interviewed by Diane Sawyer, *60 Minutes,* April 20, 1986.

285   " 'It's hard to understand' ": PB, p. 225, from "The Man from At-
            lanta," May 1982, BBC-TV.
285   *"Mein Kampf"*: RVTT, p. 147; Cable executives interviews, Sep-
            tember 29, 1992, and April 4, 1994.
285   " 'Castro's not a communist' ": PRR-2, p. 63.
286   " 'We couldn't operate' ": HW, p. 234.
286   " 'I'm gonna kick' ": Taylor, "The American Hero," p. 102.
286   "it was 156 to 53": HW, p. 234.
286   "undue pressure": Taylor, "The American Hero," p. 102.
287   " 'one speck of trash' ": HW, pp. 164–165.
287   "to solidify its reputation": Ibid., p. 231.
287   " 'I called the bureau' ": Reese Schonfeld interview; HW, p. 230.
288   " 'the kind of management style' ": HW, p. 241.
288   " 'Schonberg' or 'Schonstein' ": Don Farmer interview.
288   "Schonfeld felt that CNN should have pushed on": Reese Schonfeld
            interview.
289   " 'Ted felt he should have been talked to' ": HW, p. 238.
289   " 'You know, Reese' ": Ibid.
289   "Carnegie's chapters": Dale Carnegie, *How to Win Friends and Influ-
            ence People* (New York: Pocket Books, 1981).
289   " 'It's not your air' ": Reese Schonfeld interview.
290   "Schonfeld would linger": Ibid.
291   " 'I must respectfully disagree' ": Daniel Schorr interview; PB,
            p. 231.
291   " 'To my knowledge' ": HW, p. 245.
291   " 'turns into a battle' ": Ibid., p. 247.
293   "her own personal checkbook": Mary Alice Williams interview.
293   " 'the fight got uglier' ": HW, p. 253.
293   "to lend TBS $50 million": Ingham and Feldman, *Contemporary
            American Business Leaders,* p. 713.
293   " 'I'm going to collapse' ": Taylor, "The American Hero," p. 62.
294   " 'They'll have to kill me first!' ": "Ted Turner and the News War,"
            1984, WGBH.
294   " 'in the other hand' ": HW, p. 247.
294   "the idea for Turner Educational Services": Ira Miskin interview.
294   "Turner loved this war": Reese Schonfeld interview.
295   " 'with a lot of charisma' ": *New York Times,* December 18, 1982.
295   " 'a damn good chance' ": HW, p. 253.
295   "a $60-million loss": PB, p. 266.
295   "closer to $100 million": Robert Wussler interview.
295   "Turner Broadcasting stock": PB, p. 227.
295   "ABC-Westinghouse had caved in": HW, p. 254.
296   " 'a level playing field' ": Ingham and Feldman, *Contemporary Amer-
            ican Business Leaders,* p. 713.

Chapter 17

Page 297   " 'Hurricane Turner' ": James Dodson, "Teddy Comes About," *At-
            lanta,* May 1993, p. 108.
297   " 'was often away' ": Ibid.

298   " 'So you think I'm crazy' ": *Atlanta Constitution,* June 23, 1982.

298   " 'I remember' ": PB, p. 250.

299   " 'We used to say' ": Former CNN employee interview, December 11, 1992.

299   "nothing more than a towel": Cable CEO interview, March 18, 1993.

299   "bringing his girlfriends": Bob Hope interview.

299   " 'He and Elmer Gantry' ": Waters et al., "Ted Turner Tackles," p. 65.

299   "Dick Ruthven became so incensed": Ibid.

299   "A Turner employee": Former TBS employee interview, July 29, 1992.

300   "made up in volume": Judy Nye Hallisy interview.

300   " 'I don't listen to women' ": Zeke Siegel interview.

300   "According to witnesses": Former CNN staffer interview, August 29, 1992.

301   "Turner was known to employ": TBS staffer interview, August 12, 1992.

301   "Rare was the interview": *Savannah News Press,* November 30, 1982.

301   "the 'stupid sleazy sex' ": *Atlanta Constitution,* February 20, 1982.

301   " 'off-color humor' ": Ibid., February 20, 1982.

301   " 'terrible role models' ": Ibid.

301   "He strictly forbade": Mike Gearon interview.

301   " 'I love pictures' ": PRR, p. 84.

301   " 'I photograph nudes myself' ": Ibid.

301   "Some sources claim": Female friend interview, May 19, 1994; Former TBS staffer interview, August 12, 1992; Peter Ross Range, "The Demons of Ted Turner," *Playboy,* August 1983, p. 63.

301   " 'If you ever' ": TBS executive interview, April 4, 1994.

302   " 'No mas' ": *Savannah Morning News,* February 26, 1985.

302   "light up joints": Former senior TBS executive interview, March 28, 1994.

302   " 'the most interesting guy' ": J. J. Ebaugh interview, December 28, 1993.

302   " 'Our first real date' ": Kathy Leach interview.

303   " 'Her teddy' ": *Playboy,* April 1981, p. 7.

303   "Turner had gone ballistic": PRR-2, p. 163.

303   " 'had to be a pro' ": PB, pp. 238–239.

304   " 'to work with Liz' ": Zeke Siegel interview.

304   "where she accompanied": J. J. Ebaugh interview.

304   " 'J.J. was the best-looking thing' ": Vincent Coppola, "JJ: Today Ted Turner, Tomorrow the World," *Atlanta,* November 1988, p. 114.

304   "skiing and climbing": J. J. Ebaugh interview.

305   "she had become": Coppola, "JJ," p. 113.

305   " 'a wonderful invitation' ": Ibid.

306   "a story that became legend": Former TBS vice president interview, August 2, 1992.

306   "'Jane wasn't supposed to know'": Turner family friend interview,
          March 22, 1994. In fact, Robert Wussler was once shocked when
          Ted began itemizing his own extracurricular love affairs in front
          of Jane and Wussler. Robert Wussler interview.
306   "'One time'": Ibid.
306   "'Fishermen found her'": Ibid.
307   "offered Jane a job": Reese Schonfeld interview.
308   "to baby-sit his dates": Cable executive interview, December 20,
          1993.
308   "'I've never been in love'": Mike Gearon interview.
308   "to see a marriage counselor": Robert Wussler interview; Mike Gea-
          ron interview.
308   "Dr. Frank Pittman": *Atlanta Constitution,* January 25, 1989.
308   "'are attracted to philanderers'": Frank Pittman, M.D., *Private Lies:
          Infidelity and the Betrayal of Intimacy* (New York: W. W. Norton,
          1989), p. 178.
309   "'Philandering is addictive behavior'": Ibid., p. 181.
309   "'I don't wish'": Smith, "What Makes Ted Run?" p. 88.
309   "not uncommon": "Two million to three million adults in the
          United States, or about 1.6 percent of the nation's population,
          suffer from the disorder. Without treatment, about one-fifth of
          those with the illness will commit suicide." *New York Times,*
          December 8, 1994.
309   "was taking lithium": Robert Wussler interview; Gerry Hogan in-
          terview; Painton, "The Taming," p. 36.
309   "On the negative side": Martin Lafferty interview, August 3, 1992.
310   "'it was pretty scary'": Painton, "The Taming," p. 36.
310   "'We all worked'": Dodson, "Teddy Comes About," p. 108.
310   "earning a dollar an hour"; Ibid., p. 110; CW, p. 28.
310   "expected to pay rent": Mike Gearon interview.
310   "'It's the only thing'": *Atlanta Constitution,* September 6, 1985.
311   "'away from their mother'": CW, pp. 24–25.
311   "'a fortress of duty'": Rick Reilly, "What Is the Citadel?" *Sports
          Illustrated,* September 14, 1992, p. 70.
311   "'harder for my sons'": Smith, "What Makes Ted Run?" p. 84.
311   "'have to get whipped'": Dodson, "Teddy Comes About," p. 110.
311   "'it was a nightmare'": Ibid.
312   "'who had everything'": Painton, "The Taming," p. 38.

Chapter 18

Page 313   "a small crowd": R. Serge Denisoff, "Ted Turner's Crusade: Eco-
              nomics versus Morals," *Journal of Popular Culture,* Summer
              1987, p. 27.
313   "Turner was not as out of place": *New York Times,* October 25,
          1984.
314   "'Turner went into the music business'": Cable CEO interview,
          March 18, 1993.
314   "'The company anticipated'": Denisoff, "Ted Turner's Crusade,"
          p. 29.

314   "the MTV stock would plummet": Ibid.

314   " 'a typically competitive Turner ploy' ": *New York Times,* October
        10, 1984.

314   " 'I was really disturbed' ": Denisoff, "Ted Turner's Crusade,"
        p. 36.

315   "He attacked the video": *New York Times,* October 25, 1984.

315   " 'that would not be damaging' ": Denisoff, "Ted Turner's Cru-
        sade," p. 28.

315   "80 percent of CMC's selections": *New York Times,* October 25,
        1984.

315   " 'the hero of my country' ": Waters et al., "Ted Turner Tackles,"
        p. 66.

315   " 'what I want to do' ": Taylor, "The American Hero," p. 57.

315   " 'just ambling through life' ": Smith, "What Makes Ted Run?"
        p. 88.

315   " 'I needed to find out' ": "Sign-off," *New Yorker,* September 12,
        1988, p. 25.

316   "his four great ambitions": Irwin Mazo interview.

316   " 'any man except Jimmy Carter' ": PRR, pp. 86–87.

316   "talk among CNN staffers": Bill Zimmerman interview.

316   " 'The country needs me' ": PRR-2, p. 63.

316   " 'They're all my friends' ": Reese Schonfeld interview.

317   "out-of-work blacks": *Atlanta Constitution,* May 12, 1985.

317   "Other times he joked": *60 Minutes,* April 20, 1986.

317   " 'Turner's no racist' ": *Atlanta Constitution,* May 12, 1985.

318   " 'TURNER HOPING TO SAVE THE WORLD' ": *Savannah
        Morning News,* August 5, 1984.

318   " 'The culmination of his life' ": Painton, "The Taming," p. 39.

318   " 'some little Timbuktu' ": DF.

318   " 'will put a bullet in me' ": Waters et al., "Ted Turner Tackles,"
        p. 66.

319   " 'a pretty terrific species' ": DF.

320   " 'no such thing as good news' ": Zeke Siegel interview.

320   "on a visit to Moscow": Smith, "What Makes Ted Run?" pp. 78–
        80.

322   " 'they'd like to do something in '86' ": Robert Wussler interview.

323   "involved in national conservation groups": Jim Kennedy (CEO of
        Cox Enterprises) interview, October 7, 1992.

324   " 'nightmarish evocations of a ruined earth' ": *Washington Post,* Feb-
        ruary 11, 1979.

324   "Turner had read such books": J. J. Ebaugh interview.

324   "a somber message it was": Gerald O. Barney, ed., *The Global 2000
        Report to the President: Entering the 21st Century* (Washington,
        DC: Seven Locks Press, 1980, revised 1988), pp. 2–3.

325   " 'if present trends continue' ": Ibid., p. 1.

325   " 'his darkest fears' ": Turner friend interview, January 4, 1993.

325   "Turner would hand out copies": Mary Anne Loughlin interview.

325   " 'Population growth' ": Taylor, "The American Hero," p. 64.

326   "appointed his friend": July 22, 1980; Public Papers of the President,
        Carter Library.

326   "he quit smoking": In the mid-1980s, Ted and David McTaggart, the head of Greenpeace, had a $100,000 bet on who could quit smoking first. Ted won. Robert Wussler interview.

326   "to ban air-conditioning": Smith, "What Makes Ted Run?" p. 86.

327   "flew down to South America": Taylor, "The American Hero," p. 61.

327   "he and Turner were talking": DF.

327   " 'I gave him four million' ": PRR-2, p. 64.

328   " 'You're going to come down to Atlanta' ": "Barbara Pyle: TBS' Environmental Conscience," *Atlanta Business Chronicle,* June 12, 1989, p. 31.

329   " 'She tried real hard' ": Turner friend interview, January 6, 1994.

330   " 'that we had in common' ": J. J. Ebaugh interview.

330   " 'As systems start to collapse' ": Coppola, "JJ," p. 120.

330   " 'Thanks for not coming sooner' ": DF.

330   "death threats": Robert Wussler interview.

330   "to protect their legacy": Turner friend interview, January 4, 1993.

331   " 'so-called larger-agenda issues' ": PB, pp. 285–286.

331   "Ted had recruited": Julie Lanham, "The Greening of Ted Turner," *Humanist,* November 1989, p. 6.

331   " 'We are very disappointed' ": *New York Times,* November 29, 1984.

332   " 'There is no explanation' ": Denisoff, "Ted Turner's Crusade," p. 40.

332   " 'unwanted, unneeded' ": Ibid., p. 33.

332   " 'He lost some money' ": Cable company president interview, April 1, 1993.

Chapter 19

Page 333   "a loss on the books": "Neither Broke Nor Broken," *Broadcasting,* August 17, 1987, p. 63.

333   "thirty-four million U.S. households": *Wall Street Journal,* April 19, 1985.

334   " 'a few major companies' ": "Neither Broke," *Broadcasting,* p. 50.

334   "75-percent unrented": Paul Beckham interview, March 19, 1994.

334   " 'you don't buy real estate' ": Mike Gearon interview.

334   "he would pay $64 million": Smith, "What Makes Ted Run?" p. 77.

334   " 'that old amusement-park space' ": "Neither Broke," *Broadcasting,* p. 46.

335   " 'just about gave me a hug' ": Ibid.

335   " 'a hell of a lot cheaper' ": Bill Leonard, *In the Eye of the Storm: A Lifetime at CBS* (New York: Putnam, 1987), p. 220.

336   "an odd meeting": Ibid., pp. 220–222; Robert Wussler interview.

337   "The second meeting": Robert Wussler interview.

337   " 'some sort of collaboration' ": *New York Times,* October 16, 1981.

338   "In the first two weeks": *Wall Street Journal,* April 18, 1985.

338   "According to Bill Bevins": Ibid., April 19, 1985.

338   "Turner saw CBS": Gerry Hogan interview.

339   " 'On Wall Street' ": *Wall Street Journal,* April 14, 1985.

339   " 'liberal bias in news reporting' ": Peter Boyer, *Who Killed CBS?* (New York: Random House, 1988), p. 197.

339   " 'become Dan Rather's boss' ": Ibid.

339   " 'I do have qualifications' ": *Atlanta Constitution,* March 22, 1985.

339   "he'd be making changes": Ibid.

340   " 'You've got such good friends' ": Ibid.

340   "Wyman shot off a memo": Ed Joyce, *Prime Times, Bad Times* (New York: Doubleday, 1988), p. 448.

341   "Thomas Wyman himself called": *Wall Street Journal,* April 12, 1985.

342   " 'Don't underestimate this Turner' ": Joyce, *Prime Times,* pp. 450–452.

342   " 'a scene from *Day of the Locust*' ": Robert Wussler interview.

342   " 'We have been very interested' ": *New York Times,* April 19, 1985.

342   " 'I'll never call them junk bonds' ": HW, p. 264.

342   " 'the first leveraged buyout' ": *New York Times,* April 19, 1985.

343   " 'It's all paper' ": Stephen Koepp, "Captain Outrageous Opens Fire," *Time,* April 29, 1985, p. 60.

343   " 'the probability is zero' ": *Wall Street Journal,* April 19, 1985.

343   "an 'old friend of Jesse Helms' ": Koepp, "Captain Outrageous," p. 60.

343   " 'totally repugnant' ": *Wall Street Journal,* April 19, 1985.

343   "a grand adventure": Martin Lafferty interview.

344   " 'Paley was an outsider' ": *Atlanta Constitution,* May 29, 1985.

345   "Letters to the shareholders": *New York Times,* June 21, 1985.

345   "he didn't think Ted Turner": Ibid., May 3, 1985.

345   " 'high risk of financial ruin' ": *Wall Street Journal,* June 4, 1985.

345   " 'You could gather the news' ": HW, p. 265.

346   " 'I needed distribution' ": "Neither Broke," *Broadcasting,* p. 50.

346   "continued to lobby": *Wall Street Journal,* June 11, 1985.

346   "And in Albany": Ibid., June 27, 1985.

346   " 'There's a lot of money' ": Joyce, *Prime Times,* p. 454.

346   "the so-called poison pill": *Wall Street Journal,* July 5, 1985.

347   "He attacked the CBS strategy": *Atlanta Constitution,* July 10, 1985.

347   "he challenged": *Wall Street Journal,* July 17, 1985.

347   " 'It's dead in the water' ": Ibid., July 31, 1985.

347   " 'Why are you doing this?' ": Mike Gearon interview.

347   " 'It's a long shot' ": Ibid.

347   "easier for TBS to sell ads": *Wall Street Journal,* July 26, 1985.

Chapter 20

Page 349   " 'I admire what you've tried' ": Robert Wussler interview.

350   "was full of ideas": Ibid.

350   "MGM was scaling back production": *New York Times,* March 30, 1986.

350   "Kerkorian told Turner": Ibid.

351   " 'The directors' meeting' ": TBS board member interview, March 12, 1994.

351   " 'We gotta do this' ": Mike Gearon interview.

351   " 'Movie fees' ": *New York Times,* March 30, 1986.

351   "'the big are getting bigger'": Ibid.

352   "'a tremendous business opportunity'": *Wall Street Journal,* August 6, 1985.

352   "regarded as 'a pigeon'": Bill Powell et al., "Turner's Windless Sails," *Newsweek,* February 9, 1987, p. 46.

352   "'Turner's found oil'": Ingham and Feldman, *Contemporary American Business Leaders,* p. 714.

353   "'It's worth a fortune'": Reese Schonfeld interview.

353   "sold ten days later": Mike Gearon interview.

354   "'the deal began unravelling'": *Variety,* December 9, 1991.

354   "'we've got a problem here'": Robert Wussler interview; *New York Times,* March 30, 1986.

355   "'a brilliant piece of financing'": *Wall Street Journal,* October 3, 1985.

355   "'an atmosphere of urgency'": Ibid., October 31, 1985.

355   "'to squeeze blood'": *New York Times,* March 30, 1986.

355   "he didn't want to give up control": Robert Wussler interview; Gerry Hogan interview.

356   "Murdoch kept his focus": Gerry Hogan interview.

357   "'the safe course to take'": Mike Gearon interview.

357   "'my bosses'": Reese Schonfeld interview.

358   "'Ted Turner will go'": *Wall Street Journal,* November 27, 1985.

358   "'You put them in business'": "Neither Broke," *Broadcasting,* p. 60.

358   "'I have always bought troubled things'": *New York Times,* March 30, 1986.

358   "'In twenty-twenty hindsight'": Ibid.

359   "'certain difficulties'": *Wall Street Journal,* January 14, 1986.

359   "their share of conflicts": TBS board member interview, March 12, 1994.

359   "to clasp hands": Michael Milken story as cited in Jesse Kornbluth interview, January 5, 1993.

360   "Turner publicly announced": *Wall Street Journal,* March 26, 1986.

360   "'and left in a barrel'": Bill Powell et al., "Turner's Windless Sails," *Newsweek,* February 9, 1987, p. 46.

360   "the combined TBS and MGM": *Wall Street Journal,* March 21, 1986.

360   "'put a gun to Turner's head'": *New York Times,* March 30, 1986.

360   "'The whole empire'": Smith, "What Makes Ted Run?" p. 82.

361   "'I'm pretty proud of that'": *Wall Street Journal,* February 10, 1986.

361   "'Two billion dollars I owe'": Smith, "What Makes Ted Run?" p. 77.

361   "One night, with his mother": Peter Manigault interview.

362   "One afternoon, he was talking": Robert Wussler interview.

362   "'anything like this before'": *New York Times,* March 30, 1986.

362   "to convince Steven Spielberg": Ted Turner interview.

362   "mesmerized by Hollywood": Robert Wussler interview.

362   "'one of the most incredible'": Cable CEO interview, March 18, 1993.

363   "'the way our company would do it'": Cable executive interview, April 1, 1993.

363   "'too shallow'": In the fifty-three weeks Turner ran MGM, the
             studio brought out eleven movies that he inherited, and lost
             about $65 million.
363   "'had taken him bad'": Ibid.
364   "And he put his head down": Ted Turner interview.

Chapter 21

Page 365   "'You don't look back'": Cable CEO interview, March 18, 1993.
       365   "the stock had plummeted": *Wall Street Journal,* August 5, 1986.
       366   "In one seven-day period": Smith, "What Makes Ted Run?" p. 77.
       366   "Responding to accusations": *Wall Street Journal,* August 28, 1986.
       366   "'the same thing his daddy did'": Smith, "What Makes Ted Run?"
                   p. 84.
       366   "'I've begged him'": Ibid.
       367   "'We can best achieve peace'": *Wall Street Journal,* July 3, 1986.
       367   "'What we need'": *New York Times,* July 13, 1986.
       368   "Turner invited Gorbachev": Robert Wussler interview.
       369   "'I don't wanna breathe'": *Wall Street Journal,* July 8, 1986.
       369   "'As a business venture'": Ibid., July 17, 1986.
       369   "'I don't want you to cut'": Robert Wussler interview.
       370   "'I'm so happy'": *Wall Street Journal,* July 8, 1986.
       371   "'a nation of *winners*'": Ira Miskin interview.
       372   "he would joke about": Ibid.; Don Farmer/Chris Curle interview,
                   August 20, 1992.
       372   "'It was a friendship'": Coppola, "JJ," p. 115.
       372   "'never any ultimatums'": J. J. Ebaugh interview.
       372   "he drove across the African plains": Robert Wussler interview.
       373   "'he fell apart emotionally'": Turner friend interview, March 22,
                   1994.
       373   "obsessed": Ibid.
       373   "'the most aggressive campaign'": Painton, "The Taming," p. 38.
       373   "'I can't live without you'": Coppola, "JJ," p. 115.
       373   "She put the odds": Ibid.
       373   "'Ted had told Janie'": Ibid.
       374   "moving into two new homes": Robert Wussler interview.
       374   "An IRS audit": *Wall Street Journal,* June 5, 1987.
       374   "'I think the accident'": Dodson, "Teddy Comes About," p. 112.
       375   "a specially convened Washington hearing": *Wall Street Journal,* De-
                   cember 24, 1986.
       375   "'That's coloring, isn't it?'": Ingham and Feldman, *Contemporary
                   American Business Leaders,* p. 715.
       376   "'My lifestyle's changed a lot'": "Neither Broke," *Broadcasting,*
                   p. 66.
       376   "the main concessions": Turner family friend interview, January 4,
                   1993.
       376   "they traveled extensively": Coppola, "JJ," p. 120.
       376   "a rustic lodge": PB, p. 310.
       376   "'He determined'": Coppola, "JJ," p. 118.
       377   "'A reasonable compromise'": Ibid.

377   "'I've changed quite a bit'": Ibid., p. 88.

377   "'I learned to give and take'": Painton, "The Taming," p. 38.

378   "'I counseled Ted'": Scott Ticer, "Captain Comeback," *Business Week,* July 17, 1989, p. 104.

378   "selling the entire CNN Center complex": "Neither Broke," *Broadcasting,* p. 46.

378   "'I think the cable industry'"; Cable CEO interview, December 20, 1993.

378   "perhaps *the* most powerful force": *Wall Street Journal,* January 27, 1992.

379   "Turner's financial picture": Ibid., January 12, 1987.

379   "'If Turner's at risk'": Ibid., January 16, 1987.

379   "the agreement was announced": Ibid., May 8 and June 4, 1987.

380   "'Ted no longer runs the company'": Ibid., June 5, 1987.

380   "'I don't consider it a rescue'": "Neither Broke," *Broadcasting,* p. 60.

380   "'I do have some negative covenants'": Ibid., p. 54.

380   "'that was Ted being rescued'": Cable CEO interview, March 18, 1993.

381   "Turner was nervous": Ibid.; Robert Wussler interview.

381   "They were the role players": Paul Beckham interview; Martin Lafferty interview; Robert Wussler interview.

381   "The best work": William Sanders interview.

382   "Bevins 'wanted to be the successor'": *Wall Street Journal,* November 4, 1987.

382   "'It's like slaves on the plantation'": *New York Times,* October 12, 1990.

382   "'This is our first meeting'": Mike Gearon interview.

382   "had publicly declared": Bill Powell et al. "Turner's Windless Sails," p. 46.

383   "'If we don't buy it'": TBS board member interview, March 12, 1994; TBS board member interview, March 18, 1993.

383   "'For a number of different reasons'": Ibid.

384   "'I can remember personally feeling'": Ibid.

384   "'Ted once told me'": Ibid.

384   "'When you've got partners'": "Neither Broke," *Broadcasting,* p. 66.

384   "both NBC and Time": *Wall Street Journal,* November 5, 1987.

385   "'There had to be a restructuring'": Ibid., December 29, 1987.

385   "'he wasn't going to live with that'": TBS board member interview, March 18, 1993.

385   "'wildly enthusiastic'": *Wall Street Journal,* February 29, 1988.

386   "'I've got him'": Jolie Solomon, "Big Brother's Holding Company," *Newsweek,* October 25, 1993, p. 41.

386   "lose every one of them": *Wall Street Journal,* April 15, 1988.

386   "the country music disease": Cable executive interview, April 1, 1993; Robert Wussler interview. Turner himself worried that he had yellow tick problems.

387   "'I'm divorcing her'": Mike Gearon interview.

387   " 'like someone coming out of a prison camp' ": Turner friend interview, March 22, 1994.

388   " 'Our destiny' ": Coppola, "JJ," p. 120.

389   " 'a very, very rough time' ": Publishing figure interview, December 10, 1993.

Chapter 22

Page 390   "On a bitterly cold December night": *Atlanta Constitution,* December 15, 1989; *New York Times,* December 16, 1989; Kathy Leach interview.

391   " 'a noble movie' ": Ted Turner interview.

391   "an article comparing the two": *Atlanta Constitution,* December 10, 1989.

391   "he had fashioned his mustache": *Bozeman Daily Chronicle,* September 22, 1991.

391   "considered calling his first son": Judy Nye Hallisy interview.

392   " 'a whole series of things' ": *Wall Street Journal,* June 5, 1987.

392   "worth at least $1.76 billion": "Forbes 400," *Forbes,* October 23, 1989.

392   " 'He's on a roll' ": *Wall Street Journal,* July 31, 1989.

392   " 'He's on a roll' ": Ticer, "Captain Comeback," p. 98.

392   "actually showing a small": *Wall Street Journal,* August 3, 1989.

392   " 'We're like the Allies' ": Ticer, "Captain Comeback," p. 98.

392   "it had doubled in value": Ibid., pp. 100–101.

392   "broadcast standards": *Wall Street Journal,* July 31, 1989.

393   " 'I was seeing things' ": Ibid., July 17, 1989.

393   "Turner was able to ride": Ibid.

394   "One of the current directors": TBS board member interview, March 18, 1993.

395   " 'this has worked out better' ": Ibid.

396   " 'on a more solid financial footing' ": *Wall Street Journal,* November 13, 1987.

397   "Over the first year": *Wall Street Journal,* October 3, 1988.

397   "the most successful launch": Ticer, "Captain Comeback," p. 98.

397   " 'Ted and the board' ": TBS board member interview, March 18, 1993.

398   "been proved right": Ticer, "Captain Comeback," p. 104.

398   "to refinance his debt": PB, p. 330.

399   "CNN had a global reach": *Wall Street Journal,* February 1, 1990.

400   " 'Karl Marx, meet Marshall McLuhan' ": Jonathan Alter, "Karl Marx, Meet Marshall McLuhan," May 29, 1989, p. 28.

400   "Foreign Ministry's first call": *Wall Street Journal,* February 1, 1990.

400   " 'to tie the world together' ": Ibid.

400   " 'to have my father come back' ": Ted Turner interviewed by Diane Sawyer, *60 Minutes,* April 20, 1986.

401   "a small group of his close friends": Robert Wussler interview; Gerry Hogan interview.

401   " 'his duty to die tragically' ": Painton, "The Taming," p. 35.

401   "worked it into conversations": Gerry Hogan interview.

402   "he thought of Leach": Kathy Leach interview.

402   "'gave up a whole harem of women'": Coppola, "JJ," p. 88.

403   "'go and go until you drop'": Turner friend interview, May 19, 1994.

403   "They traveled together": Kathy Leach interview.

403   "wasn't enough land there": Turner friend interview, January 4, 1993.

403   "to stay out in the country": Kathy Leach interview.

403   "'I like these ranches'": Peter Manigault interview.

404   "'Hell, I love rattlesnakes'": Ibid.

404   "the old Sixteen Mile Ranch": PB, p. 381.

404   "'I'll give you one day'": Kathy Leach interview.

405   "both men steered clear": Ibid.

406   "'needs to be peace here'": Ibid.

406   "bought a second property": PB, p. 382.

407   "It cost Turner $22 million": Ibid.

408   "'I want buffalo'": Kathy Leach interview.

Chapter 23

Page 409   "Ted and Jane would meet": Jane Fonda friend, May 17, 1994; Jane Fonda friend, September 1, 1994.

409   "For Jane Fonda": Vincent Coppola, "Jane," *Atlanta*, February 1994, p. 33.

409   "the marriage of Hayden and Fonda": *Washington Post*, May 13, 1991.

409   "had mixed feelings": Adler, "Jane and Ted's," p. 72; Coppola, "Jane," p. 107.

410   "'Tom couldn't acknowledge her'": *Washington Post*, May 13, 1991.

410   "Hayden began a public romance": Adler, "Jane and Ted's," p. 72; Coppola, "Jane," p. 107.

410   "At Christmas in Aspen": Ibid.

410   "during this unhappy period": PB, p. 338.

410   "One night at a screening": Jane Fonda friend interview, May 17, 1994.

410   "wanting to date an actress": Kathy Leach interview.

411   "Ted discovered that her brother": PB, p. 387.

411   "remaking her life": *Washington Post*, May 13, 1991; PB, p. 346.

411   "in love with Lorenzo": Kathy Leach interview.

412   "lunching with Mikhail Gorbachev": *Wall Street Journal*, June 4, 1990.

412   "'Give us a tour'": Bill Tush interview.

412   "he burst into song": Ibid.

413   "'demonstrated through his life'": Video of Tougaloo College Spring Commencement, Tougaloo College, May 20, 1990.

413   "'schools don't teach you about marriage'": Ibid.

414   "'a male chauvinist pig'": Jane Fonda interviewed by Nancy Collins, *Primetime Live*, September 9, 1993; Painton, "The Taming," p. 38.

414   "her all-consuming amatory voracity": Adler, "Jane and Ted's,"
        p. 72.

414   " 'powerful parental figures' ": Jane Fonda, "Remembering Dad,"
        TV Guide, January 11, 1992, p. 6.

414   " 'As a child' ": Christopher P. Andersen, *Citizen Jane: The Turbulent
        Life of Jane Fonda* (New York: Henry Holt, 1990), p. 46.

415   " 'empty a refrigerator' ": "A Debilitating Obsession," *People,* Au-
        gust 3, 1992, p. 68.

415   " 'Jane was a superachiever as well' ": DF.

415   " 'Miss Fonda has enough influence' ": *Wall Street Journal,* July 24,
        1990.

415   " 'he was courting' ": Peter Dames interview.

415   " 'By any definition' ": Coppola, "Jane," p. 29.

416   " 'In previous relationships' ": Peter Dames interview.

416   " 'a trophy mate' ": Adler, "Jane and Ted's," p. 72.

416   " 'Mr. Fonda' ": Bill Tush interview.

417   "oft-proclaimed motto": Maria Shriver, *First Person,* May 6, 1992.

418   "merge their stables of horses": Adler, "Jane and Ted's," p. 74.

418   " 'a community of interests' ": Peter Boyer, "Taking on the World,"
        Vanity Fair, April 1991, p. 104.

418   "to address a celebrity crowd": Adler, "Jane and Ted's," p. 68.

419   "might lose up to $20 million": *Wall Street Journal,* June 26, 1990.

419   "Rising early in the morning": *Larry King Live,* October 8, 1990.

419   "Ted was jumping up and down": Paul Beckham interview.

419   "the Seattle Goodwill Games": *Wall Street Journal,* July 30, 1985.

420   "with a few provisos": Ibid., October 12, 1990.

421   " 'a matter of survival' ": Lanham, "Greening," p. 6.

421   " 'Ted sees the angles' ": Tom Belford interview, June 23, 1994.

421   "to all facets of his life": Boyer, "Taking on the World," p. 104;
        New York Daily News, June 21, 1990.

423   " 'As an American' ": HW, p. 304.

423   "Americans, he added": PB, p. 335.

424   " 'We'll give the other bozos' ": *New York Daily News,* July 14, 1989;
        New York Times, July 15, 1989.

424   " 'a religion for losers' ": Tim Powis, "Life at the Top," *Maclean's,*
        June 18, 1990, p. 54.

424   " 'The great minds of today' ": *Wall Street Journal,* November 21,
        1989.

424   "the prizewinning book": Ibid., June 4, 1991.

425   " 'done what we set out to do' ": Tom Belford interview.

425   "the produced programs 'mediocre' ": TBS programming executive
        interview, January 7, 1993.

425   "talked privately of closing it down": Kathy Leach interview.

426   " 'We need the most major' ": Video of Jane Fonda and Ted Turner
        at Harvard's John F. Kennedy Institute of Politics, October 9,
        1992.

427   " 'We were like pirates before' ": *New York Times,* October 12, 1990.

427   " 'the emphasis less on innovation' ": *Wall Street Journal,* October 12,
        1990.

428   "the defection of": Ibid.

428   "there were many": Cable executive interview, April 1, 1993; For-
      mer TBS executive interview, April 4, 1994.

428   "prenuptial agreement": *National Enquirer,* November 5, 1991. Oth-
      ers said that the real problem was that Ted had gone out with
      other women, and had told Jane about it. TBS executive inter-
      view, April 4, 1994.

429   " 'Ted became depressed' ": Ibid.

429   "the pair had agreed": Ibid.; "I Do, I Do, I Do, I Do," *People,*
      January 13, 1992, p. 36.

429   "wedding": Peter Dames interview; Bill Tush interview; "I Do,"
      People, p. 36.

Chapter 24

Page 431   "On May 4": Emma Edmunds, "Scenes From a Marriage," *Atlanta,*
           July 1991, p. 41; Judy Nye Hallisy interview; Emma Edmunds
           interview, May 20, 1994.

432   " 'Dysfunctional?' ": Emma Edmunds interview.

433   " 'very manic without the lithium' ": TBS executive interview, April
      4, 1994.

433   " 'not a very warm person' ": Irwin Mazo interview.

433   " 'Jane was Dad's salvation' ": Dodson, "Teddy Comes About,"
      p. 108.

434   " 'He does laugh a little more' ": Painton, "The Taming," p. 38.

434   "exactly thrilled with him": Dan Burkhart interview, June 1, 1994.

434   " 'I spend more time here' ": *Bozeman Daily Chronicle,* September
      22, 1991.

434   "The Montana house that Ted and Jane built": Ibid.; *Atlanta Con-
      stitution,* November 12, 1991; Dan Burkhart, "Turner and
      Fonda: The View from the Flying D," *Montana,* May 1992,
      p. 62.

435   "had a more aggressive quality": *Atlanta Constitution,* November 12,
      1991.

435   "Indian arrowheads": Dan Burkhart interview.

436   "Jimmy Carter helicopters in": Peter Manigault interview.

436   " 'with the new thing' ": Dan Burkhart interview.

436   " 'the new Wowuka' ": Burkhart, "Turner and Fonda," p. 67.

436   "to host Indian game wardens": Peter Manigault interview.

437   "is being funded by Turner": Ibid.

437   "both are vocally committed to causes": Dan Burkhart interview.

437   "When Jane gave up alcohol": Gerry Hogan interview.

437   " 'She's good eyes' ": Entertainment industry CEO interview, March
      18, 1993.

437   "interest in the visual arts": Mike Gearon interview.

437   "started to collect Western art": Peter Manigault interview.

438   "she has mirrored the man": Jane Fonda friend interview, May 17,
      1994.

438   " 'People say to me' ": Jane Fonda interviewed by Nancy Collins,
      Primetime Live, September 9, 1993.

438 "'roles left for Jane'": Entertainment industry CEO interview, March 18, 1993.

438 "'Cut what you're doing in half'": Jane Fonda interviewed by Nancy Collins, *Primetime Live,* September 9, 1993.

438 "never being away from him": Ibid.

438 "'She's going at it'": Turner friend interview, March 22, 1994.

439 "'they're not going to last'": Emma Edmunds interview.

439 "it paid dividends": TBS annual report, 1992; *Wall Street Journal,* February 3 and May 8, 1992.

439 "a very real explosion of joy": Robert Wussler interview.

439 "who had been cheated": Mike Gearon interview.

440 "'They grind up the feet'": Subrata N. Chakravarty, "What New Worlds to Conquer?" *Forbes,* January 4, 1993, p. 82.

440 "'an unbelievable offense'": Boyer, "Taking on the World," p. 94.

440 "'whatever you think it takes'": W. Thomas Johnson Jr. interview, May 11, 1994.

440 "'I certainly hope'": Ibid.

441 "'in serious danger'": Ibid.

441 "critics scoffed": *Wall Street Journal,* January 18, 1991.

442 "'I'm an internationalist first'": DF.

442 "a major expansion drive": W. Thomas Johnson Jr. interview.

442 "more profits than all of CBS": According to analysts at Paul Kagan Associates, CNN (and Headline News) in 1991, 1992, and 1993 had an operating cash flow of $168 million, $183 million, and $212 million, while the equivalent numbers for the entire CBS network were $32 million, $71 million, and $209 million. In fact, CNN consistently outperformed NBC and ABC as well.

442 "more homes than ever before": A. C. Nielsen data, 1993.

444 "'I can guarantee you'": *UCLA Daily Bruin,* May 28, 1991.

444 "'Visionaries are possessed creatures'": "Prince of the Global Village," *Time,* January 6, 1992, p. 22.

444 "Not quite so impressed": *Village Voice,* January 14, 1992.

445 "'Jane was married to a politician'": Ted Turner interviewed by Maria Shriver, *First Person,* May 6, 1992.

445 "too small for him": Jim Roddey interview.

445 "'I'm used to moving quickly'": DF.

445 "'I want to get my children'": Peter Manigault interview.

445 "has recently given": *New York Times,* February 8, 1994.

446 "three wishes for the new year": Reese Schonfeld interview.

446 "'We're opportunistic'": *Wall Street Journal,* February 19, 1992.

447 "an interactive unit": Ibid., February 7, 1992.

447 "talking about mergers": Ibid., February 12, 1993.

447 "what to make of Turner": Bryan Burrough, "The Siege of Paramount," *Vanity Fair,* February 1994, p. 132.

447 "'Any expenditure'": Ibid.

448 "'Turner had respect for this guy'": Ibid.

448 "'John Malone and Time'": *New York Times,* June 6, 1993.

448 "'When you have two ten-ton gorillas'": *Wall Street Journal,* April 15, 1993.

448 "let word leak to the press": *New York Times,* June 6, 1993.

448   "Time Warner advisor": *Wall Street Journal*, April 15, 1993.

449   " 'We're kind of neutral ground' ": *New York Times*, June 6, 1993.

449   " 'a trial balloon' ": Diane Mermigas, "Q & A: Ted Turner," *Electronic Media*, June 7, 1993, p. 36.

449   " 'Ted's not a seller' ": Paul Beckham interview.

449   "For approximately $600 million in cash": *New York Times*, August 18, 1993.

450   "he currently owns": Editors, "Great Rocky Mountain Land Grab," *Rocky Mountain Magazine*, Summer 1994, p. 18.

450   "the Pedro Armendaris Ranch": *Albuquerque Journal*, May 1, 1994; Leslie Linthicum interview, June 2, 1994. In recent months, Turner also purchased thousands of acres along the Platte River in Nebraska, more grasslands for his buffalo.

450   "time is running out": Ted was successfully operated on for skin cancer in August of 1994.

451   "On a warm breezy spring afternoon": Judy Nye Hallisy interview.

Epilogue

Page 453   " 'I didn't sleep at all' ": Ted Turner interview.

453   "only the previous day": *New York Times*, September 18, 1993.

453   "negotiating with Doordarshan": Ibid., April 4, 1994.

453   "two new networks": Ibid., June 6, August 27, and August 28, 1993. Apropos of TCM, Turner, in a surprising reversal, indicated in late 1994 that he was stopping the colorizing of black-and-white films.

458   "grossed about $6 million": *PR Newswire*, November 17, 1993.

458   "*Gettysburg*, the coffee-table book": James McPherson (and 65 paintings by Mort Kunstler), *Gettysburg* (Atlanta: Turner Publishing, 1993).

459   " 'We're gonna do Joan of Arc' ": Ted Turner interview.

INDEX